The Molecular and Supramolecular Chemistry of Carbohydrates

The Molecular and Supramolecular Chemistry of Carbohydrates:

Chemical Introduction to the Glycosciences

SERGE DAVID

Emeritus Professor, University of Paris-Sud, Orsay

Translated by

Rosemary Green Beau

Institute of Molecular Chemistry, University of Paris-Sud, Orsay

OXFORD NEW YORK TOKYO

OXFORD UNIVERSITY PRESS

1997

Oxford University Press, Great Clarendon Street, Oxford OX2 6DP
Oxford New York
Athens Auckland Bangkok Bogota Bombay Buenos Aires
Calcutta Cape Town Dar es Salaam Delhi Florence Hong Kong
Istanbul Karachi Kuala Lumpur Madras Madrid Melbourne
Mexico City Nairobi Paris Singapore Taipei Tokyo Toronto
and associated companies in
Berlin badan

Oxford is a trade mark of Oxford University Press

Published in the United States
by Oxford University Press Inc., New York

Published with the help of the Ministère de la Culture.

A catalogue record for this book is available from the British Library

Library of Congress Cataloging in Publication Data
(Data available)

ISBN 0 19 850047 5 (Hbk)
ISBN 0 19 850046 7 (Pbk)

Typeset by EXPO Holdings, Malaysia

Printed in Great Britain by
Bookcraft (Bath) Ltd
Midsomer Norton, Avon

I would not wish anyone so much as a tumble or a forced delay at the draw-bridge of Knippelbro etc ... like those fretting business people full of an infinite number of ventures, while we others, when the bridge is lifted, find this a good moment to sink into our thoughts.

[...] Indeed it is difficult to live in a land without ever seeing the sun or the horizon, but it is scarcely better to live in a place where the sun strikes the head so vertically that not even the slightest shadow is cast.

Søren Kierkegaard
Journal (excerpts), July 14, 1837

Preface

The outline of subject matter adopted for this work is not keeping with traditional books on organic chemistry. Handbooks and textbooks essentially describe contemporary methods for constructing covalent bonds with a few developments concerning conformation and, occasionally, a brief reference to the living world. Indeed there has been considerable progress in the synthetic organic chemistry of carbohydrates during the past decades. The optimizing of new techniques and the introduction of new concepts have permitted most of the important reactions of organic chemistry to be extended to this family of compounds. Much intense effort has greatly improved the outcome of the glycosidation reaction, which was often inefficient using older methods. The author has devoted half of this work to these synthetic aspects. However, with the current evolution of research ideas, limiting a book on carbohydrates to the description of the best methods for constructing carbon–carbon and carbon–oxygen covalent bonds boils down to dropping half the subject. One of the most important topics of contemporary organic chemistry happens to be the study of the associations between molecules which, while being relatively stable, do not involve covalent bonds. Some of this research has developed in a totally autonomous fashion with respect to the living world. However, in the chemistry of oligosaccharides (see Chapter 9) a great number of associations of this type are encountered, not only with the macromolecular receptors present in living cells, but with inorganic structures as well. Of course the complexity of natural organic receptors makes the analysis of association types rather conjectural in the majority of cases, but the importance of the mechanisms of the living world upon which they rely justifies, in the author's mind, devoting nearly half of this work to them. Without this, a rapidly expanding scientific field would have been ignored.

After all, from time immemorial, carbohydrate chemists have been quite concerned with physical chemistry. Let us consider the attempt to understand optical rotation using the means of that·time, the Hudson rules. Although they have been nearly forgotten—we will not discuss them—the spirit remains and most carbohydrate chemists, although mainly concerned with synthetic problems, find it necessary to examine carefully the physicochemistry of their molecules using modern methods, possibly using *ab initio* and MM calculations ... The reader will find, here and there, and more particularly in Chapters 1, 2, and 9, a selection of recent results.

The author hopes that the general panorama laid out in this book faithfully reflects the ambiance of the major carbohydrate chemistry laboratories and the atmosphere at the meetings and symposiums specializing in this field. We are in the presence of a science which was not built up from a particular technical arsenal, but from a rather homogeneous family. The term *glycoscience* has been proposed. This is an interesting scientific step in itself, perhaps a model for other families, regardless of its obviously anthropocentric repercussions. One could argue with the author that a multi-author text would have better met his goal. However, this is not a textbook, but rather an attempt to place a collection of work within a precise perspective using characteristic examples. The references do not constitute a prize list of discoveries, but point to an easily accessible supplementary document, or to experiments having pedagogical value. We will add that the editing by a single author makes a homogeneous treatment possible. The framework of this book is organic chemistry which seems to us justifiable since, sooner or later, all interactions will be described at the molecular level. There may be more practical applications. For example, according to one report (Raugel 1994), a top American biotechnology company was obliged to contact a large industrial group capable of helping on an organic chemistry level.

For anything having to do with the relationship of carbohydrates to the living world, we have put a great deal of importance on molecules with a *very large distribution*, often universal, with special attention to general mechanisms. This is the viewpoint of biochemistry, which has led us to exclude enticing areas such as aminoglycosidic antibiotics. But we did not want to produce a manual on the biochemistry of sugars and we will not discuss the elaboration, so characteristic, of certain oligosaccharide units of glycoconjugates (Shibaev 1986). We were above all interested in the problems concerning man and higher animals. This restriction had led us to neglect, barring some exceptions, the very complex structure and problems of practical importance concerning microbial oligosaccharides. The two references cited are to articles among the most recent from two European schools active in this field (Kenne *et al.* 1993; Auzanneau *et al.* 1992). We shall end these general comments with a practical warning: the drawings are most often schematic and should not be used as a source of quantitative data. For the latter, the reader should refer to the numerous tables in this work.

The biosynthesis of proteins follows the genetic code. The analogues in the carbohydrate field to proteins are oligosaccharides. The junctions between the monosaccharide units are catalyzed by enzymes, the glycosyltransferases, obviously being coded. Contrary to what happens with amino acids, at the present time we have no indication that a code exists which organizes the sequences of monosaccharides in the oligosaccharides. Simply (if it is possible to use the word!), the glycosyltransferases must appear at the right moment and in the right place. Does this cause a certain fuzziness in the synthesis? The vague opinion circulating in the community of specialists is that a certain disorder could well be advantageous for an organism by tempering the excessive rigour of the genetic code. To our knowledge, this idea has not yet been worked out. The fact remains that many of these sequences have a rather forbidding appearance, and the reader coming from the vivid world of the chemistry of natural products would have the impression of entering an arid and disorderly land, but this is because the meaning of these structures is only revealed slowly, and that is enough to make them fascinating.

The author is grateful to Professor André Lubineau for his collaboration in the editing of Section 17.6, concerning selectins and their ligands, and Sections 11.2 and 11.6, dealing with the intimate relationship, so obvious and yet still mysterious, between sugars and water. The help of Doctor Claudine Augé, director of research at the CNRS, for all the questions on enzymic preparative chemistry was immensely appreciated. In general, being immersed in the center of an active group greatly facilitated the author's collection and verification of information. Finally, the author thanks Doctor Ten Feizi of the Medical Research Council in Harrow (England) for her help in editing Sections 17.4 to 17.6, and Professor Sen-itiroh Hakomori for his assistance in the elaboration of Chapter 16.

Paris S. D.
Jan 1997

References

Auzanneau, F. I., Mondange, M., Charon, D., and Szabo L. (1992), *Carbohydr. Res.*, **228**, 37–45.

Kenne, L., Lindberg, B., Matibubur Rahman, M., and Mosihuzzaman, M. (1993), *Carbohydr. Res.*, **243**, 131–138.

Raugel, P. J. (1994), *La Recherche*, **262**, 224–233.

Shibaev, V. N. (1986), *Adv. Carbohydr. Chem. Biochem.*, **44**, 277–339.

Contents

1 Configuration of monosaccharides

1.1 Glucose

Glucose is extremely soluble in water: 0.5 kg can be dissolved in 250 mL of hot water. The addition of acetic acid to this solution brings about a slow precipitation of crystals. This is one of various tautomers, referred to in the official nomenclature as 'α-D-glucopyranose', a word whose exact meaning will be defined later in this chapter. The absolute configuration of this solid is known through the association of X-ray and neutron diffraction analyses which give the 23 bond lengths, the 42 valency angles, and the 69 torsion angles of this molecule (Brown and Levy 1979). In the schematic representation **1.1** of this configuration, carbons 2 and 3 of the chain are assumed to be in front of the molecule and carbons 1 and 4 in the plane of the drawing. The other carbons and the ring oxygen are at the back of the molecule.

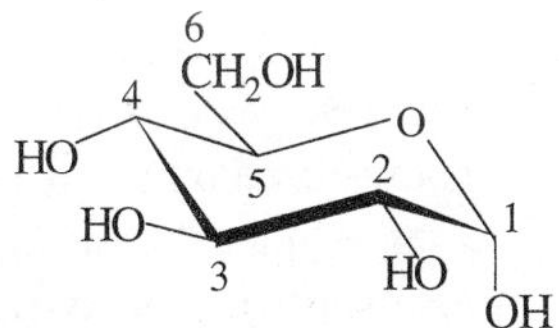

1.1

One recognizes an oxane ring (tetrahydropyran) substituted by three secondary alcohol functions in an equatorial orientation, a side chain carrying a primary alcohol function and finally a hemiacetal hydroxyl carried by carbon 1. This intramolecular hemiacetal is derived from the addition of the oxygen carried by C-5 to an aldehyde function.

Starting from any glucose sample, an isomer of compound **1.1** can be prepared by the following protocol: the sample is recrystallized in acetic acid, crystals are then dissolved in ice water (100 mL for 100 g), filtered, ethanol (0.5 L) is added to the filtrate to bring about a rapid precipitation. The obtained compound has the configuration **1.2** in the solid state (Chu and Jeffrey 1968).

1.2

The only difference with molecule **1.1** is in the hemiacetal hydroxyl orientation. All substrates of molecule **1.2** are equatorial. There is a great underlying simplicity in the D-glucose configuration in spite of its forbidding aspect for a beginner. This observation may be a useful starting point for memorizing carbohydrate structures. Molecule **1.2** is called 'β-D-glucopyranose'.

Isomers **1.1** and **1.2** are in tautomeric equilibrium in aqueous solution according to equation (1.1).

(1.1) α-D-glucopyranose $\rightleftharpoons$ β-D-glucopyranose

Thus the optical rotation of an aqueous solution of the α-D-isomer, which corresponds to $[\alpha]_D^{20}$ + 112° immediately after dissolution, decreases to 52.7° in a few hours. Conversely, the optical rotation of the β-D-isomer increases from 18.7°, the value at dissolution, to the same equilibrium value. This allows the following calculation: $[\alpha]/[\beta] = 38/62$. The all-equatorial compound dominates, but we will see in Section 2.6 that we must avoid seeing here the classical rules of conformational analysis. These are the experiments which allowed the tautomeric equilibrium (1.1) to be observed for the first time, and for this reason, it has kept the name of mutarotation.

The proton NMR spectrum in D_2O gives similar results. The H-1 proton carried by C-1 shows a downfield signal, because of the two geminal oxygens, separated from the group of other protons and easy to spot. Immediately after dissolution, a ^3J 4 Hz doublet is observed on the α-D-glucopyranose spectrum, due to an axial-equatorial coupling. Under the same conditions, a large ^3J 8 Hz doublet on the β-D-glucopyranose spectrum is observed immediately after dissolution, because of a *trans*-diaxial coupling. At equilibrium both signals are observed (Fig. 1.1).

In fact, this aqueous solution contains other tautomers but in concentrations much too weak to show up during routine NMR studies. For the time being we will disregard their existence. It must be clear that tautomers **1.1** and **1.2** are two chemically distinct molecules whose differences are not only revealed by their physical characteristics, but also by their chemical and enzymic reactivity. However, one observes that the C-1 carbon is distinguished from others by its unstable configuration, hence its particular name of an *anomeric carbon*. Traditionally, glucose has been represented by the aldehyde parent **1.3** in which only stable configurations are found. *However, this tautomer is only present, under any circumstance, in a very small concentration.*

Aldehyde **1.3** is drawn using the Fischer projection formula. The hydroxyls located below the average plane of the oxane are to the right, the hydroxyl situated above is to the left. The correspondence for carbon 5 linked to the side chain is more difficult. The reader should remember that, using the Fischer projection, the vertical valencies recede from and the horizontal valencies project towards the viewer. The viewer may then check that the heavy atoms of D-glyceraldehyde **1.4** can be superimposed on the portion corresponding to carbons 4, 5, and 6 of oxanes **1.1** and **1.2**.

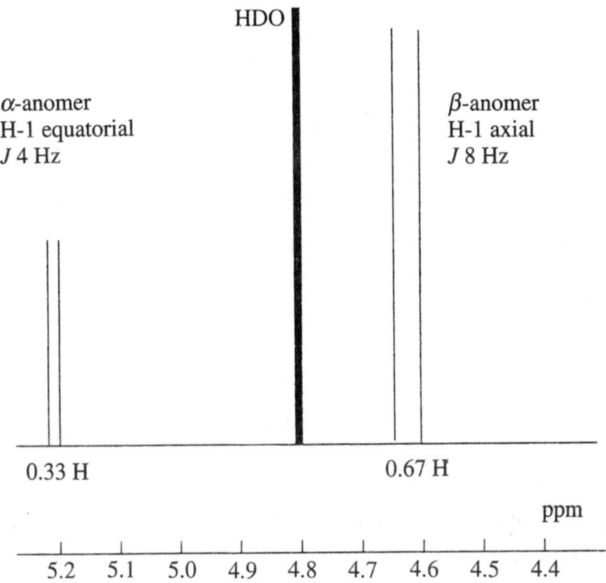

Fig. 1.1 ^{1}H NMR signals of anomeric protons of α- and β-D-glucopyranoses.

CHO
H—C—OH
HO—C—H
H—C—OH
H—C—OH
CH_2OH

1.3

CHO
H—C—OH
CH_2OH

1.4

1.2 Other carbohydrate configurations

There are four other asymmetric carbons in the configuration **1.3**, and thus $2^4 = 16$ isomers, each having its own name. The reader will find a table of these sugars in Chapter 4, which deals with nomenclature. The majority of these configurations are found in derivatized forms in living cells. To confine ourselves to the general universally known constituents, we will cite D-mannose **1.5** and D-galactose **1.6**, epimers at C-2 and C-4 of D-glucose, respectively. We will encounter, just as frequently, three sugars in which the hydroxyl at C-2 has been replaced by an acetamido group, called *N*-acetylglucosamine **1.7**, *N*-acetylmannosamine **1.8** and *N*-acetylgalactosamine **1.9**. Partially deoxygenated molecules are also observed, such as L-fucose **1.10**. All of these sugars with a latent aldehyde function are

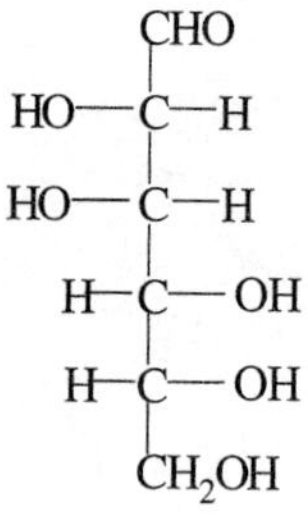

CHO
HO—C—H
HO—C—H
H—C—OH
H—C—OH
CH_2OH

1.5

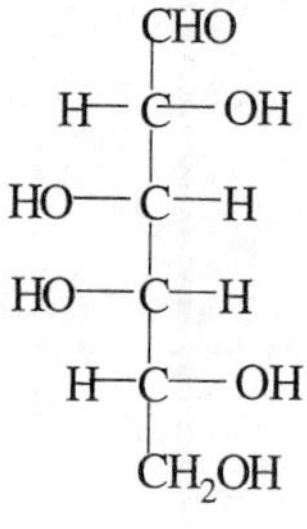

CHO
H—C—OH
HO—C—H
HO—C—H
H—C—OH
CH_2OH

1.6

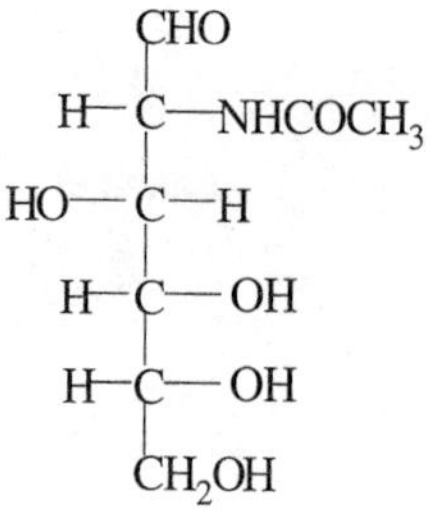

CHO
H—C—$NHCOCH_3$
HO—C—H
H—C—OH
H—C—OH
CH_2OH

1.7

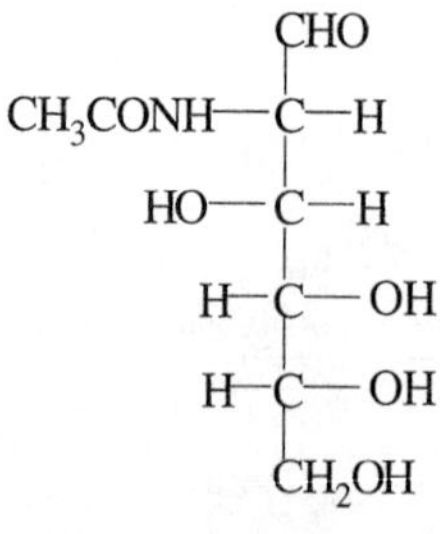

CHO
CH_3CONH—C—H
HO—C—H
H—C—OH
H—C—OH
CH_2OH

1.8

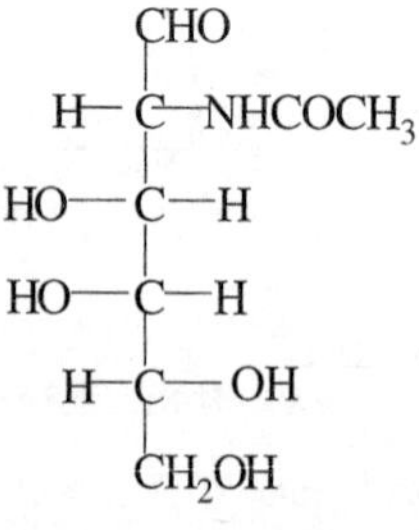

CHO
H—C—$NHCOCH_3$
HO—C—H
HO—C—H
H—C—OH
CH_2OH

1.9

CHO
HO—C—H
H—C—OH
H—C—OH
HO—C—H
CH_3

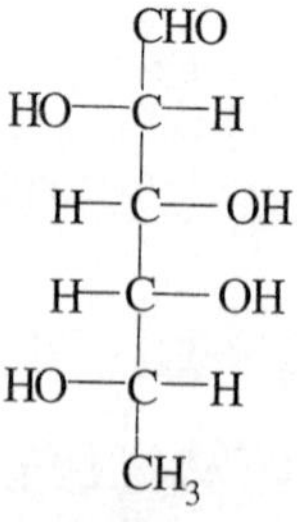

1.10

called *aldoses*. However, the latent carbonyl can also be a ketone, hence we have *ketoses* such as the fructose **1.11**. All sugars comprising a six-carbon non-branched chain have been given the general name of *hexoses*.

There are also five-carbon sugars, the *pentoses*, of which two representatives, the D-ribose **1.12** and the deoxyribose (using the correct nomenclature, 2-deoxy-D-*erythro*-pentose) **1.13**, are infinitely more important than the others. A sugar with nine carbons, the sialic acid **1.14**, gathers on the same chain a carboxyl, a ketone carbonyl, five alcohol hydroxyls and one amide function. The carbohydrate chains are numbered by giving the lowest number to the carbonyl carbon. All of these molecules belong to the group called *monosaccharides*.

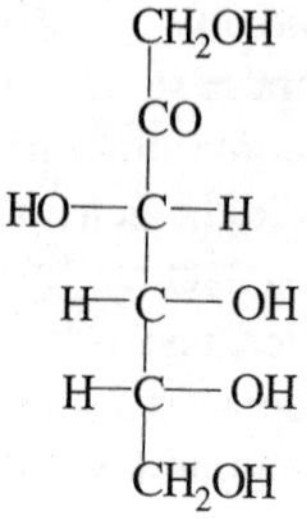

CH_2OH
CO
HO—C—H
H—C—OH
H—C—OH
CH_2OH

1.11

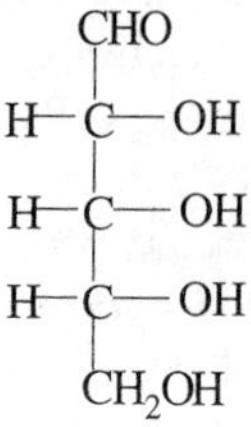

CHO
H—C—OH
H—C—OH
H—C—OH
CH_2OH

1.12

CHO
H—C—H
H—C—OH
H—C—OH
CH_2OH

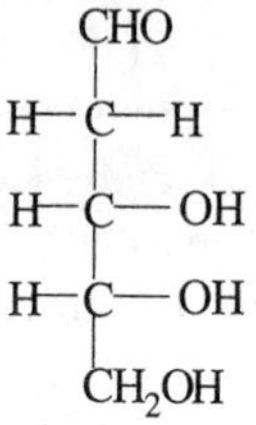

1.13

$$
\begin{array}{rcl}
 & \mathrm{COOH} & \\
 & | & \\
 & \mathrm{CO} & \\
 & | & \\
\mathrm{H}- & \mathrm{C} & -\mathrm{H} \\
 & | & \\
\mathrm{H}- & \mathrm{C} & -\mathrm{OH} \\
 & | & \\
\mathrm{CH_3CONH}- & \mathrm{C} & -\mathrm{H} \\
 & | & \\
\mathrm{HO}- & \mathrm{C} & -\mathrm{H} \\
 & | & \\
\mathrm{H}- & \mathrm{C} & -\mathrm{OH} \\
 & | & \\
\mathrm{H}- & \mathrm{C} & -\mathrm{OH} \\
 & | & \\
 & \mathrm{CH_2OH} &
\end{array}
$$

1.14

With the exception of fucose, all these sugars have the same configuration on the penultimate carbon as does the central carbon of D-glyceraldehyde. This can easily be explained because living cells produce all sugars from D-glyceraldehyde, and the biosynthetic pathway does not involve, at any step, a cleavage between the central carbon of D-glyceraldehyde and one of its four substituents. Figure 1.2 shows the 'genealogical tree' of these monosaccharides. D-Fructose results from the aldol condensation of dihydroxyacetone (nucleophilic partner) on D-glyceraldehyde. This leads to either D-glucose or D-mannose by modifications at C-1 and C-2. D-Glucose is epimerized at C-4 to give D-galactose. The same D-glucose loses C-1 and undergoes some transformations at C-2 and C-3 to give D-ribose (there is another biosynthetic pathway, the *pentose–heptose cycle*, which is more complicated but does not involve the penultimate carbon). Deoxyribose is produced by deoxygenation of D-ribose at C-2. The amination of D-fructose

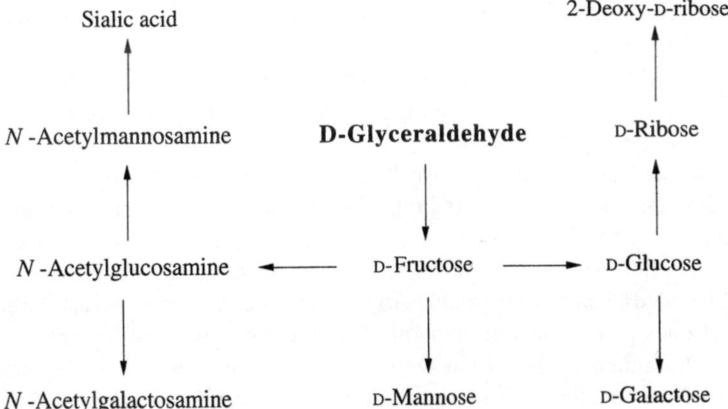

Fig. 1.2 'Genealogy' of the major sugars of the D-series.

followed by acetylation gives *N*-acetylglucosamine, epimerized to *N*-acetylgalactosamine and *N*-acetylmannosamine. The aldol condensation of pyruvic acid with *N*-acetylmannosamine gives sialic acid.

Among the sugars **1.5** to **1.14**, only fucose has the L-glyceraldehyde configuration at its penultimate carbon. The biological precursor is D-mannose which is converted to a derivative of intermediate structure **1.15**. The latter undergoes epimerizations at C-3 and C-5, in agreement with the organic chemist's intuition since these carbons are adjacent to a carbonyl, and undergoes carbonyl reduction to an alcohol function. It is perhaps noteworthy that the L-glyceraldehyde configuration is observed at the penultimate carbon of other deoxygenated natural sugars at C-6. To avoid any error of interpretation, we must point out that the real substrates of enzymes in biosynthetic pathways *are not* free sugars but rather phosphates or complex phosphates. However, in no way does this invalidate our deductions.

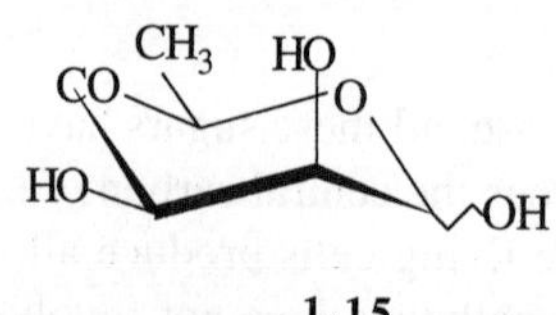

1.15

Let us leave biochemistry for geometry. We have taken the habit of classifying D-glucose and its 15 isomers with a stable configuration into two groups of eight, the D-series where C-5 has the D-glyceraldehyde configuration and the L-series where C-5 has the L-glyceraldehyde configuration. The names of the sugars of the D-series are preceded by the prefix D such as D-glucose, D-mannose, etc. The enantiomers of these hexoses belonging to the L-series are called L-glucose, L-mannose, etc. Finally, whenever the configuration rather than the molecule is to be designated, the words are written in italics as in D-*manno*, D-*gluco*, D-*galacto*, L-*manno*, etc., derived from the current names of sugars. This holds true for words appearing within a text or for a sugar named according to the official nomenclature (see Chapter 4). These rules can be applied without difficulty to the pentoses. There are eight pentoses, pairs of enantiomers, divided into two series, D and L, depending on whether the penultimate carbon has the D-glyceraldehyde or opposite configuration. The configuration of the pentose D-ribose is designated as D-*ribo*.

The words 'D-series' and 'L-series' do not have the same biological meaning, depending on whether we are looking at sugars or amino acids The amino acids of proteins, 20 in all, belong exclusively to the 'L series of amino acids'. The oligosaccharide sequences, parallel structures to polypeptides in tissues (but not directly coded), can be constructed from D-sugars as well as L-sugars, although the first ones predominate in general. The following explanation may be suggested: the amino acids carbon whose configuration determines the series is linked to two functional groups, amino and carboxyl, involved directly in the peptidic bond. On the other hand, the penultimate carbon of sugars, whose configuration determines the series is rather passive in the creation of the glycosidic bond between monosaccharides.

1.3 Tautomerism

1.3.1 General

Examining the formula of aldehyde **1.3** points to, *à priori*, the presence of six tautomers. We have already looked at pyranoses **1.1** and **1.2** resulting from attack by oxygen O-5 of one of the two faces of the prochiral carbonyl of aldehyde **1.3**. But we have no reason to exclude the possibility of attack by oxygen O-4 with the formation of two tautomers having a five-membered ring, **1.16** and **1.17**. The stability of a five-membered ring is not very different from that of a six-membered one. At this point, we will complete our explanation of the terms dogmatically introduced at the beginning of the chapter. Oxane ring sugars (tetrahydropyran) are *pyranoses*, while those with an oxolane ring (tetrahydrofuran) are known as *furanoses*. The symbol α is defined with reference to the penultimate carbon. A simple explanation for this convention is based on the carbonyl hydrate such as **1.18** corresponding to D-glucose. On paper, it is transformed into furanose or pyranose by replacing one of the hydroxyls carried by C-1 by the oxygen of an alcohol function carried by C-4 or C-5. If the remaining hydroxyl is strictly *trans* with respect to the oxygen of the penultimate carbon, the anomer is called α. Thus **1.16** and **1.17** are α- and β-D-furanose, respectively. Six-membered sugars are dealt with in Section 4.2.6.

We also have to consider the presence of a free aldehyde in solution and its hydrate **1.18** because it is known that hydrates of α-hydroxylated aldehydes are relatively stable.

In fact, all of these tautomers exist in aqueous solution, but usually some are present in concentrations too low to be visible without sophisticated techniques. For example, in the case of D-glucose, tautomers other than α- and β-D-pyranoses are only present in insignificant amounts. The problem of the tautomeric content of sugar solutions at equilibrium, generally aqueous, has stimulated a great deal

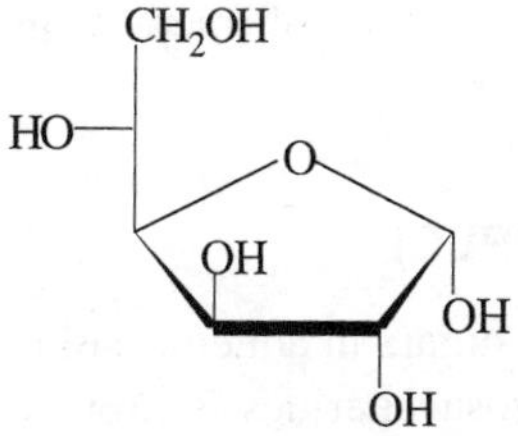

1.16

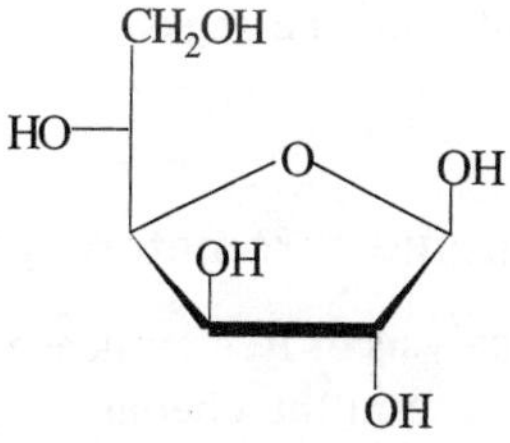

1.17

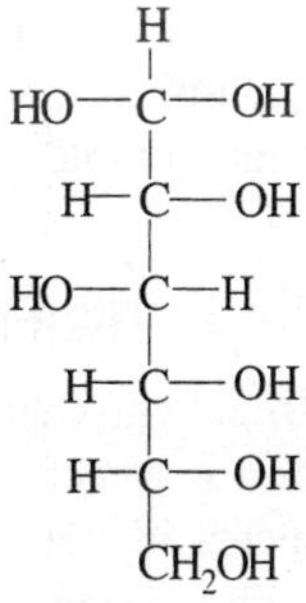

1.18

of interest (Angyal 1984, 1991). Contrary to tradition, which gives precedence to physical measurements rather than attempting isolation, we will first discuss separative methods because they call for very basic techniques of carbohydrate chemistry. Let us acknowledge, however, that they are not the most powerful in the present context. Of course isolation of a tautomer must not noticeably modify the equilibrium. Solvents in which mutarotation is slow, low temperature reactions, and the fastest possible derivatization reactions will be used.

1.3.2 Gas chromatography

Sugars are extremely stabilized in solid and liquid phases by a network of hydrogen bonds, since it is possible for each hydroxyl to be a donor or an acceptor. The total destruction of this network is not possible at temperatures below that when the molecule begins to decompose. Sugars cannot be distilled. Substitution of all acidic hydrogens by $Si(CH_3)_3$ suppresses any possibility of hydrogen bonding. The accumulation of methyl groups on the outside—15 in the case of glucose—gives the molecule the approximate form of a sphere limited by 45 neutral hydrogen atoms with minimal cohesion. Although rendered considerably heavier, the molecule becomes volatile. Thus the derivative of α-D-glucopyranose in which every OH is replaced by $OSi(CH_3)_3$ boils at 107°–110°C under 0.1 mm of Hg. Thus the persilylated sugar derivatives can be rapidly separated by gas chromatography. In a classical silylation procedure, sugar (10 mg) is dissolved in pyridine (1 mL), then hexamethyldisilazane ($Me_3SiNHSiMe_3$, 0.1 mL) and cholorotrimethylsilane (0.1 mL) are added. Each hydroxyl is silylated according to equation (1.2). The reaction is normally completed in 5 min at room temperature.

$$3ROH + ClSiMe_3 + Me_3SiNHSiMe_3 \rightarrow 3ROSiMe_3 + NH_4Cl \tag{1.2}$$

In order to follow the mutarotation, the sample (5 μL) is quickly dissolved in *N,N*-dimethylformamide and the solution is cooled in liquid nitrogen. The silylating mixture is added, allowed to warm to room temperature, then applied to a column. Used at 150–200°C, the column contains a liquid with a high boiling point, adsorbed on a powdery solid phase. Utilizing this method, it is possible to observe as many peaks on the chromatogram as there are tautomers in noticeable quantities in solution. Identifying the peaks requires the isolation of fractions in a measurable quantity.

1.3.3 High-pressure liquid chromatography (HPLC)

Direct chromatographic analysis of free, or nearly free, sugars in aqueous solution is a technique whose use in the chemistry of oligosaccharides is rapidly increasing (Hicks 1988). The adsorbant is in the form of monodispersed particles with varying dimensions from 3 to 15 μm in 10 to 15 cm analytical columns. Preparative columns (2.5 × 30 cm) are also used. Elution is only possible under pressures varying from 1 to 30 atmospheres, and the procedure takes around 1 h. In the eluent, sugars are determined by measuring the change in conductance, the refractive index, or the ultraviolet absorption. This requires absolute stability of all operational conditions, so much so that the column is surrounded by particularly expensive regulators.

Among other stationary phases adapted for these separations, the use of sulfonated divinylbenzene polystyrene beads of strictly controlled diameters (6 μm) filling a 6 × 150 mm column has been described. This cation exchanger is used in the Ca^{2+} form with a flow rate of 0.5 mL min^{-1}. In these experiments (Honda *et al.* 1984), the authors determined sugars in an eluent by transformation into a derivative absorbing at 280 nm, but this would certainly not be necessary today using modern equipment. Figure 1.3 shows the data obtained with D-glucose, D-galactose, and D-mannose (Honda *et al.* 1984).

The α- and β-furanoses equilibrate too quickly to be separated at 0–4°C, but this can be carried out between –25 and –45°C using special solvents with D-galactose and D-fucose.

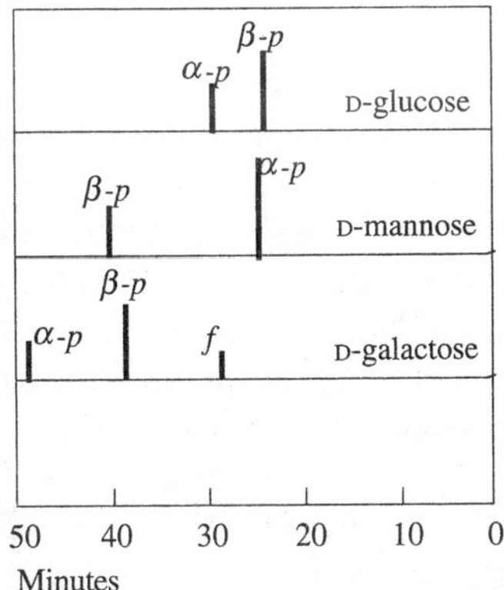

Fig. 1.3 Separation of anomers by HPLC at 4°C, eluent, water–acetonitrile (20:80, v/v); ordinate, absorption at 280 nm. For other conditions see text. Adapted from Honda *et al.* (1984).

1.3.4 Circular dichroism

Circular dichroism measurements have been considered in order to observe the carbonyl tautomers which are particularly quickly fading. Carbonyl compounds absorb near 280 nm due to the nπ* transition of the C=O double bond. The beginning of this band is visible in aqueous solutions of sugars, but because of its weak intensity, it is largely masked by the shoulder of a more intense one. For this reason, extinction cannot be directly determined. But what is characteristic is the presence of a circular dichroism in this region, that is to say an extinction difference of $\varepsilon_L - \varepsilon_R = \Delta\varepsilon$ between left- and right-circulatory polarized light. This is a result of the presence of an asymmetric carbon adjacent to the carbonyl. Table 1.1 gives a selection of results.

Table 1.1 Circular dichroism of sugars in aqueous solution at 20°C (from Hayward and Angyal 1977) (reproduced with kind permission from Elsevier Science).

Sugar	$10^3\ \Delta\varepsilon$	λ (nm)	α-Carbon configuration
D-ribose	–0.469	285	*R*
D-galactose	–0.170	287	*R*
D-glucose	–0.0222	285	*R*
D-mannose	+0.0535	292	*S*
5,6-di-*O*-methyl-D-glucose (**1.19**)	–9.57	289	*R*
D-fructose	+6.72	273	*S*
1-deoxy-D-fructose (**1.20**)	+138	274	*S*

Table 1.1 first shows that $\Delta\varepsilon$ is positive when the configuration of the adjacent chiral center is *S*, and negative when this configuration is *R*. This is a general rule, verified by 33 examples. Next one notes the considerable difference in the order of magnitude between the non-substituted aldehyde sugars and the ketone sugars, suggesting that the concentration of the carbonyl tautomer is much higher in the latter. A higher value is also noted with 5,6-di*O*-methyl-D-glucose **1.19**. This derivative cannot exist in the form of a pyranose and it essentially has the furanose form in solution, but the considerable increase of $\Delta\varepsilon$ indicates that the difference of free enthalpy between aldehyde and furanose is less than between aldehyde and pyranose. At the moment the deoxy sugar **1.20** holds the absolute record for these values. Probably the greater natural stability of the ketone function is reinforced by suppressing the inductive effect of the alcohol function. These values cannot be used to determine carbonyl tautomers exactly because we do not have access to the $\Delta\varepsilon$ values of pure compounds. Taking the unit as an approximately plausible value, we obtain concentrations of the same order by this calculation as by other methods.

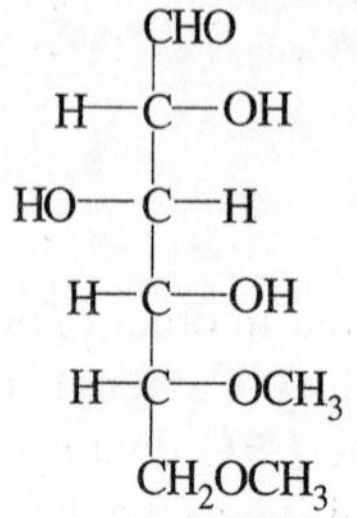

CHO
H—C—OH
HO—C—H
H—C—OH
H—C—OCH_3
CH_2OCH_3

1.19

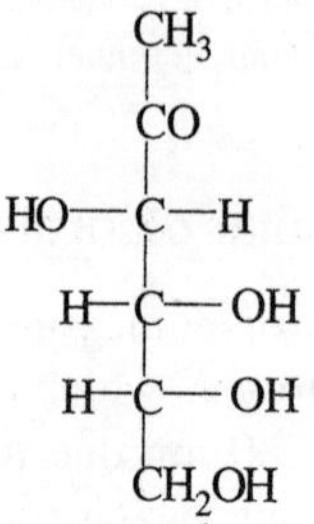

CH_3
CO
HO—C—H
H—C—OH
H—C—OH
CH_2OH

1.20

1.3.5 Nuclear magnetic resonance

Everything that has been said about glucose can be applied to other aldoses. The H-1 proton signals of different tautomers in solution in deuterium oxide appear at low field, clearly separated from the others. The proton NMR technique, as it is routinely used in most laboratories, reveals only the α- and β-pyranose signals in solutions of D-glucose and D-mannose, while in solutions of D-galactose, two other very weak peaks show the presence of two furanoses. The carbonyl signal is visible in the ^{13}C NMR spectrum in a 4M solution of fructose (Angyal 1984).

Further, ^{13}C NMR is used with sugars labelled at C-1 by a synthetic method which multiplies the signal intensity by 100. Using special accumulation and prolonged procedures, the six tautomers of D-glucose at 37°C can be observed (Fig. 1.4). Because of the disproportion of concentrations a quantitative presentation is not possible and the reader should consult Table 1.2.

The utilization of sugars labelled at C-1 has another equally important advantage in that we can observe the coupling of C-1 with the ^{13}C nuclei present in very small amounts at other positions of the sugar, which in general only show ^1J couplings. Thus we have a tool to facilitate the interpretation of the spectrum (Barker and Serianni 1986; King-Morris and Serianni 1987). A supplementary simplification is carried out using the INADEQUATE technique which only records the signals due to carbons coupled at C-1. The parameters can be

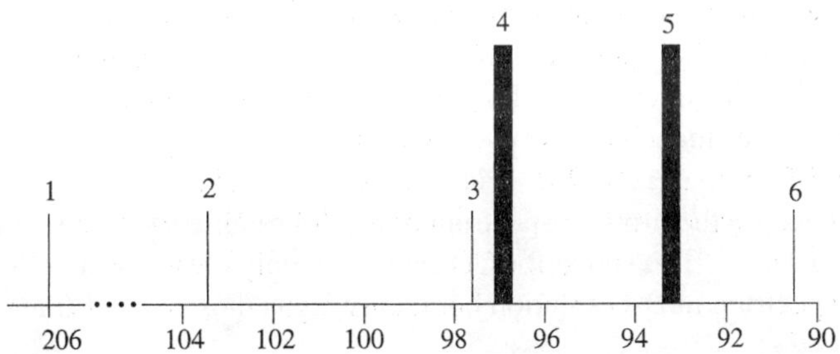

Fig. 1.4 ^{13}C NMR spectrum of [1-^{13}C]-D-glucose in water at 37°C. Abscissae: displacements in ppm from Me_4Si. Ordinate: qualitative intensities. Signal attributions: (1) aldehyde; (2) β-D-glucofuranose; (3) α-D-glucofuranose; (4) β-D-glucopyranose; (5) α-D-glucopyranose; (6) *gem*-diol (from Maple and Allerhand 1987) (reproduced with kind permission from the American Chemical Society).

Table 1.2 Tautomeric composition of sugars in D_2O according to Angyal (1984; 1991) (reproduced with kind permission from Academic Press).

Sugar	I (°C)	Pyranose α	Pyranose β	Furanose α	Furanose β	Aldehyde	Aldehydrol
D-Glucose	27	38.8	60.9	0.14	0.15	0.0024*	0.0045
D-Mannose	21	68.0	32.0				
D-Galactose	31	30	64	2.5	3.5	0.02	
D-Ribose	31	21.5	58.5	6.5	13.5	0.05	
2-Deoxy-D-erythro-pentose	30	40	35	13	12		
Fructose	31	2.5	65	6.5	25	0.8	
Fructofuranose-1,6-diphosphate (**1.21**)	6			13	86	0.9	
1-Deoxy-fructose (**1.20**)	37	4	75	6	9	6**	

*at 37°C; **ketonic tautomer.

adapted so that one can retain only the signals of carbons separated from C-1 by either one, two, or three bonds, etc. This technique removes the signals of the carbons not coupled to C-1.

1.3.6 Results and discussion

Table 1.2 displays the tautomeric composition of pentoses and hexoses in aqueous solution as well as 1-deoxyfructose **1.20**, and fructofuranose-1,6-diphosphate **1.21**. The predominance of pyranoses is observed. The galactofuranoses **1.22** are relatively more stable than the glucofuranoses **1.23**, perhaps due to the *trans* arrangement of the hydroxyl at C-3 and the side chain in the former. We also see that there are ten times as many β- as there are α-pyranoses in the fructose solutions. At this point we are jumping ahead to Chapter 2, which is devoted to problems of conformation. We can draw β-D-fructopyranose as conformation **1.24** where only one unfavorable interaction takes place between H-3 and OH-5. The exchange of substituents at C-2 leads to an eminently unfavorable conformation which shifts to **1.25**, the lesser of two evils, but where strong 1,3-diaxial interactions remain.

We draw attention to the remarkable stability of the recently confirmed (Kennedy *et al.* 1995) open chain form of 1-deoxy-D-*threo*-pentulose **1.26**. This pentulose, first discovered in culture broth of *Streptomyces hygroscopicus*, is a precursor of the thiazole moiety of thiamine, and the pyridine of pyridoxine (David and Estramareix 1996). It crystallizes, undoubtedly, as the open chain tautomer because the infrared spectrum of a solid sample shows a strong ketonic band at 1710 cm^{-1}. The signal for $COCH_3$ is not only clearly visible in a routine ^1H NMR spectrum in D_2O solution but it outweighs those α- and β-anomers, the

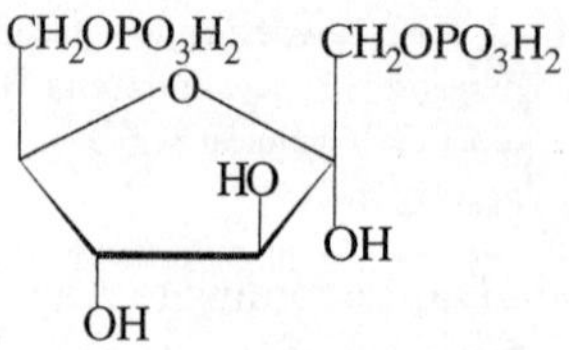

1.21 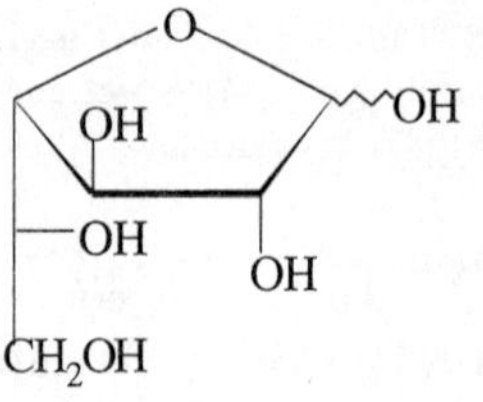 **1.22**

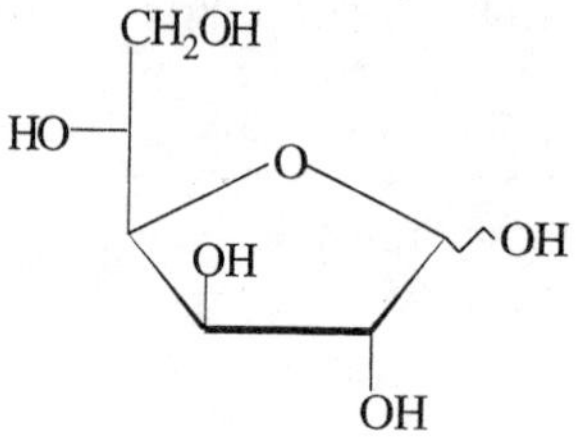

1.23 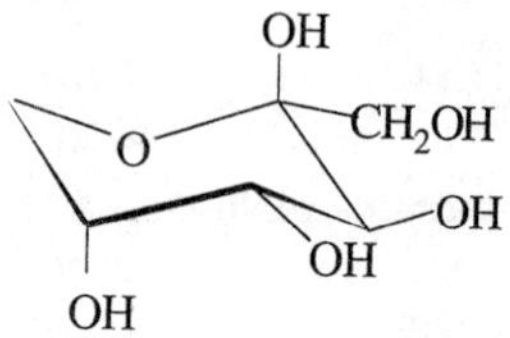 **1.24**

1.25 **1.26**

proportions being 16:15:67. Thus the *keto* form is more stable than the rings by *c.* 0.9 kcal mol^{-1} at room temperature. The conjunction of deoxygenation at C-1 and the impossibility of building a pyranose may be responsible for this.

In dimethyl sulfoxide solution, some of the sugars have nearly the same composition as in aqueous solution. Others, particularly fructose, galactose, and talose, have very different compositions. There seems to be no obvious explanation for this behaviour (Angyal 1994).

1.4 Kinetics of mutarotation

In the case of D-glucose, there are practically only pyranoses in solution. Their interconversion can be formulated as a reversible reaction with a first-order law (1.3), $C\alpha$ and $C\beta$ being concentrations (activities) of each anomer. The rate of disappearance is given by equation (1.4) which is integrated in the usual manner. A practical approach is to convert the concentration variations into optical rotation variations with a set wavelength, giving equation (1.5) where r_0 and r_∞ represent the rotations measured for $t = 0$ and $t = \infty$.

(1.3) $$\alpha\text{-D-glucopyranose} \underset{k_2}{\overset{k_1}{\rightleftharpoons}} \beta\text{-D-glucopyranose}$$

(1.4) $$-\frac{dC\alpha}{dt} = k_1[\alpha] - k_2[\beta]$$

(1.5) $$k_1 + k_2 = \frac{1}{t}\log\frac{r_0 - r_\infty}{r_t - r_\infty}$$

This formula has been checked in a good number of cases (Isbell and Pigman 1968). The rate is multiplied by a factor of about 2.5 for a temperature of 10°C which corresponds to an activation energy close to 17 kcal mol^{-1}. Sometimes, as in the case of D-galactose, we can observe an appreciable discrepancy and even, with D-ribose, a variation which is not at all linear. These abnormalities can be easily explained by the presence of more than two tautomers interconverting in solution. Mutarotation is catalysed by acids and bases and is slowest between pH

3.0 and 7.0. This is represented by a function of $[H^+]$ and $[OH^-]$ of the form $A + B[H^+] + C[OH^-]$ which gives, for example, equation (1.6) for glucose at 20°C.

$$k_1 + k_2 = 0.0060 + 0.18[H^+] + 16\,000\,[OH^-] \tag{1.6}$$

The observation of mutarotation only allows the sum $k_1 + k_2$ to be known. At any rate $k_1 + k_2$ are composite constants. We have every good reason to believe that the anomeric equilibrium occurs via the transient cabonyl (Fig. 1.5). Among other indications, the oxygen carried by the anomeric carbon does not exchange with water during the process.

Each of the partial equilibria of Fig. 1.5, represented by a formula of the type (1.7) brings about two rate constants k_o and k_c corresponding to the opening and closure of the ring, respectively. Their ratio $K = k_c/k_o$ is the equilibrium constant, measurable in the NMR spectra if the carbonyl signal can be seen with enough precision.

$$\text{cyclic tautomer} \underset{k_c}{\overset{k_o}{\rightleftharpoons}} \text{carbonyl tautomer} \tag{1.7}$$

We can measure k_o and k_c in a certain number of cases (Barker and Serianni 1986). When the rate constants are in the order of 10–200 s^{-1}, the method for measuring line broadening in the NMR spectra, according to Gutowsky and Holm, is applicable. For lower values from 0.05 to 10 s^{-1}, another method can be applied, usable in proton NMR as well as ^{13}C NMR, that is the transfer of saturation between two nuclei. One irradiates to saturation at the frequency of the carbonyl of the acyclic tautomer. The ring-closure to the hemiacetal changes its environment and it becomes the hemiacetal carbon but does not contribute to the intensity of the carbon signal. Experimentally, by prolonging the irradiation at the carbonyl frequency resonance, we observe a lowering in the intensity of the hemiacetal resonance. The latter becomes stabilized at a final level which depends on the relaxation of the hemiacetal site and the rate of the ring opening. One formula allows the opening rate to be extracted and the method is usable in the 0.05–10 s^{-1} zone. By way of example, we have shown in Fig. 1.6 the variations with pH of k_o for **1.27** and **1.28**, the two anomers of a very important metabolite, D-ribose 5-phosphate, obtained with the labelled molecule D-[1-^{13}C]ribose 5-phosphate. The reader will observe the order of magnitude of k_o. Moreover, we can observe much higher values with other sugars, even at pH

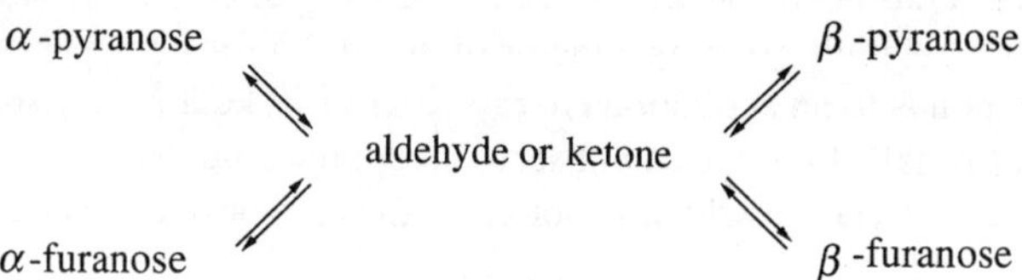

Fig. 1.5 Tautomeric equilibrium in solution.

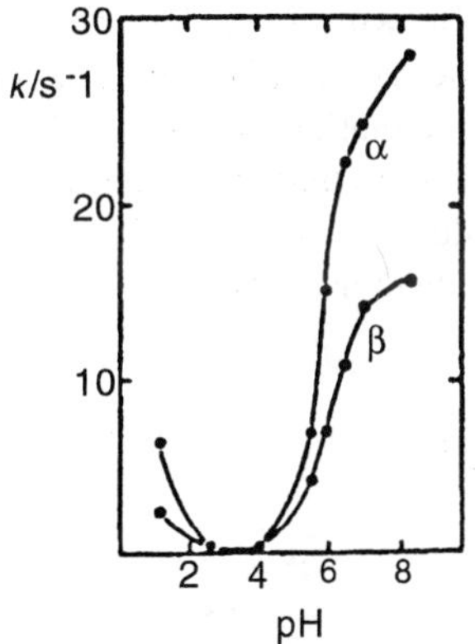

Fig. 1.6 Variation of the opening constant of α- and β-furanoses, **1.27** and **1.28**, as a function of pH in 0.3 M solution of 15% ^2H_2O at 24°C (Barker and Serianni 1986) (reproduced with kind permission from the American Chemical Society, © 1986).

$CH_2OPO_3H_2$ — O — R' — R — OH — OH

1.27 R = OH, R' = H

$CH_2OPO_3H_2$ — O — R' — R — OH — OH

1.28 R = H, R' = OH

7.5. As to the constant k_c, it is obviously much higher since K is generally much higher than 1.

1.5 General remarks

The reader may be tempted to think that the experimental details described in the preceding paragraphs requiring difficult techniques are essentially of academic concern. In fact, in the field of organic chemistry, a number of sugar reactions are most easily explained by assuming that the carbonyl tautomer is in rapid equilibrium with the dominating rings. Typical carbonyl reactions are observed. Moreover, even approximate measures of aldehyde or ketone concentrations allow, by applying Gibbs' expression, the magnitude order of the excess free enthalpy to be estimated in relation to the cyclic forms, that is around 6 kcal mol^{-1} for glucose.

As regards the situation in living cells, the anomeric configuration of free sugars is probably not unimportant since Nature foresaw an enzyme, the *mutarotase*

(Aldose-1-*epimerase*), quite widespread in animal tissue and bacteria, which catalyses mutarotation. The *Escherichia coli* enzyme has a maximum activity close to neutral pH. The activation energy $\Delta G^{\neq} = 11.9$ kcal mol^{-1} is greatly lowered, as usual with respect to that of the non-enzymically catalysed reaction, close to 17 kcal mol^{-1}. D-Glucose, D-galactose, and D-fucose are substrates but not D-mannose (Hucho and Wallenfels 1971).

(1.8)

$$\begin{array}{c} CH_2OPO_3H_2 \\ | \\ CO \\ | \\ HO-C-H \\ | \\ H-C-OH \\ | \\ H-C-OH \\ | \\ CH_2OPO_3H_2 \end{array} \rightleftharpoons \begin{array}{c} CH_2OPO_3H_2 \\ | \\ CO \\ | \\ CH_2OH \\ + \\ CHO \\ | \\ H-C-OH \\ | \\ CH_2OPO_3H_2 \end{array}$$

Table 1.2 shows that the proportion of the carbonyl form, in this case ketonic, is much greater with D-fructose and its diphosphate **1.21**. In the acyclic form both have a characteristic difunctional group, the β-hydroxy carbonyl (aldol or ketol). One of the features of this group is its cleavage by a reversible reaction in the presence of purely chemical catalysts. The reaction of the diphosphate **1.21**, written according to equation (1.8), is catalysed by the enzyme *aldolase*, and this is a major pathway to creating carbon–carbon bonds in cells.

References

Angyal, S. J. (1984), *Adv. Carbohydr. Chem. Biochem.*, **42**, 15–68; (1991), **49**, 19–35.

Angyal, S. J. (1994), *Carbohydr. Res.*, **263**, 1–11.

Barker, R. and Serianni, A. S. (1986), *Acc. Chem. Res.*, **19**, 307–313.

Brown, G. M. and Levy, H. A. (1979), *Acta Crystallogr.*, **B35**, 656–659.

Chu, S. S. C. and Jeffrey, G. A. (1968), *Acta Crystallogr.*, **B24**, 830–838.

David, S. and Estramareix, B. (1997), *Adv. Carbohydr. Chem. Biochem.*, submitted.

Hayward, L. D. and Angyal, P. J. (1977), *Carbohydr. Res.*, **53**, 13–20.

Hicks, K. B. (1988), *Adv. Carbohydr. Chem. Biochem.*, **46**, 17–72.

Honda, S., Suzuki, S., and Kakehi, K. (1984), *J. Chromatogr.*, **291**, 317–325.

Hucho, F. and Wallenfels, K. (1971), *Eur. J. Biochem.*, **23**, 489–496.

Isbell, H. S. and Pigman, W. (1968), *Adv. Carbohydr. Chem.*, **23**, 11–57; (1969), **24**, 13–65.

Kennedy, I. A., Hemscheidt, T., Britten, J. F., and Spenser, I. D. (1995), *Can. J. Chem.*, **73**, 1329–1337.

King-Morris, M. J. and Serianni, A. S. (1987), *J. Am. Chem. Soc.*, **109**, 3501–3508.

Maple, S. R. and Allerhand, A. (1987), *J. Am. Chem. Soc.*, **109**, 3168–3169.

2 Conformation of monosaccharides and their derivatives

2.1 Conformation symbols: pyranoses

The conformations of the oxane ring (tetrahydropyran) of pyranoses are the same as those of cyclohexane. The carbons are numbered starting with the hemiacetal carbon, referred to as anomeric. This convention is not in keeping with the rule for numbering heterocycles where number one is assigned to the heteroatom (in this case, the oxygen). The oxane is represented with carbons 1, 3, and 5 in the horizontal plane, carbons 1 and 4 in the plane of the vertically positioned paper, and the cyclic oxygen behind the paper. To the viewer situated above the ring the numbers appear clockwise. In a pyranose sugar, all the carbons, or nearly all of them, are substituted, but for practical purposes, one only needs to introduce a substituent R to an arbitrary site. Equation (2.1) then represents the extension of the classic conformational equilibrium of cyclohexane to pyranoses.

(2.1)

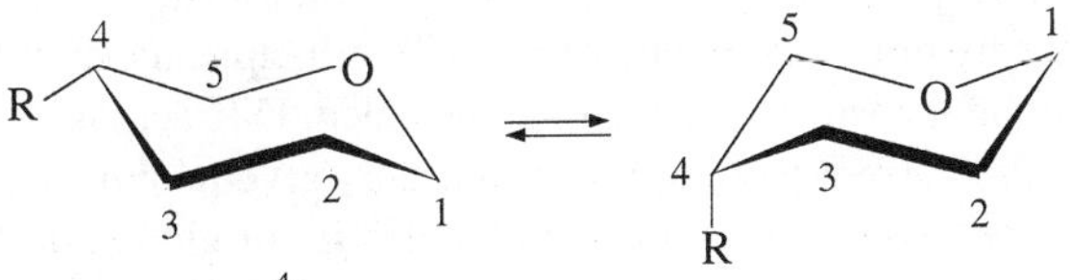

2.1 ($^4C_1$) ⇌ **2.2** ($^1C_4$)

Conformation **2.1** is symbolized by $^4C_1$ which indicates that in the conventional representation, carbons 1 and 4 are below and above the average reference plane of the molecule, respectively. The symbol for conformation **2.2** is thus $^1C_4$. Likewise, the enantiomer **2.3** of pyranose **2.1** gives rise to a conformational equilibrium (2.2), symmetrical to the preceding one, to which the symbols $^1C_4$ and $^4C_1$ correspond according to our convention.

This leads us to a surprising result: the same symbol $^4C_1$ is attributed to conformations **2.1** and **2.4**, neither superimposable nor symmetrical. The symbols

(2.2)

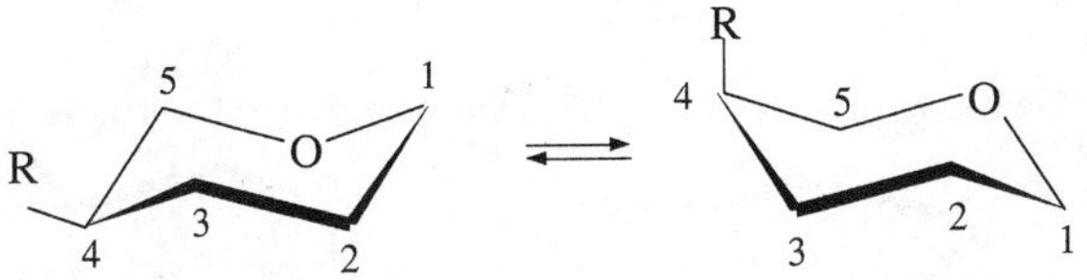

2.3 ($^1C_4$) ⇌ **2.4** ($^4C_1$)

^iCj have no meaning if we do not know the D- or L-series of the pyranose, which must be introduced to avoid any ambiguity. If the schematic sugar **2.1** belongs to the D-series, the correct symbols for conformations **2.1**, **2.2**, **2.3**, and **2.4** are thus D-$^4C_1$, D-$^1C_4$, L-$^1C_4$, and L-$^4C_1$, respectively. Note that while the enantiomer of a molecule in the D-$^4C_1$ conformation is a molecule in the L-$^1C_4$ conformation, they both behave identically in any achiral environment.

The pyranoses containing a double bond in the ring or a fused oxirane ring, important intermediates in synthesis, exist in half-chair conformations. We will give their symbolic descriptions as they are dealt with.

Finally, there are intermediary conformations between the chair and the skew. Their symbolic descriptions will be given by using the most important example, L-iduronic acid (see Section 2.8).

2.2 Conformations in solid state

It is obvious that the elucidation of the structure of a crystalline sugar gives, at the same time, its absolute configuration and its conformation. Considerable technical progress carried out on the construction of diffraction analysers have made structure determination by X-ray not only faster and faster but also more easily accessible to the non-specialist. Occasionally, X-ray spectra have been associated with neutron diffraction spectra (X, N) which give geometric sizes with more precision, allowing the hydrogens to be located and, in principle, the distribution of the electronic density in the valence layers to be revealed. Diffraction methods are the most precise of the current techniques in that they give lengths, torsion angles, and valency angles. Nonetheless, only molecules in a rigid crystalline network can be observed. A good number of sugar structures in solid state are known and have been regularly compiled in the periodical *Advances in Carbohydrate Chemistry and Biochemistry*. There the reader will find a critical investigation of results from the crystallographer's point of view (Jeffrey and Sundaralingam 1974; 1975; 1976; 1977; 1980; 1981; 1985).

However, the preparation of an appropriate crystal can prove to be more difficult than the spectroscopy itself. Naturally, sugars are typically a 'highly crystalline' family. But in order to purify them, contemporary chemists would sooner rely on the more systematic and powerful chromatographic methods than on the uncertain search for the ideal solvent for crystallization. There is also a more fundamental problem in that a conformation in the crystal may not be the

(2.3)

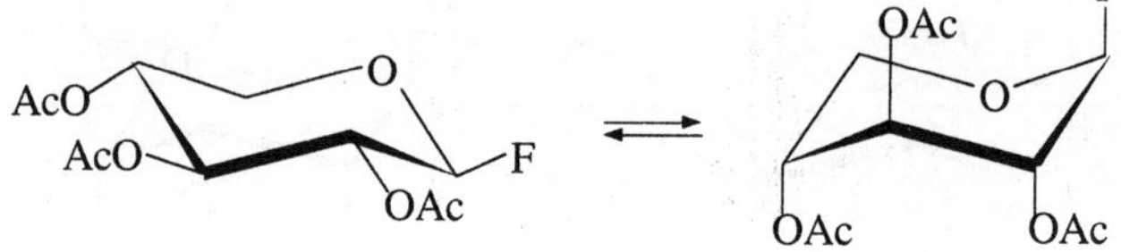

AcO AcO O F OAc ⇌ OAc O F OAc OAc

2.5 **2.6**

same as in solution. For example, let us consider the conformational equilibrium (2.3) of an acetylated derivative of β-D-xylopyranosyl fluoride.

This 2,3,4-tetra-*O*-acetyl-β-D-xylopyranosyl fluoride adopts the tetra-equatorial conformation **2.5** in the crystal, whereas in solution it gives rise to a conformational equilibrium in which the tetra-axial conformation **2.6** is strongly predominant (80–90%) (Paulsen 1979). The tetra-equatorial conformation **2.5** thus has an energy difference over conformation **2.6** (which can be calculated by employing the Gibbs equation) equal, at least, to 0.8 kcal mol^{-1}. This slight difference is compensated for when the crystal develops whereby it selects the tetra-equatorial conformation from the solution and displaces the equilibrium totally towards the left. The greater planarity of conformation **2.5** favours, perhaps, compact stacking.

Nevertheless, when only one conformation of a molecule appears in solution, in the majority of cases using contemporary investigating methods, this is what is found in the crystalline form.

2.3 Conformation in solution: proton nuclear magnetic resonance

In the first chapter we have already discussed the use of NMR for the investigation of anomeric equilibria by analysing that part of the spectrum related to anomeric protons. In general, the analysis of a 250 MHz spectrum of a monosaccharide presents no difficulty. The vicinal coupling constant values between axial and *gauche* protons are in the order of 8–11 and 1–3 Hz, respectively. The axial–equatorial values are higher than equatorial–equatorial ones, which are often close to zero. When the configuration of a pyranosyl derivative is known, we can generally find a pair of *trans*-related vicinal protons. Whenever their coupling constant value can be measured in the spectrum, the conformation is determined without ambiguity if it is close to one of the extreme values given above. It is pointless to calculate the torsion angles with high precision using the Karplus relationship, and this kind of information is not necessary to predict reactivity, for example.

A few examples will indicate the characteristics detectable in the spectra of the monosaccharides reproduced in Figs 2.1, 2.2, and 2.3. Methyl β-D-galactopyranoside **2.7** is representative of an internal galactose residue of the glycolipid chains. Methyl α-L-fucopyranoside **2.8** in the L-$^{1}C_{4}$ conformation represents an epitope branch (see Chapter 16) of antigens of the major blood groups. Sialic acid **2.9** in the D-$^{2}C_{5}$ conformation plays an important role in recognition phenomena. The numerical values of chemical shifts and coupling constants are grouped together in Table 2.1.

In certain cases we observe intermediate values of the coupling constants for various reasons.

One reason is that the conformation moves significantly away from the classic chair shape. This is what we see with the bis-ketal **2.10**, 'diacetone-galactose' (1,2:3,4-di-*O*-isopropylidene-α-D-galactopyranose). It has been drawn as an ordinary alicyclic compound in order not to prejudge its conformation.

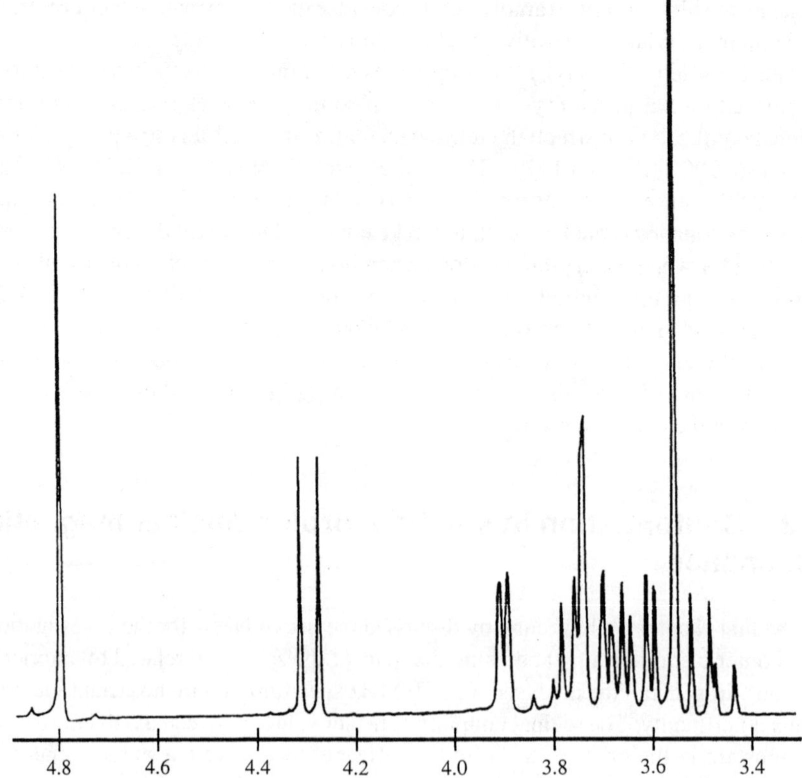

Fig. 2.1 Proton NMR spectrum at 250 MHz of methyl β-D-galactopyranoside in D_2O.

2.7

2.8

2.9

2.10

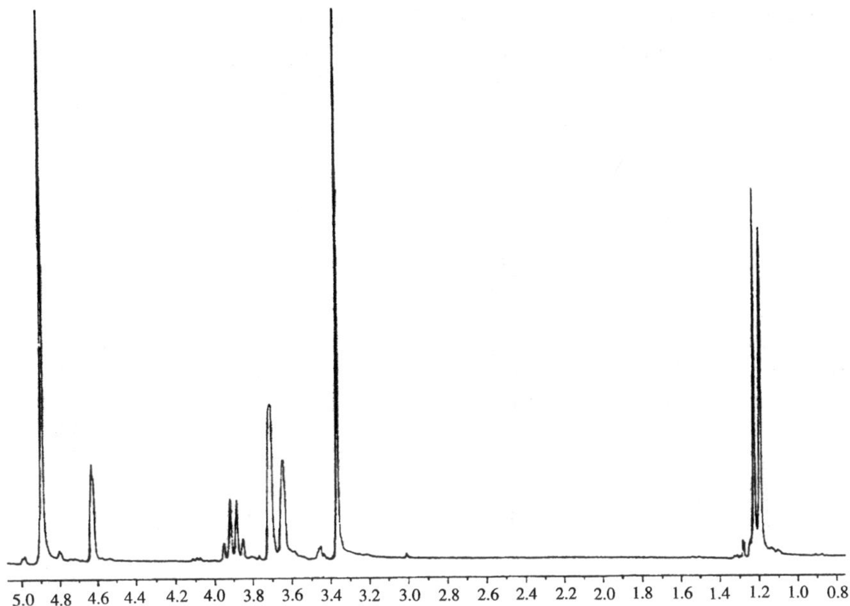

Fig. 2.2 Proton NMR spectrum at 250 MHz of methyl α-L-fucopyranoside in D_2O.

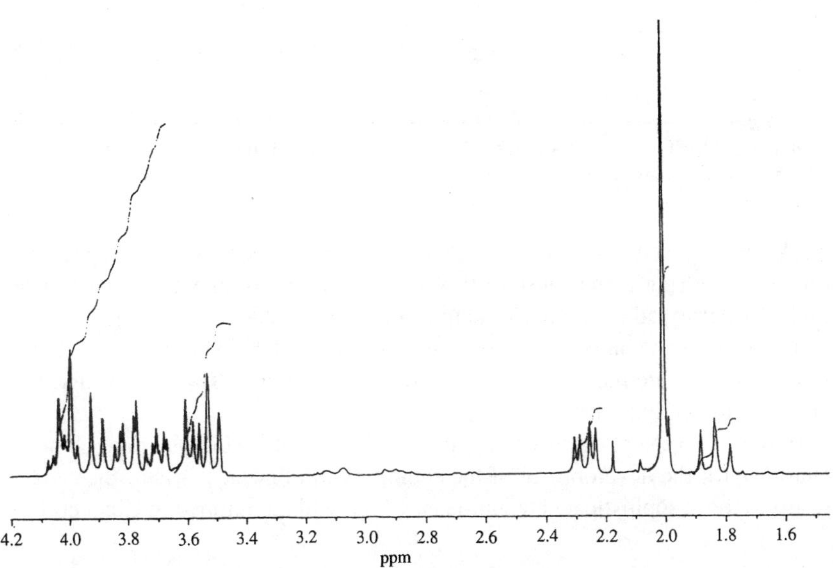

Fig. 2.3 Proton NMR spectrum at 250 MHz of *N*-acetylneuraminic acid.

In $CDCl_3$, the coupling constant values are $J_{1,2}$ 5.0, $J_{2,3}$ 2.4, $J_{3,4}$ 8.0, and $J_{4,5}$ 1.4 Hz. The reader may check that these values are not compatible with the D-$^4C_1$ conformation. An intermediate conformation between the skew and boat forms

Table 2.1 NMR data for **2.7, 2.8**, and **2.9**. Below δ (in ppm from Me_4Si), the $^2J, J_{n,n+1}$ coupling constants are given in this order between parentheses.

Protons	**2.7***	**2.8****	**2.9***
H-1	4.30 (8)	4.60 (1)	
H-2	3.47 (10)	3.71	
H-3ax	3.62 (3.5)	3.71	1.88 (13) (12)
H-3eq			2.32 (13) (4.5)
H-4	3.90 (1)	3.65	4.07 (10)
H-5	~3.75 (4.9) (7.4)***	3.85 (6.5)	3.93 (10)
H-6	3.75 (–12)***	1.20	4.07 (1)
H-7			3.55 (9)
H-8			3.75 (6) (2.5)
H-9			3.62 (11.5)
H-9′			3.84 (11.5)
N-acetyl			2.05
O-methyl	3.55	3.38	

*solvent D_2O, HOD peak at 4.8 ppm; **solvent CD_3OD, HOD peak at 4.85 ppm; ***calculated values (from Welti 1977).

has been proposed (Cone and Hough 1965). The fusion of two pentagonal rings on the oxane ring is the cause of this distortion. This is an extreme case as less radically deformed chair conformations are also noted.

The second reason is that there is an equilibrium between several conformations. What we are measuring is thus a weighted average. This will be discussed at length in Section 2.6.

Here we have spoken only about the application of NMR spectra to monosaccharides; for the development of the chemistry of oligosaccharides one needs to resort to more sophisticated techniques, which will be outlined in Chapter 9.

2.4 General comments on the conformational features of monosaccharides

As shown in example **2.10** of the preceding section, the mechanical constraints introduced by the fusion of oxane with other rings have a dominating influence on the conformation. We will deal with similar cases as they appear along in this work. In the rest of this chapter we will only look at monocyclic compounds.

When two chair conformations of a substituted cyclohexane are at equilibrium, the excess free enthalpy of the less stable conformation is calculated as the difference between the sum of two terms, that is to say, the 1,3-diaxial and 1,2-*gauche* interactions of each conformer. We assume, therefore, a law of additivity of steric crowding. This semi-quantitative treatment loses part of its meaning with pyranoses. All the carbons are functional which should facilitate the diffusion of interactions from one end of the molecule to the other and diminish the plausibility of using the addition of independent contributions. Moreover, with non-branched sugars, those with which we will deal essentially, the substituents are most often hydroxy, acetoxy, and benzoyloxy groups. The bulkiness of the hydroxyl varies with the degree of solvation. Acylation diminishes its volume in an unpredictable fashion by displacing the electronic density towards the carbonyl. Finally, practically every position has, to a certain extent, a particular status. The anomalies are very pronounced at position 1 of all pyranoses and at position 5 of hexopyranoses. These will be discussed in detail in Sections 2.5 and 2.6. Positions 2, 3, and 4 remain to be seen. In 4-acetoxyoxane **2.11**, the axial substituent undergoes the traditional steric strain due to two axial C–H bonds at positions 2 and 6. On the other hand, in 3-acetoxyoxane **2.12** (and in the 5-acetoxy compound), the substituent interacts with only one axial C–H bond, the other position being occupied by the endocyclic oxygen. Compound **2.12** leads to a conformational equilibrium in which there is nearly the same proportion of conformations **2.12** and **2.13**. The conformational energy value is close to zero for this position.

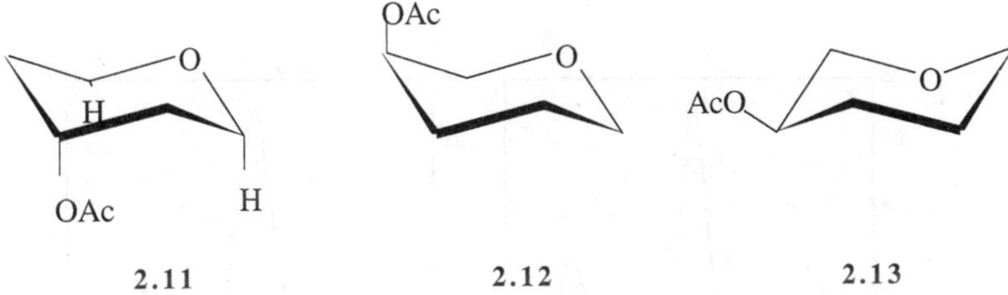

In fact, the conformation of pyranoses is dominated by two effects, not present in the cyclohexane, which appear at positions 2 and 6 of the oxane. One of them is characteristic of hexopyranoses and I propose that we call this the 'coplanar effect' in order not to imply a particularly restrictive structure by using the name of an effect already present in methoxyethane. The other effect, present in all pyranoses, is referred to as anomeric. This name, taken from the nomenclature of sugars because it was first recognized in this family, in fact disguises its general nature since it is also present in methyl chloromethyl ether. The consequences of these effects can be modulated by cyclohexane-type interactions, but not to the point where more than a qualitative discussion is necessary.

2.5 Coplanar effect

It is well known that butane has two favoured conformations represented as **2.14** (*anti*) and **2.15** (*gauche*) according to a Newman projection perpendicular to the C-2–C-3 bond.

2.14 **2.15**

In Fig. 2.4, the population of the corresponding conformations (part I*a*) and the energy variation of the molecule as a function of the MeC–CMe torsion angle (part I*b*) are given by *ab initio* calculations (Jorgensen *et al.* 1981). Figure 2.4 (part II) shows the corresponding variations for methoxyethane (Jorgensen and Ibrahim 1981), whereby we can anticipate, by analogy, two preferred conformations, **2.16** and **2.17**. What is important to recognize in the curves of Fig. 2.4 is that the excess energy of the *gauche* form over the *anti*, around 0.70 kcal mol^{-1} in butane, increases to 1.96 kcal mol^{-1} in methoxyethane. Consequently, the population of the *gauche* conformation is extremely low. Here we have the simplest possible example of the considerable stability of the *anti* conformation due to the presence of oxygen. The name 'coplanar effect' given to this phenomenon reminds us of the exaggerated tendency of the CMe bond to remain in the C–O–C plane.

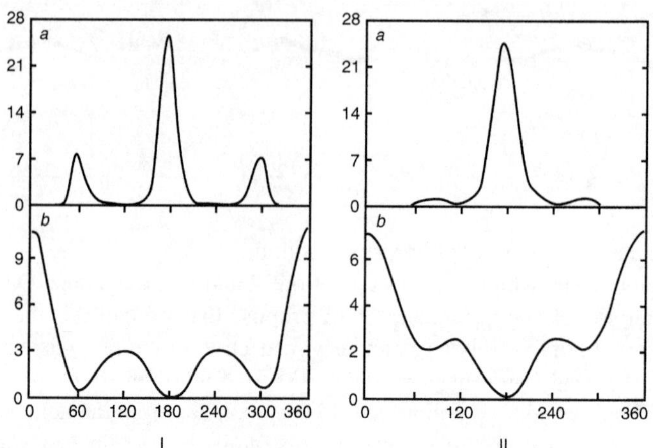

Fig. 2.4 Theoretical quantitative data on the conformational equilibrium of butane (I) and methoxyethane (II). Abscissae: MeC–CMe or MeC–OMe torsion angles; ordinates: (a) 10^3 molar fractions by angle degree; (b) kcal mol^{-1} (from Jorgensen *et al.* 1981; Jorgensen and Ibrahim 1981) (reproduced with kind permission from the American Chemical Society).

Let us move on to the case of oxane (Eliel *et al.* 1982), which gives three types of monomethylated derivatives, **2.18**, **2.19**, and **2.20**.

For a monosubstituted oxane, the excess free enthalpy of the conformation with an axial substituent over the conformation with an equatorial substituent is, as we know by definition, the conformational free energy (CFE) of the substituent in oxane at this position. These values, possibly measured indirectly by utilizing intermediate compounds, are shown in Table 2.2. Equatorial conformations correspond to *anti* conformations in butane and methoxyethane, and the axial conformations (not represented) to *gauche* conformations. We can observe that the environment of derivative **2.20** is closest to that of the cyclohexane and that the CFE is of the same order. On the other hand, the presence of the cyclic oxygen lowers notably the CFE of derivative **2.19**. The important point is the noteworthy increase in the CFE of compound **2.18**, where the methyl group is close to the cyclic oxygen and possesses, on one side, an environment similar to that of methoxyethane. Let us look at the equilibrium (2.4) of the dimethylated derivative **2.21**.

(2.4)

Table 2.2 Conformational free energies of substituted oxanes (between 163–183 K in chlorinated solvents) (from Eliel *et al.* 1982) (reproduced with kind permission from the American Chemical Society).

Substituent	$-\Delta G°$/kcal mol^{-1}
2-CH_3	2.86
2-CH_2OH	2.89
3-CH_3	1.43 ± 0.04
4-CH_3	1.95 ± 0.05

This compound adopts nearly exclusively conformation **2.21b**. We find that

$$k = \frac{[2.21b]}{[2.21a]} = 86.0$$

which corresponds to $-\Delta G^{\ominus} = 1.62$ kcal mol^{-1}. If we assume that $-\Delta G^{\ominus}$ represents the conformational energy difference of the methyl groups at positions 2 and 4, we find that the CFE of the methyl group at position 2 is 1.62 + 1.43 = 3.05 kcal mol^{-1}. The lower value found in Table 2.2 results from an indirect calculation using more measurable equilibria, for here the k value is very high inducing an unprecise measure for the concentration of **2.21a**. The CFE of a CH_2OH side chain at position 2, the usual arrangement of the hexopyranoses, was calculated in the same way.

The coplanar effect did not create as much excitement with the theoreticians as did the anomeric effect and its cause is not known with certainty. The simplest explanation is that the 1,3-diaxial interaction of a methyl group at position 2 of the oxane with the C–H bond at position 6 is increased because these two substituents are closer than if they were separated by $-CH_2-$ instead of –O–. The calculation for the crystalline α-D-glucopyranose, with carbon–oxygen bonds of 1.439 and 1.427 Å making an angle of 113.7° between them, gives 2.400 Å for the C-1–C-5 distance, whereas the corresponding value for cyclohexane is at least equal to 2.5 Å. It is a well-known fact that steric strain increases rapidly as the internuclear distance decreases.

2.6 Anomeric effect

2.6.1 Experimental data

The anomeric effect, still an object of active research (Kirby 1983; Juaristi and Cuevas 1992; Thatcher 1993), was first observed as a property of pyranose sugars (Edward 1955; Lemieux 1964). However, the anomeric effect emerges in a pure state on a very simple molecule, methyl choromethyl ether (CH_3OCH_2Cl), represented as projected along the $O–CH_2Cl$ bond of formula **2.22**.

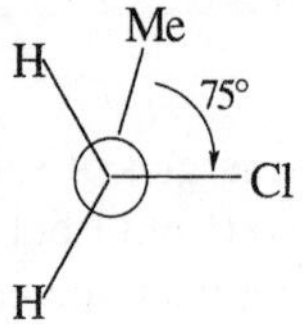

Me
H
75°
Cl
H

2.22

The known conformation is that of the molecule in the gas phase, thus isolated, as determined by electron diffraction (Planje *et al.* 1965). Instead of adopting the favored *anti*-position of butane, the carbon–chlorine bond defines a torsion angle of 75° with the OMe bond. It is nearly coplanar to the orbital axis of the 2p lone pair of the oxygen, the gap (15° in the projection) probably being due to a non-bonding interaction between methyl and hydrogen. The carbon–chlorine bond (1.813 Å) is longer than with the chloroalkanes, and the O–CH_2Cl bond (1.368 Å) is shorter than with the aliphatic ethers and the CH_3–O bond (1.414 Å). Finally, with methyl chloromethyl ether in solid state we observe a ^{35}Cl quadrupolar resonance frequency that is exceptionally low (29.817 MHz) compared to that of 1-chloropropane (32.968 MHz), which indicates an increase in the 3p orbital population in the direction of the bond or, in less precise terms, an increase in the chlorine ionicity.

The same conformational effect is found in 2-halooxanes (**2.23**, X = Cl, Br, I). These compounds only exist in the conformation **2.23a**, where the halogen is axial. This corresponds to the *gauche* conformation of methyl chloromethyl ether, of which they are the cyclic analogues, taking into account the constraints exerted by the ring according to equation (2.5).

(2.5)

2.23a **2.23b**

Until the present we have become accustomed to the idea that a bulky substituent imposes a six-membered ring conformation where this substituent is equatorial. The tendency is thus opposite in the α-position of an ether. This phenomenon can be observed on sugar derivatives with a halogen or, more generally, an oxygen atom at C-1. We will try to evaluate the anomeric effect starting from an equilibrium such as (2.5), relative, this time, to the most common pyranose.

If A_x is the conformational free energy of X

$$\text{anomeric effect} = \Delta G^{\ominus}_{x} + A_x$$

Unfortunately, A_x at position 2 of an oxane is not measurable for a substituent with an anomeric effect because steric repulsion cannot be separated experimentally from this effect. Let us look at the CFE in cyclohexane. The example of the methyl group, without the anomeric effect (see Section 2.5), leads us to suppose that the repulsion is greater at this position. We obtain values lower than they are in reality. There are other definitions, but none can escape criticism.

The anomeric effect of halogens is too powerful for a conformation other than **2.23a** to be observed in 2-halooxanes. The evaluation is not based on a conformational equilibrium but rather on an equilibrated chemical reaction (2.6), the inversion of configuration at C-1 of the *cis*- (**2.24c**) and *trans*- (**2.24t**) 2-halo-4-methyloxanes catalysed by HCl (Anderson and Sepp 1967).

A mixture (97:3) in which the *trans*-derivative **2.24t** (X = Cl) with the axial chlorine atom predominates, is always observed ($-\Delta G^{\ominus} = 2.15$ kcal mol^{-1}). The

(2.6)

2.24c ⇌ 2.24t

numerical value of the anomeric effect is obtained by adding the CFE of chlorine (0.5 kcal mol^{-1}), which finally gives 2.65 kcal mol^{-1} for the pure liquid. Likewise, the value of the anomeric effect was estimated from glycosidation equilibria (Bishop and Cooper 1963). Table 2.3 shows a few results (Durette and Horton 1971; Aebischer *et al.* 1983). The order given by these values is, without a doubt, quite important.

As just mentioned, a more significant value for the anomeric effect of a polar substituent could be calculated if the *A* value at position 2 of oxane were known. But this can be measured only for weakly polar substituents such as methyl, hydroxymethyl, vinyl, and ethynyl, which are supposed to exhibit no anomeric effect. For such substituents, the *A* value at position 2 of oxane correlates fairly well with the conformational free energy in cyclohexane.

The relationship, A(oxane) = 1.53 A(cyclohexane) + 0.02, should also be valid for polar substituents if it expresses only an effect of bulkiness, due to the greater steric constraints at position 2 of oxane. It may be used to calculate their correct *A* value at this position and therefore derive a more significant value for the anomeric effect. Thus the minimum value of *O*-methyl is recalculated as 2.1 kcal mol^{-1}, a figure 60% higher than that in Table 2.3 (Franck 1983).

Some basic nitrogen substituents at C-2 of oxane show a tendency to adopt the equatorial position on protonation which, on the basis of the cyclohexane *A* values, looked abnormally high. This was called the *antianomeric effect.* However, when the corrected *A* values are considered, the shift toward the equatorial position is a perfectly normal consequence of the extra bulk introduced by the hydrogen atom (Franck 1983). The experimental proofs for the antianomeric effect could not be confirmed using contemporary techniques (Fabian *et al.* 1994).

Table 2.3 Numerical evaluation of the anomeric effect

Substituent	kcal mol^{-1}
Hydroxy	0.9–1.35
Methoxy	1.3
Acetoxy	1.4
Fluorine	?
Chlorine	2.7
Bromine	3.2
Iodine	3.1
Nitro*	3.4

(*from Aebischer *et al.* 1983).

2.6.2 Origin of the anomeric effect

In chloro, bromo, and iodo compounds, we observe one or several absorption lines in the Hertzian spectrum whose frequencies are characteristic of the bonding state of the halogen. This comes from the fact that atomic nuclei, ^{35}Cl for example, have a quadrupolar momentum which can have several energy levels in an electric field gradient. The Townes and Dailey equation links this resonance frequency directly to parameters describing the bond in molecular orbital terms, the *a* population of the p_z orbital of the halogen invovled in the carbon–halogen bond and the average *b* populations of the p_x and p_y perpendicular orbitals, here roughly equal to 2. For the ^{35}Cl nucleus, this equation can be written

$$\nu/\text{MHz} = 55(2 - a)$$

The resonance frequency decreases as the $3p_z$ population in the direction of the bond increases. Finally, no resonance is observed with an ionic compound ($a = 2$). To explain it more loosely, the more ionic the bond, the lower the resonance frequency.

The exceptionally low resonance of methyl chloromethyl ether was explained as a consequence of the delocalization of the $2p_z$ orbital of the high-energy electron pair at oxygen in the antibonding orbital σ^*_{CH} of the carbon–chlorine bond (Lucken 1959) (Fig. 2.5).

This implies a quasi-parallelism between the axes of the two orbitals. An angle, $\theta = 15°$, $\cos \theta = 0.97$, is observed. In the *anti*-conformation, the two orbitals would be orthogonal with no interaction. The introduction of electrons furnished by oxygen makes the $3p_z$ orbital more populated and hence lowers the resonance frequency, but since it involves antibonding electrons, the carbon–chlorine bond is weakened and lengthened. On the other hand, the carbon–oxygen bond, in which two p orbitals with parallel axes participate, takes on a certain π character which shortens it. The delocalization hypothesis explains satisfactorily all the behaviour of methyl chloromethyl ether.

Subsequent studies are more directly related to pyranose sugars. An important family of derivatives is known, the pyranosyl halides, in which the alcohol hydroxyls are acylated (generally acetylated) and the hemiacetal hydroxyl is

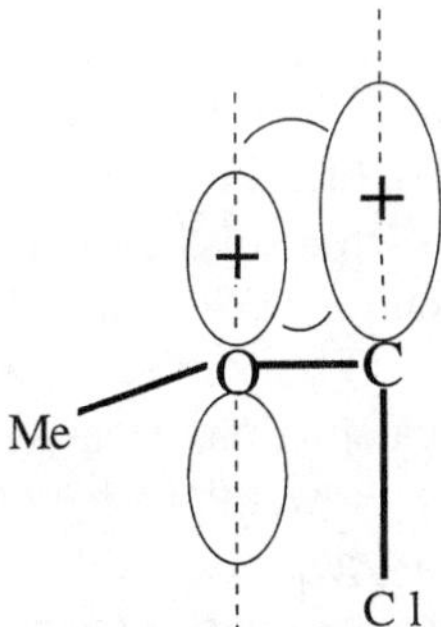

Fig. 2.5 Orbital delocalization in methy chloromethyl ether.

replaced by fluorine, chlorine, or bromine. Formulas **2.25** and **2.26** give prototypes of the D-*gluco* series.

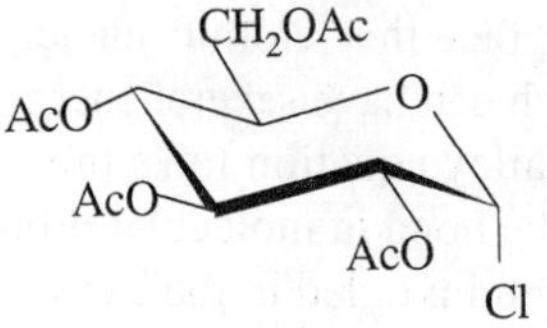

2.25 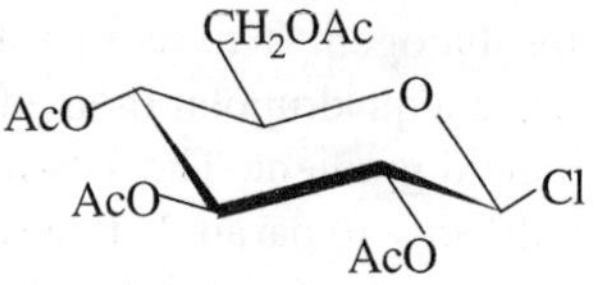 **2.26**

Starting from pentoses and hexoses having varied conformations, we can prepare two collections of pyranosyl chlorides, with axial or equatorial chlorine atoms, the analogues of **2.25** and **2.26**, respectively. As with 2-chlorooxane, the axial orientation of the chlorine in **2.25** corresponds to the stable conformation of methyl chloromethyl ether. A comparison of the geometrical data in solid state, when available, shows that the axial carbon–chlorine bond is invariably longer than the same bond in an equatorial position. Finally, Fig. 2.6 shows that, according to quadrupolar resonance spectroscopy, these compounds are divided into two groups.

The resonance frequency of the axial chlorine is always lower than that of the equatorial chlorine (David 1979). Figure 2.6 suggests another point of view: the dispersion of equatorial resonances, close to 0.5 Hz, has an order of magnitude, called 'the crystal effect' by the specialists, of intermolecular origin. When first analysed, they should not be considered significant. On the other hand, the variation range of the axial resonances, 1.7 MHz, is quite superior to the crystal effects. This dispersion expresses the before-mentioned fact that the anomeric effect of chlorine (as with any other substituent) in a pyranose is not independent of the configuration of the rest of the molecule. Thus, the resonance of the axial D-*manno* chloride is by far the lowest and it is well known that the anomeric effect is intensified in α-D-*manno* derivatives.

The theory also explains the increase in the effect of the order of chlorine, bromine, and iodine as well; the σ^* orbital is more and more diffuse and the overlap with the $2p_z$ orbital of oxygen is more and more efficient. The atomic polarizabilities of halogens are as follows: fluorine, 0.557; chlorine, 2.18; bromine, 3.05; iodine, 4.7. The anomeric effect of fluorine should be the weakest because of the compact nature of its orbitals. It has not been measured but it is unquestionable. It imposes 85% of the tetraaxial conformation on derivative **2.6** in solution. The comparison of the solid structures **2.5** and **2.27** is indicative. The length of the axial C–F bond is 1.386 Å and that of the equatorial C–F bond, 1.367 Å. The length of the C-1–O bonds in derivatives **2.5** and **2.27** are 1.406 Å and 1.339 Å, respectively. A calculation (Tvarosvka 1989) leads to an anomeric effect of 1.85 kcal mol^{-1}, indeed lower than that of chlorine.

For sugars substituted by oxygen at C-1, delocalization is more difficult to prove because the quadrupolar resonance spectrum cannot be observed. The atomic polarizability of oxygen, 0.802, places it between fluorine and chlorine, so it seems very unlikely that the anomeric effect would comes from a radically

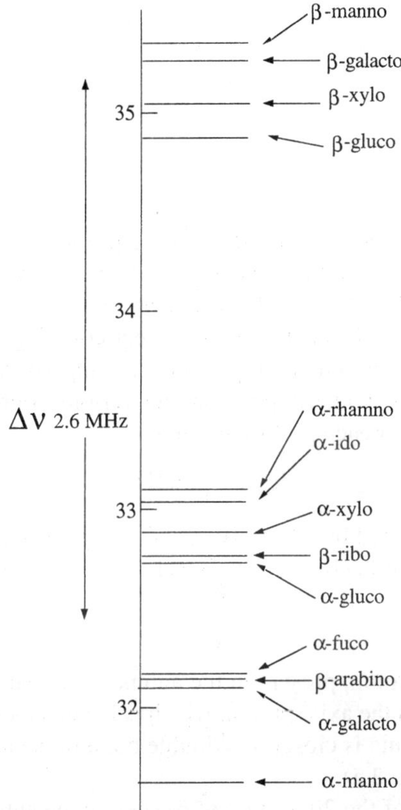

Fig. 2.6 Quadripolar resonance frequency of peracetylated hexopyranosyl chlorides having configurations as indicated.

OBz F O OBz OBz

2.27

different mechanism. Dimethoxymethane ($MeOCH_2OMe$) has a *gauche* conformation **2.28** in the gas phase (Astrup 1971) (projected along the CH_2–O bond) which corresponds to that of methyl α-hexopyranosides in which the methoxy group is axial. Nonetheless, it is noteworthy that the two oxygens play the same role. Delocalization can happen in the two directions with an appropriate geometry. It is here where the origin of the exo-anomeric effect was first seen. Since this effect is especially interesting in the chemistry of oligosaccharides, it will be discussed in Chapter 9.

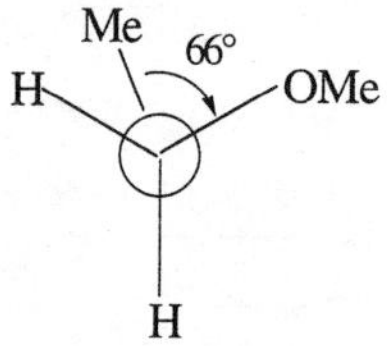

Me 66° H OMe H

2.28

Take note that this effect would stabilize the equatorial anomers, hence diminishes the anomeric effect of oxygenated substituents. However, there are other physical indications of the favoured delocalization of the cyclic oxygen towards the exocyclic oxygen as furnished by the direct coupling constant value $^{1}J_{CH}$ between the anomeric carbon and hydrogen. A comparison of the $^{1}J_{CH}$ values, measured on about 20 anomeric pairs having various configurations and substitutions, shows that for each pair we observe

$$J_{eq} - J_{ax} \cong 10 \text{ Hz}$$

Moreover, $^{1}J_{CH}$, measured by ^{13}C NMR spectroscopy, is related to the percentage of the s character of the bond, namely ρ, by the equation

$$^{1}J_{CH} = 500\,\rho$$

Thus the equatorial proton, present in the anomer with the axial oxygen, has a higher s character than the axial proton, which is in agreement with the idea that the C–O bond of the ring is closer to a double bond in the axial anomer than in the equatorial one (Fig. 2.5).

The delocalization of the $2p_z$ orbital of oxygen in the antibonding C–X axial bond is indisputable but the latter does not resolve, in any way, the question concerning the 'cause' of the anomeric effect. One of the dogmas of the qualitative electronic theories states that delocalization is stabilizing, but theoreticians are already beginning to contest this, even with the traditional benzene. Another objection is that these stabilizations are calculated from non-delocalized configurations which are conceptual monsters. Here the theoretical study has been carried out within the framework of the theory of molecular orbitals, for **2.23a** and **2.23b** (X = Cl). In Fig. 2.7, the cyclic oxygen is at the origin of the coordinates, and the p and sp^2 orbitals of the pairs are directed following Oz and Oy, respectively. The C-5–O–C-1 bonds are in the xOy plane, and the C–Cl bond makes a 30° angle with Oz.

The stabilization of a filled molecular orbital with an energy of E_0 in interaction with an empty orbital with an energy of E_1 is given by the classical equation

$$\Delta E = \frac{2\beta C_0 C_1}{E_0 - E_1}$$

The β value depends on the geometrical conditions of the interaction, and C_0 and C_1 are the coefficients of the atomic orbitals in contact in the expression of

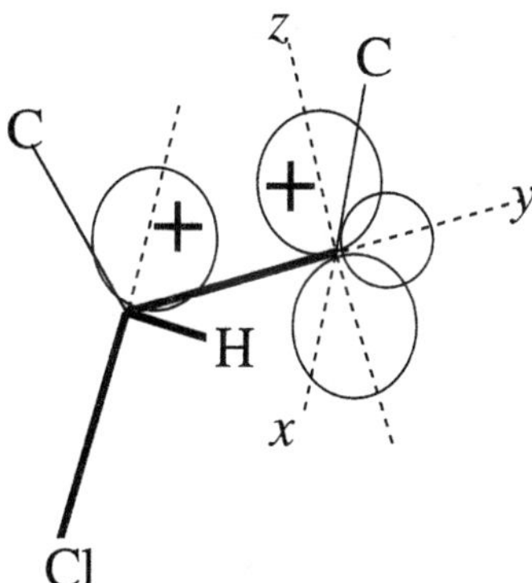

Fig. 2.7 Molecular orbitals of 2-chlorooxane involved in the anomeric effect theory.

the interacting molecular orbitals. This equation only makes sense if we can attribute energies to these two molecular orbitals, hence it is a matter of canonical orbitals.

When we study the list of molecular orbitals of 2-chlorooxane in conformations with an axial chlorine (Fig. 2.7) or an equatorial chlorine, obtained at the STO-3G level, we not only observe the interaction between $2p_z$ of the oxygen and σ^*_{CCl} on the axial conformer, but also the other interactions which, although clearly weaker, are not negligible: $2p_z\sigma^*_{CH}$ in the equatorial conformer, and interactions between the sp^2 oxygen pair (on O*y*) and the equatorial substituents in the two conformations. Figure 2.8 illustrates these interactions for the conformer with an axial chlorine.

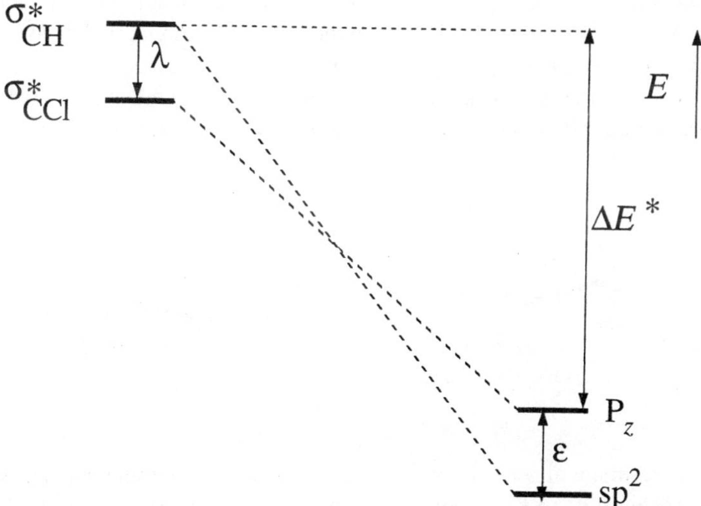

Fig. 2.8 Energy level and interactions in 2-chlorooxane with axial chlorine.

The $\sigma^*_{CCl}/p_z(O)$ interaction dominates because these two orbitals have the closest energies. For the anomeric effect of chlorine, the complete calculation gives

$$\Delta E = \frac{4h^{*2}\lambda\varepsilon}{(\Delta E^*)^3}$$

h^* being calculable on a model. This formula, which leads to a reasonable result (3.3 kcal mol^{-1}), also has the merit of underlining the fact that the anomeric effect is only observable because there is an energy difference, ε, between the two oxygen free lone pairs.

2.7 Conformation of pentopyranoses

Because of the absence of a side chain at C-5, there is frequently conformational mobility. The proton NMR spectrum can give the impression of a compound in a homogeneous conformation, while in fact, what we are observing is a time average because interconversion is rapid on an NMR time scale. In turn, the spectrum of the minor conformer can be present but escape detection.

Thus, the intermediate value of the $J_{1,2}$ coupling (4.8 Hz in deuterated acetone) of the tetra-*O*-acetyl-β-D-ribopyranose suggests an equilibrium between the $^1C_4$ and $^4C_1$ conformations **2.29** and **2.30**. Upon cooling, the signal examined at 200 MHz suddenly becomes broader towards –60°C, then settles into two signals: a narrow singlet at lower field, characteristic of an equatorial H-1 proton, and a large doublet at higher field, characteristic of an axial H-1 proton. At low temperature, the equilibrium corresponds to an excess (2:1) of the triaxial form. The determination of the coalescence temperature allows the rate constant of inversion to be calculated: it is close to 117 s^{-1} at –60°C and corresponds to an activation energy of $\Delta G^{\ddagger} = 10.3 \pm 0.3$ kcal mol^{-1} in the $^4C_1 \rightarrow {}^1C_4$ direction. For the reversed reaction, the corresponding numbers are 57 s^{-1} and 10.6 ± 0.3 kcal mol^{-1}, respectively. These values are close to those observed for cyclohexane or oxane. It is remarkable that substitution does not cause hindrance for the inversion.

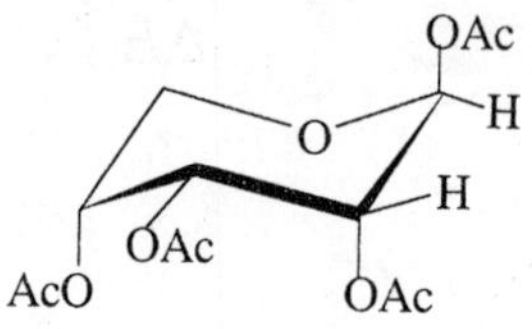

OAc H O H OAc AcO OAc

2.29

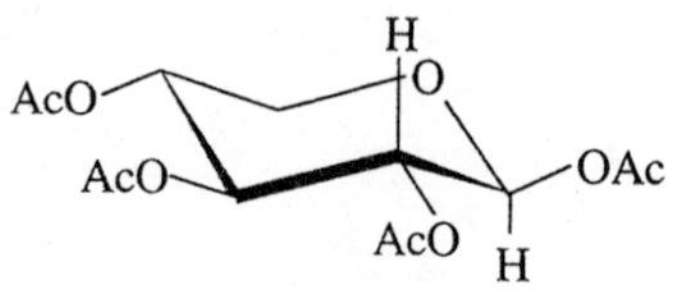

H O AcO AcO OAc AcO H

2.30

This experiment allows the 'exact' values of the $J_{1,2}$ coupling constants for the two conformations to be measured. If, at another temperature, the molar fractions of the conformers are N_e and N_a, a rule of mixing gives

$$J_{obs} = N_e J_e + N_a J_a$$

The J value allows us to calculate $k = N_a / N_e$ and the enthalpy difference, $\Delta G^{\ominus} = -RT\ln k$. Thus we find, at room temperature, 55% of conformer **2.29**.

This conformational freeze is exceptional. The only other example among the tetraacetylated pentopyranoses is the β-D-*lyxo* derivative. The calculation of the equilibrium constants from average spectra requires certain extrapolations. We cannot observe any regular effect from the nature and the polarity of the solvent. Let us now examine different types of derivatives.

The simplest case is that of peracylated halides. The conformation is dominated by the powerful anomeric effect of a halogen and all we see are conformations with an axial halogen, except for the β-D-*xylo* configuration which gives rise to equilibrium (2.7) (Table 2.4). Nevertheless, Table 2.4 shows that the tetraaxial conformations are always dominant and sometimes nearly exclusive.

(2.7)

OR X O OR OR $^1C_4$ ⇌ RO O RO X RO $^4C_1$

If we now replace the anomeric halogen by an acetoxy group, the equilibrium position is inverted, the weak anomeric effect not being able to compensate for two diaxial interactions at room temperature. The per-*O*-benzoylated derivative, however, leads to a 1:1 equilibrium. Here we recognize the limits of these analyses, as they only take into account the oxane part. With benzoate substituents, the 'essential point' is no doubt elsewhere. Another aspect of the problem is that the conformational energy differences can appear weak as compared to the stacking forces in crystals. In crystalline form, per-*O*-acetylated β-D-*xylo* chloride adopts an all-equatorial conformation, just as the peracetylated fluoride, 85% tetraaxial in solution, crystallizes in the tetraequatorial form. In these cases where there is equilibrium in solution, corresponding to free enthalpy differences close to

Table 2.4 Conformational equilibrium of β-D-*xylo* derivatives in CD_3COCD_3 (see equation 2.7).

R	X	$k = {}^4C_1/{}^1C_4$
Ac	Cl	0.26
Bz	Cl	0.19
Ac	F	0.17
Bz	F	0.05
Ac	OAc	2.60
Bz	OBz	0.98

0.8 kcal mol^{-1}, it is not surprising that the stacking forces—a tendency of compact structures—are able to dominate. More astonishing is the case of the per-*O*-benzoylated fluoride, which crystallizes in the tetraaxial form. In this case, it seems probable that the stacking of the phenyl rings is an essential factor.

In summary, these peracylated derivatives lead to conformational equilibria, except in the case where the issue is particularly obvious (α-D-*xylo*, $^4C_1$ β-D-*arabino*, $^1C_4$ configurations). The free pentoses in aqueous solution still remain to be seen. The diaxial interactions are stronger than with the acetates and the anomeric effect is weaker. Out of the eight D-pentose configurations, four of them (β-D-arabinose, α-D-lyxose, α-D-ribose, and β-D-ribose) lead to a conformational equilibrium.

2.8 Conformation of the hexopyranoses and their derivatives

The tendency of the side chain to adopt the equatorial position is an extremely dominating factor (Augé and David 1984); the only proven case where this chain adopts the axial position is that of methyl 2,4-bis(*N*-acetyl-*N*-benzoylamino)-3,6-di-*O*-benzoyl-2,4-dideoxy-α-D-idopyranoside, an exotic compound having two enormous substituents. In the preceding section, we pointed out that there is no reason that the rules applicable to simple derivatives of six-membered rings be made general for these extreme cases.

More broadly speaking, we can anticipate the following four orientations around the cyclic oxygen of a D-pyranose: I and III for *trans* derivatives, and II and IV for *cis* (Fig. 2.9). With the exception of idose (and perhaps altrose), all of the *trans* derivatives, in this case the monocyclic α-D-hexopyranoses and their derivatives, exist under the only observable conformation, D-$^4C_1$, which corresponds to the local conformation I, doubly stabilized by the anomeric and coplanar effects.

α-D-Idopyranose **2.31** is the only pyranose having two 1,3-diaxial interactions in the D-$^4C_1$ conformation. This is the moment when new conformations and their symbols should be introduced. The skew (*S*) conformation of compound **2.32** is described by taking, as a reference, the four coplanar atoms (non-consecutive). The symbol is completed by indicating the numbers of atoms located above and below the reference plane and, of course, the symbol of the series, to give, in

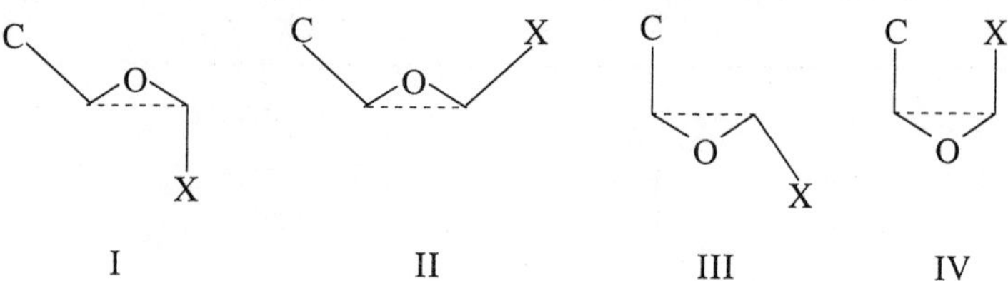

Fig. 2.9 Diverse orientations of substituents around the ring oxygen of a pyranose.

this case, D-$^0S_2$. α-D-Idopyranose in the D-$^0S_2$ conformation, **2.33**, no longer shows prohibitive diaxial interactions, and nevertheless fulfils the criteria for local stabilization around the cyclic oxygen, or at least partially. The proton NMR spectrum of α-D-idopyranose in aqueous solution corresponds to a mixture of conformations **2.31** and **2.33** in equilibrium.

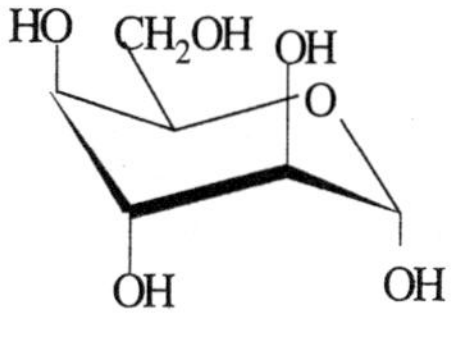

2.31 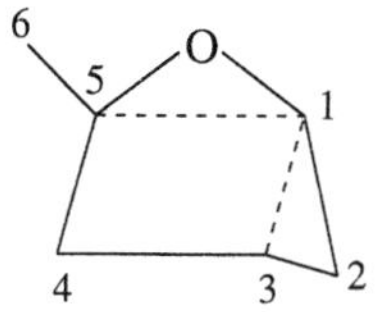 **2.32** 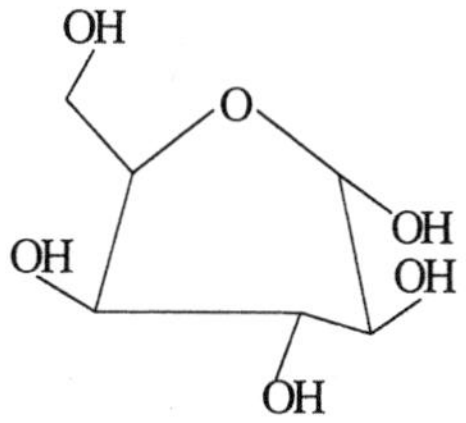 **2.33**

The *ido* configuration is present, isolated in the middle of other monosaccharide residues, in the polycondensed chains called 'glycosaminoglycans' of natural polysaccharides: dermatan sulfate, heparan sulfate, and heparin. The derivative in question belonging to the L-series is 2-*O*-sulfo-L-iduronic acid. It is represented (**2.34**, R = R′ = H) in a non-conformational manner with the α-L-*idopyrano* configuration present in these polysaccharides. The L-*ido* residue is isolated in the sequence, in the centre of the chains attached to O-1 and O-4, respectively (Section 17.3). It is found as a mixture of the L-$^1C_4$ (**2.35**) and L-$^2S_0$ (**2.36**) conformations. The proportion of the skew form varies from 40 to 60% according to the attached oligosaccharide sequences, R and R′ (Casu *et al.* 1986).

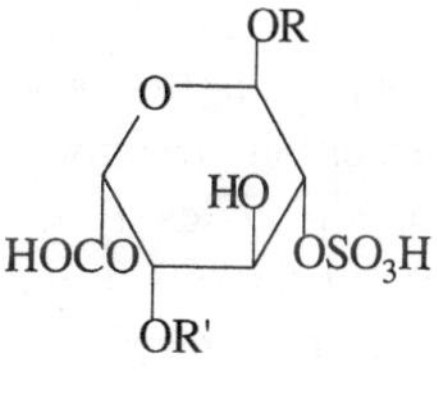

2.34 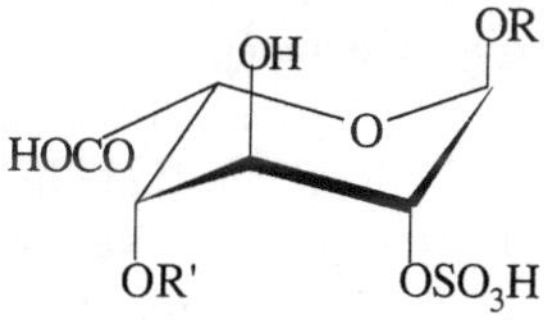 **2.35** 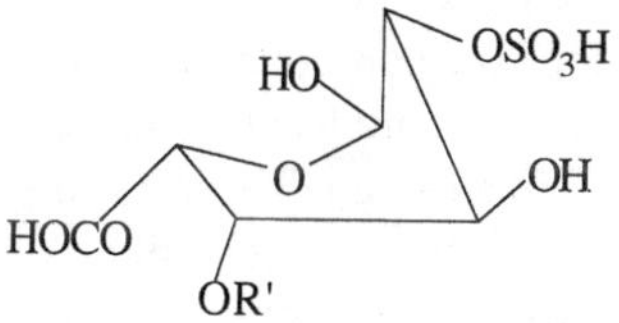 **2.36**

A novel ^{1}H NMR study suggested that α-L-iduronic acid in dermatan sulfate exists predominantly in a 'slightly distorted' L-$^1C_4$ conformation (Rao 1995).

The vigorous anchoring of nearly all the hexopyranoses in D-$^4C_1$ (L-$^1C_4$) conformations by the coplanar effect brings about a certain rigidity of the oligosaccharide chains. It is possible that the introduction of *ido* residues with a flexible conformation into certain sites creates the flexibility indispensible for certain functions.

In the peracetylated α-D-idopyranose, steric hindrance of the axial oxygens is lessened by acetylation, and the anomeric effect is increased. This ester exists exclusively in the tetraaxial D-$^4C_1$ conformation, **2.37**. (Durette and Horton 1971).

AcO CH_2OAc OAc
O
OAc OAc

2.37

2.9 Furanoses

Oxolane, as flexible as cyclopentane, will be discussed first. It is practical to assign names to certain conformations such as twist $^3T_2$ (**2.38**), envelope ^3E (**2.39**), and twist $^3T_4$ (**2.40**), whose symbols are copied from those of cyclohexane (equilibrium 2.8).

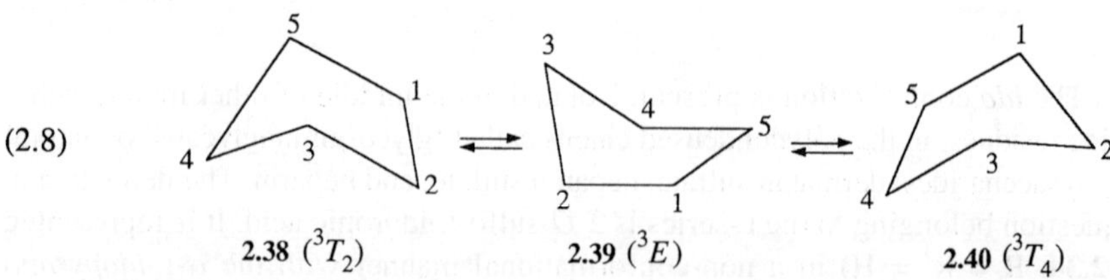

2.38 ($^3T_2$) **2.39** (^3E) **2.40** ($^3T_4$)

These have only geometrical significance. Conformations **2.38**, **2.39**, and **2.40** represent three steps of a continuous deformation which are not separated by noticeable barriers. A succession of alternate conformations, *E* and *T*, can be constructed which brings us back in a continual fashion to the starting point. For example, starting from a ^3E envelope, we go through envelopes ^5E, ^2E, ^4E, and ^1E only to come back to ^3E. From the point of view of its geometric form, each envelope is a consequence of the preceding one undergoing a 144° rotation; however, it is not the molecule which turns but rather its form. This is called pseudo-rotation.

In a frozen conformation of cyclopentane, there are five torsion angles, θ_0 ... θ_4 which are fixed (Fig. 2.10*a*). During pseudo-rotation, one of them, say θ_0, varies between two extreme values, ϕ_0 and $-\phi_0$ for example, going through θ_0, O, $-\phi_0$, ϕ_0, O, θ_0. From this came the idea of evaluating this periodical function by a Fourier transform reduced to a single term, by writing $\theta_0 = \theta_m \cos P$. The angle *P* undergoes a 360° variation when the molecule makes the complete pseudo-rotation circuit. (We thus note that a particular conformation has gone around the molecule twice, $2 \times 360°$).

The θ_0 dihedral angle has its maximum value in the $^3T_2$ conformation. Thus we have $\theta_0 = \theta_m$, hence $P = 0°$. For the following steps ^3E, $^3T_4$ of the pseudo-rotation, we have $P = 18°$ and 36°. Likewise the itinerary $^2T_1 \rightarrow {}^2E \rightarrow {}^2T_3$ corresponds to $P = 144°$, 162°, and 180°. In the $^2T_3$ conformation, the torsion angle is in the opposite direction of that of the $^3T_2$ conformation, and we have $\theta_0 = \theta_m \cos (180°) = -\theta_m$. A *P* value exists, of course, for each of the conformations *E*, *T*, *etc.*, but the novelty brought by the introduction of a *continuous* parameter, *P*, gives the possibility of characterizing the intermediate conformations between T and E which are the ones we really encounter.

Starting from θ_m and P, the other torsion angles are calculated by

$$\theta_j = \theta_m \cos(P + j\delta) \qquad j = 0, 1 \ldots 4 \; \delta = 144°$$

These derivations can be extended to the furanoses by replacing carbon 5 by oxygen (Fig. 2.10*b*). The torsion angles are written $\tau_0, \tau_1 \ldots \tau_4$. Despite the flexibility of the furanose system, there is one conformation (or two) with lower energy, determined by the site and orientation of the substituents and, in the solid phase, by the stacking forces in the crystal. In general, this does not coincide with one of the E or T conformations. Altona and Sundaralingam (1972) proposed to describe it by the parameters θ_m and P. This description found its way into the crystallography literature (Jeffrey and Sundaralingam 1974; 1975; 1976; 1977; 1980; 1981; 1985).

This notation will be explained by an example taken from the chemistry of nucleosides/nucleotides, by far the most important family of furanosides (see Section 3.4). The nucleosides are glycosidic combinations with a heterocyclic base and a β-D-ribofuranosyl or 2-deoxy-β-D-*erythro*-pentofuranosyl residue. The example chosen is the synthetic nucleoside 5-iodouridine. We can observe two different conformations of this molecule in the crystalline form, close to $^3T_2$ (**2.41**) and $^2T_3$ (**2.42**), respectively. The maximum torsion angle is between the 2′ and 3′ positions. In order for P to be close to 0, we will therefore choose $\tau_2 = \theta_0$ (Fig. 2.10) and consequently for $\theta_1 \ldots \theta_4$, the values τ_3, τ_4, τ_0, and τ_1, respectively. Knowledge of the two torsion angles allows us to calculate τ_m and P. The P angle has the character of a phase and τ_m is a measure of the flattening of the

Fig. 2.10 Convention for the representation of torsion angles of (a) cyclopentane and (b) oxolane.

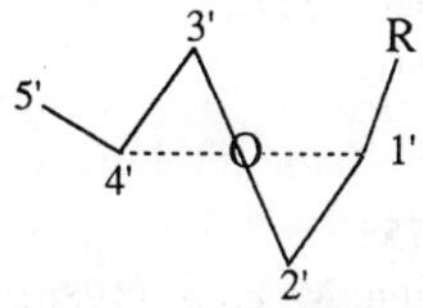

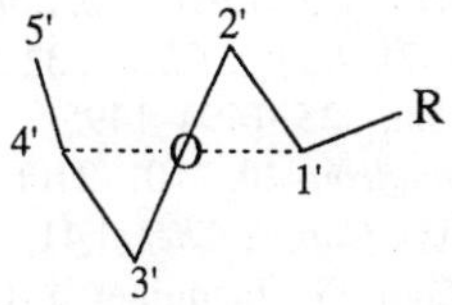

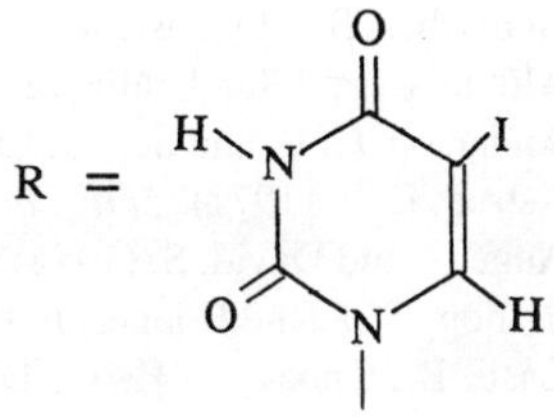

ring. Conformation **2.41**, $P = 9°$, is halfway between $^3T_2$ and ^3E. We have $\tau_m = 36°$ and the maximum torsion angle of 35° between 2′ and 3′. Conformation **2.42**, $P = 175°$, is rather close to $^2T_3$. The maximum torsion angle is still between 2′ and 3′, –42°. The fact that the same molecule can have two different conformations in the same crystal is in itself indicative of their small energy difference. All the same it is unusual and nucleosides/nucleotides are divided equally between $0° < P < 36°$ and $144° < P < 180°$ with few exceptions.

The biological functions of DNA involve flexibility of the chains. Is this the reason why nature has selected the flexible deoxyribofuranose rather than one of the rigid hexopyranoses in the construction of its building blocks? The interested reader will find a detailed investigation of this problem in papers published from 1992 by Eschenmoser and his group on the general theme 'Warum Pentose- und nicht Hexose-Nucleinsäuren?' (1992).

2.10 Non-cyclic polyols

Let us take galactitol **2.43** in the solid state as an example. The molecule has the zigzag form. This is the favorite disposition in solution, except when it leads to 1,3-eclipsed interactions of the hydroxyls. In the latter case, the molecule distorts itself into the sickle form. This conformation is derived from the zigzag plane by a 120° rotation around the internal C–C bond. The sickle conformation is observed on the diethyl dithioacetal of the peracetylated D-ribose **2.44**.

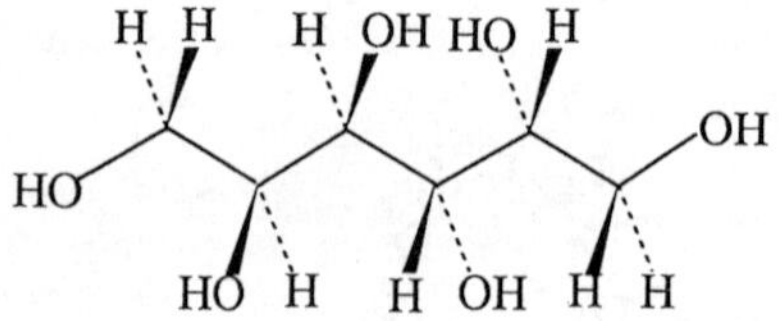

2.43

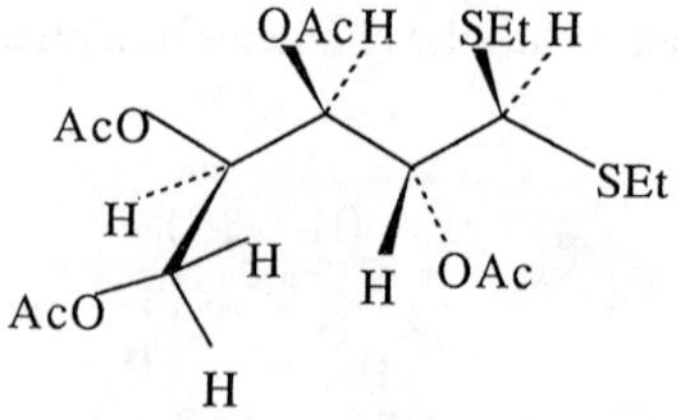

2.44

References

Aebischer, B., Hollenstein, R., and Vasella, A. (1983), *Helv. Chem. Acta*, **66**, 1748–1754.

Altona, C. and Sundaralingam, M. (1972), *J. Am. Chem. Soc.*, **94**, 8205–8212.

Anderson, C. B. and Sepp, D. T. (1967), *J. Org. Chem.*, **32**, 607–611.

Astrup, E. E. (1971), *Acta Chem. Scand.*, **25**, 1494–1495.

Augé, J. and David, S. (1984), *Tetrahedron*, **40**, 2101–2106.

Bishop, C. T. and Cooper, F. P. (1963), *Can. J. Chem.*, **41**, 2743–2758.

Casu, B., Choay, J., Ferro, D. R., Gatt, G., Jacquinet, J.-C., Petitou, M. *et al.* (1986), *Nature*, 215–216.

Cone, C. and Hough, L. (1965), *Carbohydr. Res.*, **1**, 1–9.

David, S. (1979), *ACS Symp. Ser.*, **87**, 1–16.

Durette, P. L. and Horton, D. (1971), *Adv. Carbohydr. Chem. Biochem.*, **26**, 49–125.

Edward, J. T. (1955), *Chem. Ind. (London)*, 1102–1104.

Eliel, E. L., Hargrave, K. D., Pietrusiewicz, K. M., and Manoharan, M. (1982), *J. Am. Chem. Soc.*, **104**, 3635–3643.

Eschenmoser, A. and Dobler, M. (1992), *Helv. Chim. Acta*, **75**, 218–259.

Fabian, M. A., Perrin, C. L., and Sinnott, M. L. (1994), *J. Am. Chem. Soc.*, **116**, 8398–8399.

Franck, R. W. (1983). *Tetrahedron*, **39**, 3251–3252.

Jeffrey, G. A. and Sundaralingam, M. (1974), *Adv. Carbohydr. Chem. Biochem.*, **30**, 445–466; (1975) **31**, 347–371; (1976) **32**, 353–384; (1977) **34**, 345–378; (1980) **37**, 373–436; (1981) **38**, 417–529; (1985) **43**, 203–421.

Jorgensen, W. L. and Ibrahim, M. (1981), *J. Am. Chem. Soc.*, **103**, 3976–3985.

Jorgensen, W. L., Binning, Jr., R. C., and Bigot, B. (1981), *J. Am. Chem. Soc.*, **103**, 4393–4399.

Juaristi, E. and Cuevas, G. (1992), *Tetrahedron*, **48**, 5019–5087.

Kirby, A. J. (1983), *The anomeric effect and related stereoelectronic effects at oxygen*, Springer Verlag, Berlin.

Lemieux, R. U. (1964), *Molecular rearrangements*, Part II (ed. P. de Mayo) p. 735, Interscience, New York.

Lucken, E. A. C. (1959), *J. Chem. Soc.*, 2954–2960.

Paulsen, H. (1979), *ACS Symp. Ser.*, **87**, 63–79.

Planje, M. C., Toneman, L. H., and Dallinga, G. (1965), *Rec. Trav. Chim.*, **84**, 232–240.

Rao, V. S. R., Balaji, P. V., and Qasba, P. K. (1995), *Glycobiology*, **5**, 273–279.

Thatcher, G. R. J. (1993), *ACS Symp. Ser.*, **539**.

Tvarosvka, I. (1989), *Adv. Carbohydr. Chem. Biochem.*, **47**, 45–123.

Welti, D. (1977), *J. Chem. Res.*, *M*, 3566–3587.

3 Alkyl and aryl glycosides and glycosamines

3.1 Definitions related to glycosides (*O*-glycosides)

Furanoses and pyranoses are hemiacetals. *Glycosides* are acetals. On paper, they are derived from furanoses and pyranoses by replacing the hydrogen of the hemiacetal hydroxyl group by an R group. These are thus mixed acetals, internal and external, whereby one of the acetal oxygens is derived from one of the alcohol functions of the sugar, and the other from the external hydroxylated compound, R–OH. It follows that there are four types of glycosides, corresponding to either a pentose or a hexose. Below are examples of four glycosides derived from galactose by substituting the hemiacetal hydrogen by a methyl group: methyl α-D-galactofuranoside **3.1**, methyl β-D-galactopyranoside **3.2**, methyl α-D-galactofuranoside **3.3**, and methyl β-D-galactofuranoside **3.4**.

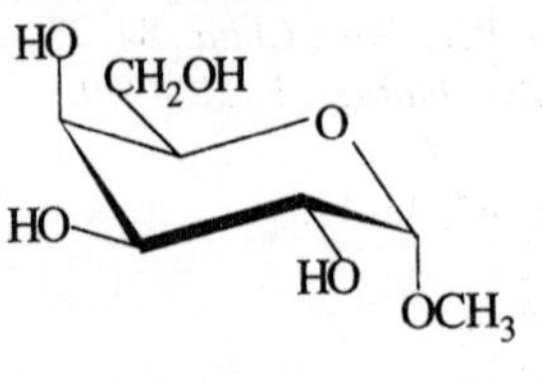

3.1 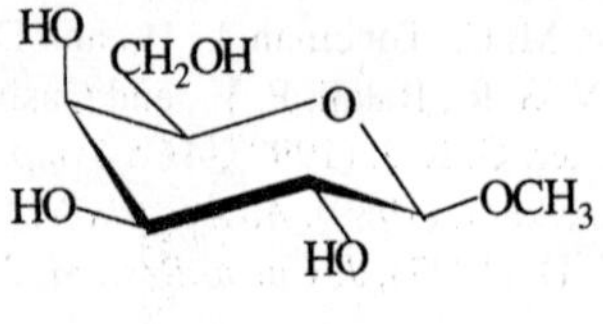 **3.2**

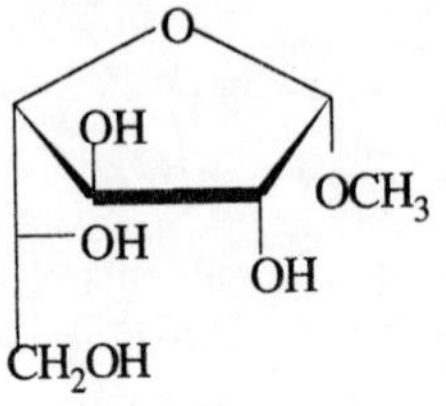

3.3 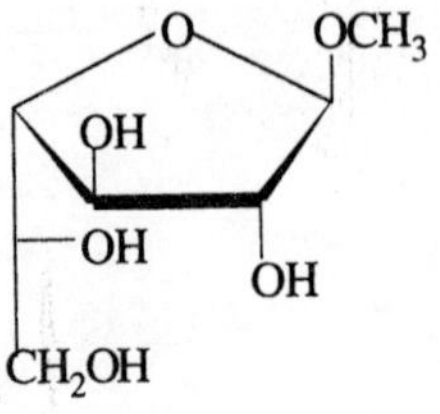 **3.4**

The definition given includes the case where the hydroxylated derivative R–OH, corresponding to the external substituent R, belongs to the sugar family and, indeed, the corresponding bond behaves exactly as the others from a chemical viewpoint; at any rate, this bond between two groups by an acetal function is called the *glycosidic bond*. In practice, these two categories of glycosides, depending on whether or not R–OH is a sugar, play very different roles which justifies their being treated separately. Glycosides in which R = methyl, ethyl, phenyl, benzyl, *etc.*, are very important as synthetic intermediates. They are placed under the general heading of glycosides and are the topic of this chapter. The 'sugar' part is called the *glycosyl unit* and the exterior R group, the *aglycon.* After examining the names of derivatives **3.1**, **3.2**, **3.3**, and **3.4**, the reader will be able to deduce the nomenclature rules. The name begins with the radical designating the aglycon, followed by the name indicating the glycosyl unit in which the suffix 'ose' of the free sugar is replaced by 'oside'.

If R–OH is a sugar, the term *disaccharide* will be used preferentially. Moreover, the association of far more than two molecules of simple sugars can be carried out by the same type of linkage. Some of these structures play a fundamental role in cell recognition phenomena and will be discussed in detail beginning with Chapter 9.

3.2 Synthesis of alkyl glycosides by the Fischer method (Ferrier and Collins 1972; Overend 1972; Ferrier 1988; Green 1966)

3.2.1 Experimental aspect

With acid catalysis, there is, on the one hand, equilibrium between a pentose or a hexose and an aliphatic or benzylic alcohol, and on the other, the corresponding mixed acetal and water, in accordance with Fig. 3.1. We will first look at the synthesis. The reaction is displaced towards the right by using a large excess of alcohol ROH, generally employed as solvent. For example, by heating under reflux a solution of galactose in methanol containing 2% HCl, we obtain, after 12 h, a mixture of the starting sugar at equilibrium with galactosides **3.1**, **3.2**, **3.3**, and **3.4**. This is a general *glycosidation* reaction. As usual, it only forms five- or six-membered rings and the only internal alcohol functions involved are those carried by C-4 or C-5. Thus, the pentoses and hexoses must show at least one free alcohol function at the γ- or δ- positions of the carbonyl (aldehyde or ketone).

However, the composition mixture at equilibrium varies greatly from one sugar to another (Table 3.1). The constituents are estimated using methods already outlined in Chapter 1 concerning the estimation of the anomers of free sugars, namely analytical HPLC or gas-phase chromatography after silylation. Here measurements are easier because, as soon as they are no longer in a relatively acidic medium, glycosides are very stable and there is no risk of the sample composition changing during the analysis. The composition of these

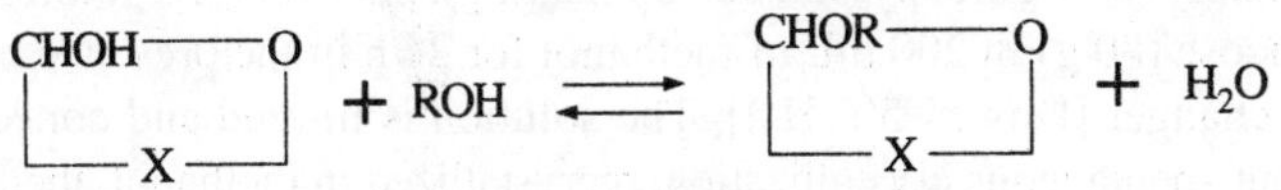

CHOH——O / X (ring) + ROH ⇌ CHOR——O / X (ring) + H_2O

Fig. 3.1 Glycosidation reaction.

Table 3.1 Acid methanolysis: composition at equilibrium of a mixture of methyl glycosides

		Furanosides		Pyranosides	
	T(°C)	α	β	α	β
glucose*	35	0.6	0.9	66	32.5
mannose*	35	0.7	0	94	5.3
galactose*	35	6	16	58	20
fucose**	65	6	13	54	27
ribose*	35	5	17	12	66
fructose***	28	25	26	3	46

*from Ferrier and Colins 1972; **from Mowery 1975; ***from Bethell and Ferrier 1973 (reproduced with kind permission from Penguin Books and Elsevier Science).

mixtures is obviously linked to the difference of the free enthalpy of each constituent. Certain consistencies in their behaviour can be observed. Pyranosides are favoured over furanosides, which are nearly absent in the case of glucose and mannose. In pyranosides having stable conformations such as gluco-, manno- and galactopyranosides, it is the axial methoxyl derivative which predominates, indicative of the anomeric effect.

If we follow the development of the glycosidation reaction over time instead of examining the composition at equilibrium, we see that the furanosides are formed at the start, only to disappear thereafter, more or less completely, to the benefit of the pyranosides.

3.2.2 Preparative usefulness and limits

As described above with galactose, we always use the alcohol as solvent. We can begin by trying to heat the solution for a few hours to 80°C in the presence of a mineral acid (~ 0.1–1 M). The best conditions must then be determined. Those which were just given would quantitatively transform deoxyribose into levulinic acid, $CH_3COCH_2CH_2COOH$. But with all 2-deoxy sugars, glycosidation is very fast under much milder conditions. For example, the conversion of deoxyribose into methyl glycosides is complete in 20 min at 27°C with hydrogen chloride in methanol (0.015 M).

If the glycoside does not crystallize directly in the reaction medium, the mineral acid catalyst must be removed, and this may cause a problem since glycosides are generally quite soluble in water and it is impossible to extract them using organic solvents. It is very practical to use a cation exchange resin (H^+ form) as a catalyst, which is separated by filtering at the end of the operation. Thus, we prepare methyl α-D-glucopyranoside by heating under reflux a solution of anhydrous glucose (80 g) in 200 mL of methanol for 24 h in the presence of 20 g of a cation exchanger [Dowex 50 (H^+)]. The solution is filtered and concentrated to bring about spontaneous crystallization, recrystallized in methanol, then ethanol to give 25 g (29%) of pure methyl α-D-glucopyranoside (Bollenback 1963).

At this stage, the idea must be firmly implanted that in these syntheses involving only a small number of steps starting from ordinary sugars, the race to obtain high yields is no longer very meaningful. What is important, however, is the simplicity of the steps. In France, a kilogram of glucose, to the degree of purity necessary for research, costs ~US$10. The purification of a derivative of this kilogram by chromatography would require a minimum of 20 kg of silica gel (~US$750) and 60 L of solvent (~US$120). It is much more economical to use a method which only requires crystallizations, even it it means increasing the work scale when the yield is low. Generally, it is the major constituent of the solution which crystallizes, namely the pyranoside whose anomeric alkoxy is axial; nonetheless, this is not an absolute rule. More often than not, the preparation of a pyranoside with an equatorial anomeric alkoxyl group using the Fischer glycosidation method involves rather tedious fractional crystallization from the mother liquors and another method is preferentially employed.

To obtain furanosides, it is possible to stop glycosidation at the beginning, the moment when concentration is at a maximum. For example, by heating under reflux a solution of galactose in methanolic hydrogen chloride (0.004 M) for 6 h, methyl β-D-galactofuranoside can be obtained in 53% yield (Augestad and Berner 1954). In this preparation, the equilibrium process has been slowed down by using a very small concentration of the catalyst, conditions which are only rarely ideal.

Replacing the protic catalyst by iron (III) chloride, a mild Lewis acid, gives exclusively a mixture of furanosides. Thus, a mixture of methyl D-glucofuranosides (α/β 3:7) can be obtained in 75% yield (Lubineau and Fischer 1991). Another pathway to a furanoside is based on intermediate complexation by Ca^{2+} ions. Such complexations will be treated in Chapter 11. By heating under reflux a solution of mannose and $CaCl_2$ in methanol in the presence of acetyl chloride as the proton source for 2 h, 53.5% of methyl β-D-mannofuranoside is obtained, isolated in 40% yield (Angyal *et al.* 1980). Finally, applying the Fischer glycosidation to a sugar whose alcohol function at C-5 is protected obviously gives only furanosides.

At this point we will look at the limits of usefulness of the Fischer method. A large excess of alcohol must be used, preferably as solvent, to displace the equilibrium in the desired direction. Equatorial anomeric alkoxyl anomers are not easily isolated. This reaction cannot be used with phenols, and aryl glycosides are not accessible.

3.3 Other methods of preparing glycosides

3.3.1 Activation of the anomeric carbon

In the Fischer glycosidation reaction, we can imagine that the role of the acid catalyst is to protonate the anomeric hydroxyl, thus transforming it into the leaving group (Fig. 3.2), which would facilitate nucleophilic substitution. In the

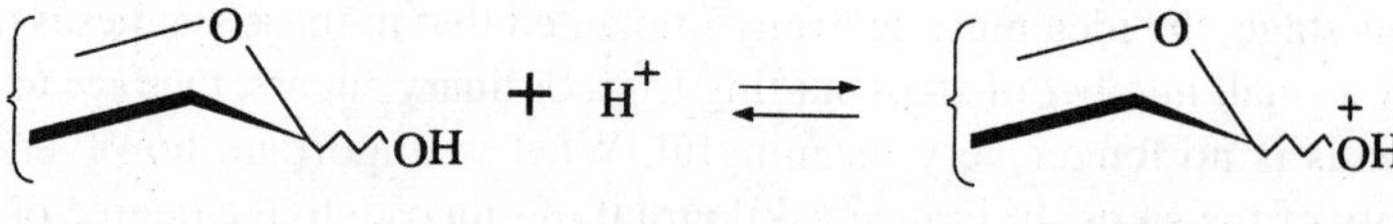

Fig. 3.2 Protonated intermediate in the glycosidation reaction.

methods we are about to describe, there is a stable leaving group at C-1, a fact which does not prevent the employment of supplementary activation reagents, the 'promoters' introduced in the medium. These methods, longer than the Fischer glycosidation and very varied, are essentially used in the synthesis of oligosaccharides. As detailed in Chapter 10, they are all usable in the synthesis of simple glycosides.

Thus, in the peracetate of β-D-glucopyranoside **3.5**, the anomeric acetoxy group is particularly labile. It is replaced by a methoxy group in the presence of $SnCl_4$ to give methyl tetra-*O*-acetyl-β-D-glucopyranoside **3.6** (Hannessian and Banoub 1980). This type of activation can prove to be insufficient. A promoter, more efficient but more expensive than $SnCl_4$, trimethylsilyl trifluoromethanesulfonate ($CF_3SO_3SiMe_3$) has been proposed. For large-scale work, it is advisable to use a bromide, readily obtained by treating a peracetate of β-D-glucopyranoside such as **3.5** with HBr. Thus, the bromide of tetra-*O*-acetyl-α-D-galactopyranoside **3.7** is prepared which leads, with methanol in the presence of a mixture of HgO (1 eq.) and $HgBr_2$ (0.04 eq.), to the peracetylated methyl β-D-galactoside **3.8**. A total yield of 24% is obtained from galactose. We should note the equatorial orientation of the incoming methoxy group in both cases, independent of the configuration of the starting product. In Chapter 10 we will have a more detailed look at how the participating acetoxy group at position 2 imposes the 1,2-*trans* configuration of the product. The yield of the actual substitution is generally excellent.

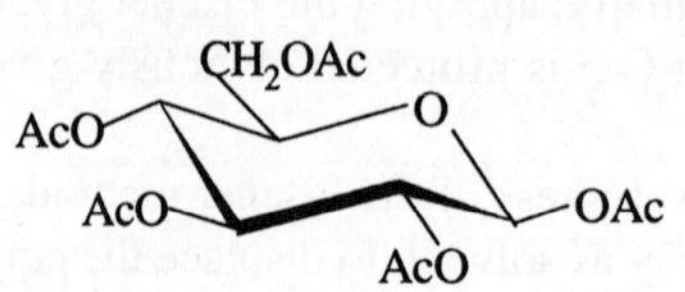

3.5

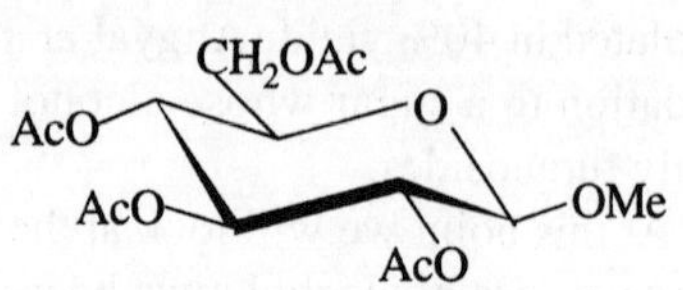

3.6

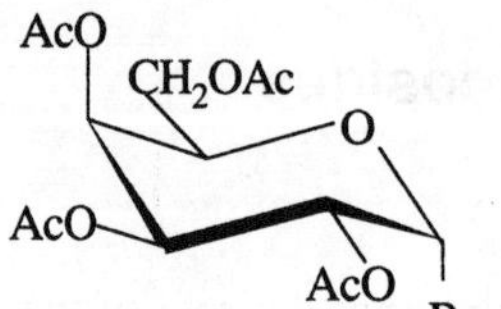

3.7

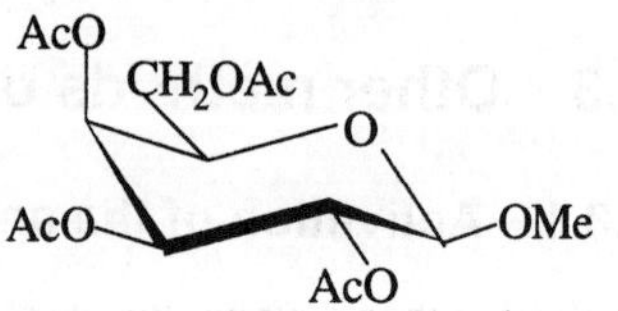

3.8

In summary, alkyl glycosides with an axial aglycon are easily obtained using the Fischer glycosidation, whether they are 1,2-*cis* or 1,2-*trans*. The 1,2-*trans* diequatorial alkyl glycosides result from participating reactions. This leaves the 1,2-*cis* with an equatorial aglycon such as β-D-mannoside **3.9**. Compound **3.9** is isolated in 30% yield by simultaneous addition of methyl sulfate and sodium hydroxide to an aqueous solution of mannose (Isbell and Frush 1940).

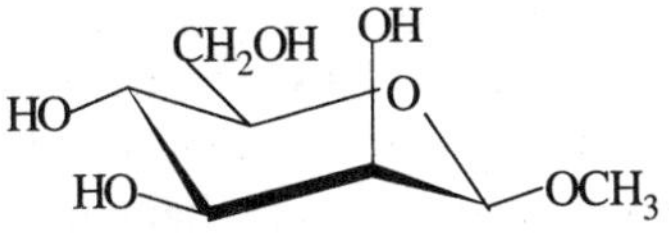

3.9

3.3.2 Aryl glycosides

Aryl glycosides cannot be obtained by Fischer glycosidation. Perhaps the phenol hydroxyl is insufficiently nucleophilic in comparison to the alcohol hydroxyl. But they can be very easily prepared from acetates in the presence of acid catalysts. The fused mixture of the peracetate of α-D-glucopyranose **3.10** and phenol gives 64% of the tetraacetylated phenyl α- D-glucopyranoside **3.11** in the presence of zinc chloride and 85% of the β-anomer **3.12** in the presence of *p*-toluenesulfonic acid (Ferrier and Collins 1972). The reaction course strongly depends on the experimental conditions. Every time an alkyl or aryl glycoside is obtained in a tetraacetylated form, it is easy, if it happens to be necessary, to deprotect the alcohol functions by alkaline methanolysis since the acetal bond is very stable with bases.

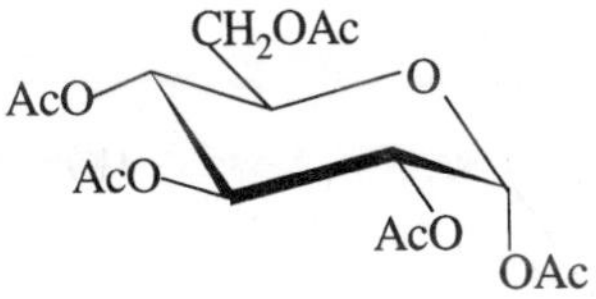

3.10

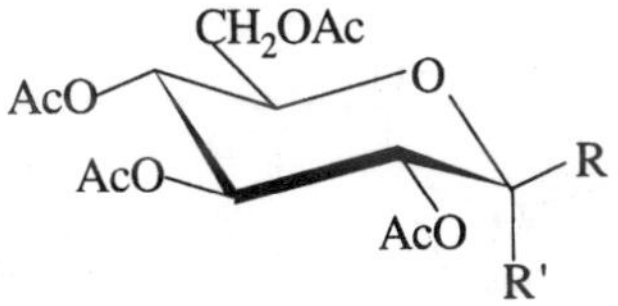

3.11 R=H, R'=OPh
3.12 R=OPh, R=H

3.4 Acetal-type anhydropyranoses and anhydrofuranoses

The reader will have perhaps wondered why, in the Fischer glycosidation reaction, there is no acetalation by *two* alcohol functions of the sugar molecule, leading to a completely internal and bicyclic mixed acetal. As a general rule, an intramolecular reaction is faster than the intermolecular analogue and this type of compound is observed in families other than sugars. The answer could be that alkyl glycosides are prepared in the presence of an excessive amount of alcohol. In fact, these internal acetal derivatives are a class of well-known compounds, perfectly stable under non-acidic conditions. They are named as derivatives of pyranoses and furanoses; for example, an *x,y-anhydro-β*-D-pyranose with an ether bridge between the carbons of symbols x and y. We will begin with 1,6-*anhydro-β*-D-pyranoses, namely the internal mixed acetal between the aldehyde function and the hydroxyl carried by carbon 6. The general skeleton of these compounds is represented by **3.13**. The pyranose ring shows the D-$^1C_4$ conformation, non-existent with free pyranoses. It displays a favourable anomeric effect but an axially disposed side chain, unacceptable in the case of a free pyranose. It is difficult to predict the free energy contribution of this side chain because of the new ring closure. At any rate, it is certain that an axial substituent at C-3 introduces strong steric compression. The most stable example of this series should be the triequatorial **3.14**, derived from D-idose **3.15**. In Chapter 2 we mentioned the conformational instability of this sugar. When heated in aqueous solution in the presence of dilute acid, it gives the 1,6-anhydride in 86% yield. It is, therefore, by far the most stable conformation in solution. On the other hand, the three other sugars, i.e. glucose, mannose, and galactose, give only traces of 1,6-anhydride under these conditions. We anticipate a powerful 1,3-diaxial interaction between CH_2 and the hydroxyl group at position 3 for these three configurations, as in the example of 1,6-*anhydro*-D-glucose **3.16**. As an exercise, the reader can interpret, in conformational analysis terms, the conversion yields of the 1,6-anhydrides of the following sugars in acid solution at 100°C (Angyal and Dawes 1968): glucose (0.2%), mannose (0.8%), galactose (0.8%), talose (2.8%), allose (14%), gulose (65%), altrose (65%), and idose (86%). The configurations of these hexoses can be found in Chapter 4.

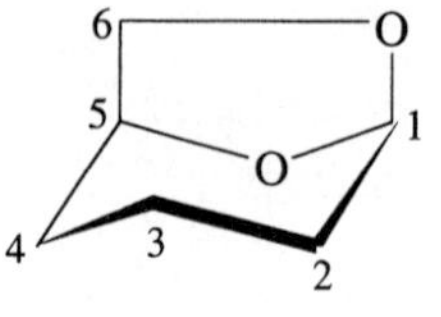

3.13

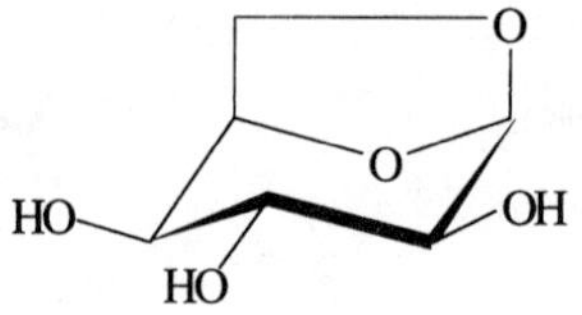

3.14

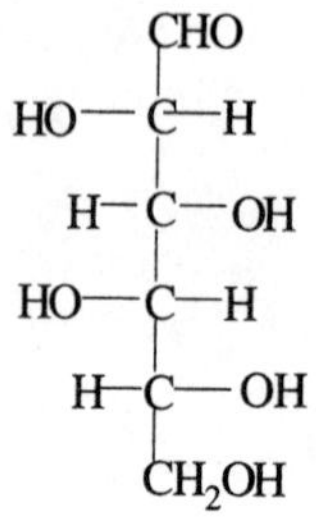

3.15

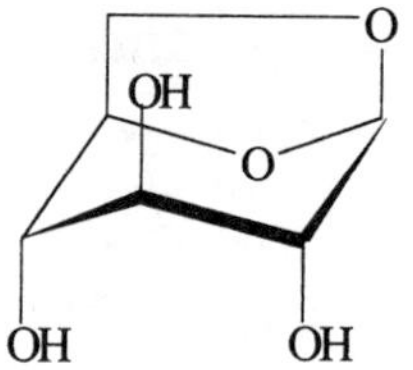

3.16

While in an acidic medium these anhydrides are at equilibrium with variable quantities of the corresponding hexoses, they are very stable in an alkaline medium, as are all glycosides and acetals in general. Those which are unstable in an acidic medium can be prepared using a novel and efficient reaction by heating at 100°C a peracetylated aryl β-D- glycopyranoside. Thus, phenyl tetra-*O*-acetyl-β-D-glucopyranoside **3.12** gives 1,6-*anhydro*-D- glucose **3.16**, easily isolated in 80% yield by peracetylation and recrystallization of the highly crystalline peracetate (Coleman 1963). This reaction apparently involves a 1,2-*anhydro* intermediate because it does not take place if the hydroxyl group at position 2 of the glucoside is protected by methylation, and the α-anomer **3.11** is not reactive.

When a pyranose is transformed into a 1,6-anhydropyranose, the axial hydroxyl groups become equatorial and vice versa. This is undoubtedly the most interesting property from a synthetic point of view because of the change in reactivity. But there are limits to the usefulness of the 1,6-anhydro; the *endo* face between the oxolane and the oxane is not very accessible and nucleophiles cannot be easily introduced there, so much so that it can be difficult to substitute activated hydroxyl groups by the S_N2 reaction involving *endo* attack. Finally, it should be noted that the preparation of certain 1,4-anhydropyranoses is also known.

1,2-Anhydrides behave as very active oxiranes. Moreover, **3.17** is prepared as an oxirane by the displacement of a chloride by a vicinal alkoxide, in this case by the action of ammonia on a glucosyl chloride with a free 2-hydroxyl group. Recently, epoxidation of *glycals* such as **3.18** by dimethyldioxirane was recommended (Halcomb and Danishefsky 1989). When the R protection is a benzyl or a *t*-butyldimethylsilyl group, the specific introduction of the oxygen *trans* to the oxygen at position 3 (equation 3.1; products **3.18** and **3.19**) is observed. These oxiranes are opened at room temperature without a catalyst by primary alcohols (CH_3OH, $PhCH_2OH$), with inversion of configuration to give, nearly quantitatively, 1,2-*trans* glycosides **3.20** (equation 3.2). Compound **3.17** allows the synthesis of saccharose **3.21** by a reaction appearing to be analogous, but obviously more complex, since there is retention of configuration, and it is necessary to react for 100 h at 100°C only to obtain a low yield (Lemieux and Huber 1956). Other derivatives of sugars are known which are transformed into glycosides at room temperature without a catalyst. The diazirines **3.22**, for example, can give glycosides, even with secondary alcohols (equation 3.3). The intermediate is probably carbene **3.23** (Brimer and Vasella 1989).

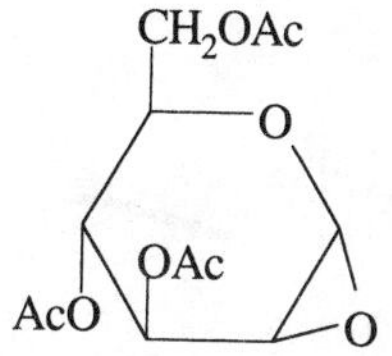

CH_2OAc

O

OAc

AcO

O

3.17

(3.1) 3.18 + $(CH_3)_2C(O_2)$ → 3.19 + CH_3COCH_3

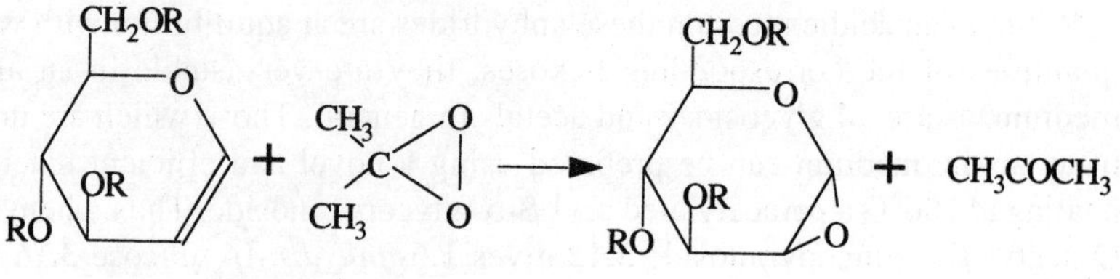

3.18 **3.19**

(3.2) + $PhCH_2OH$ →

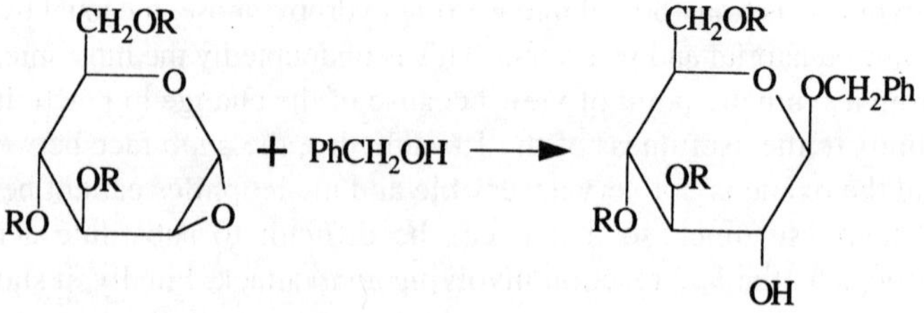

3.20

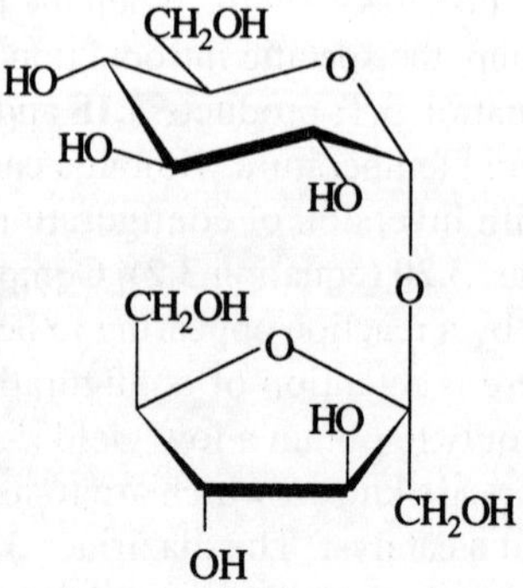

3.21

(3.3) → →

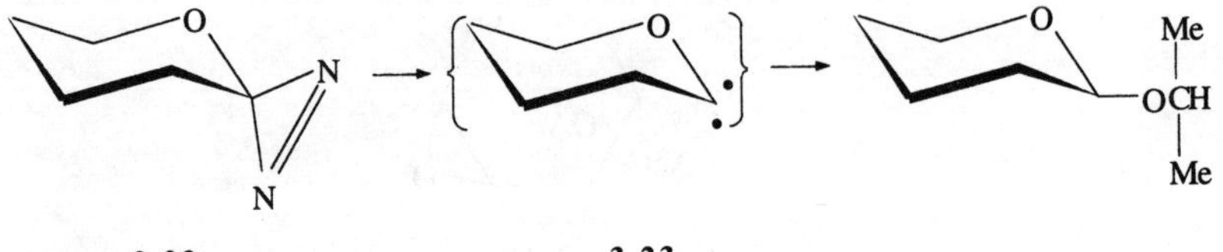

3.22 **3.23**

3.5 Chemical properties of glycosides

3.5.1 Hydrolysis in acidic medium (Bochkov and Zaikov 1979; BeMiller 1967; Szejtli 1976)

In the presence of excess water, the course of the reversible reaction shown in Fig. 3.1 obviously goes from right to left. Glycosides in aqueous solution are decomposed by acids. This reaction aroused so much interest that by 1979 there were already more than a thousand articles on the subject (Bochkov and Zaikov 1979). The most generally accepted mechanism seems to be that proposed by Edward (1955). The intermediate, reversibly protonated at the exocyclic oxygen evolves towards the carbenium ion 'glycosyl cation', or undergoes a bimolecular substitution by a water molecule (Fig. 3.3). Having a life span in water in the order of 10^{-10} to 10^{-12} s, the glycosyl cation would be 'at the threshold of actual existence' (Sinnott 1990). The environment would necessarily be involved in the reactions of such an unstable intermediate.

In a more general fashion, with S being the glycoside, the reaction pathway is shown by equation (3.4).

(3.4) $$S + H^+ \text{ (water)} \rightleftharpoons [SH^+] \longrightarrow \text{products}$$

The overall rate is proportional to $[SH^+]$ since the concentration of water stays roughly the same. As S is a very weak base, its protonation is equilibrated according to equation (3.5) and the reaction rate is thus given by equation (3.6).

(3.5) $$\frac{[S][H^+]}{[SH^+]} = K$$

(3.6) $$\nu = k'[SH^+] = \frac{k'}{K}[H^+][S]$$

In the presence of excess glycoside, the pseudo-monomolecular rate constant k is proportional to the concentration of hydronium ions (equation 3.7).

(3.7) $$k = \lambda\,[H^+]$$

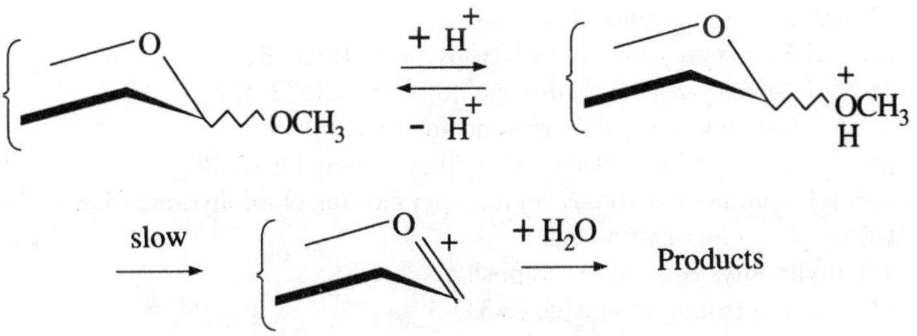

Fig. 3.3 Proposed mechanism for hydrolysis of glycosides.

For the utmost precision, it would be necessary to introduce the activity of hydronium ions because we are sometimes working with concentrated acid solutions (HCl 2.5 M, H_2SO_4 2 M, etc.). Acidity functions are also involved in the mechanistic studies, but the very basis of the theory of these functions has recently given rise to sharp criticism (Ritchie 1990). For a qualitative discussion we may consider activity and concentration as identical. The reported conditions, such as the nature and concentration of the acid and the temperature, are so variable, and the k value range so wide, that it is difficult to tabulate results in a consistent manner. One author (Szejtli 1976) chose to calculate the k_r value of the rate constant at 100°C in normal HCl starting from its k value at t°C in c molar acid concentration using equation (3.8) which supposes the activation energy $E^‡$ is known (in cal mol^{-1}).

$$\log k_r = \log k + \frac{E^‡}{4.575}\left(\frac{1}{273+t} - \frac{1}{373}\right) - \log c \qquad (3.8)$$

Equation (3.9) gives the half-life of the reaction in minutes, which is more meaningful for the synthetic chemist.

$$t_{1/2}/\text{min} = \frac{0.69}{k/\text{min}^{-1}} \qquad (3.9)$$

Table 3.2 shows the typical results for a few ordinary glucosides as well as for compounds **3.24**, **3.25**, **3.26**, **3.27**, **3.28**, **3.29**, **3.30**, and **3.31**. Certain extrapolations, far from the true measurement conditions, are undoubtedly unrealistic, especially for very labile glycosides. Nevertheless, Table 3.2 probably shows a

Table 3.2 Comparison of hydrolysis rates of glycosides

Example	Glycoside	$t_{1/2}$/min
1	Methyl α-D-glucopyranoside (**3.24**)	32
2	Methyl β-D-glucopyranoside	19
3	Ethyl β-D-glucopyranoside	13
4	Phenyl α-D-glucopyranoside	1.2
5	Ethyl β-D-glucofuranoside	0.06
6	Methyl α-D-galactopyranoside (**3.1**)	6
7	Methyl β-D-ribopyranoside (**3.26**)	4
8	Methyl 2-deoxy-α-D-*arabino*-hexopyranoside (**3.25**)	0.02
9	Methyl 2-deoxy-α,β-D-*erythro*-pentopyranoside (**3.27**)*	< 0.005
10	Methyl 6-deoxy-α-L-galactopyranoside (**3.28**)**	1
11	Methyl 2-acetamido-2-deoxy-β-D-glucopyranoside (**3.29**)	100
12	Methyl 2-amino-2-deoxy-β-D-gluco-pyranoside, chlorohydrate (**3.30**)	2800
13	t-Butyl β-D-glucopyranoside	0.075
14	Triethylmethyl β-D-glucopyranoside	0.003
15	Methyl α-D-fructopyranoside (**3.31**)	0.003

*estimation;
** measured on the enantiomer.

satisfactory basis for comparison. The reader who wishes to make use of the published data for numerical purposes is encouraged to consult the work cited (Szejtli 1976).

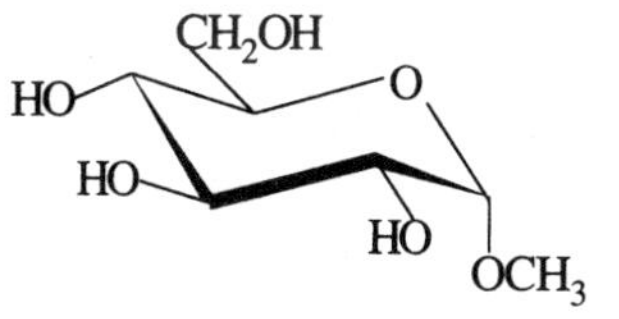

3.24 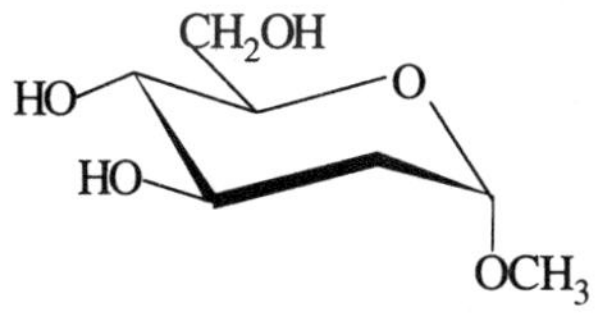 **3.25**

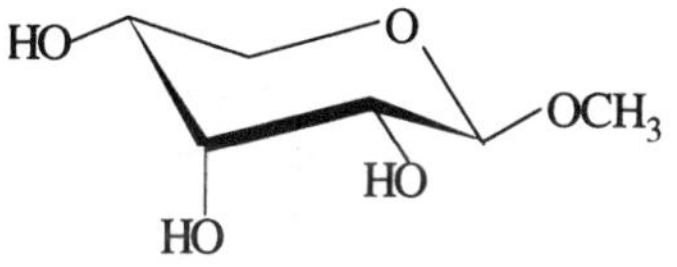

3.26 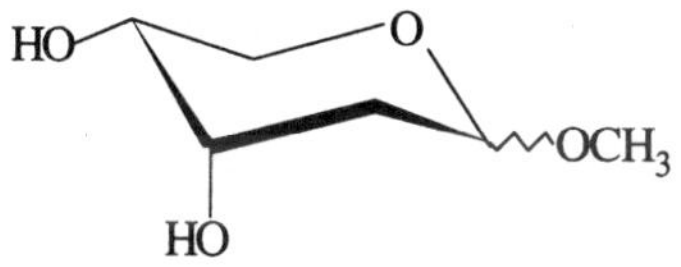 **3.27**

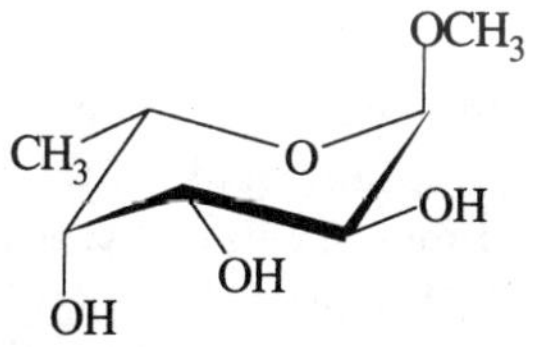

3.28 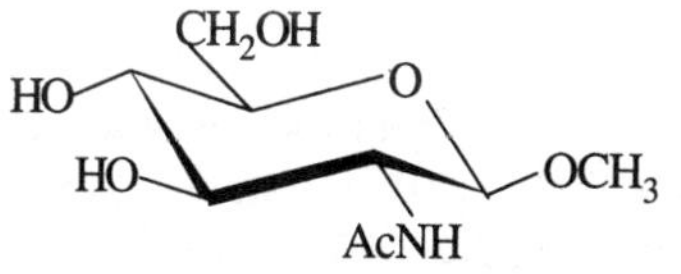 **3.29**

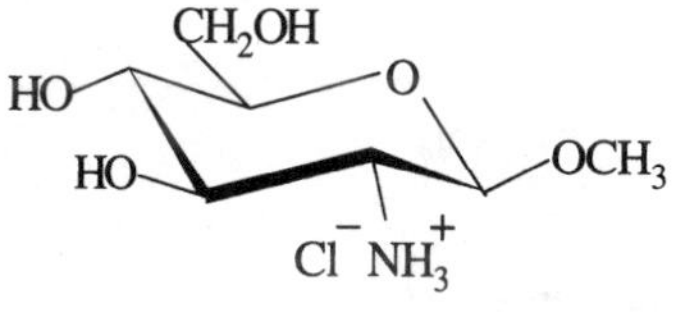

3.30 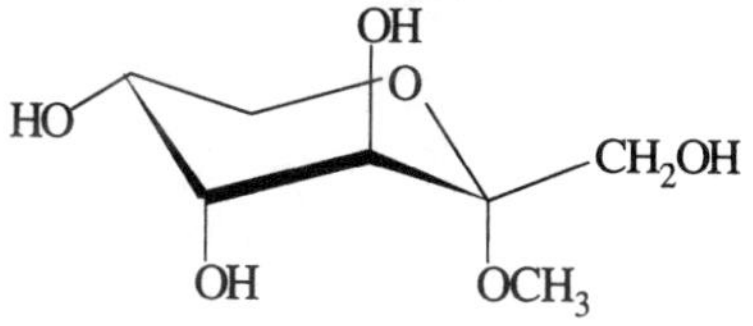 **3.31**

Examples 1 and 2 of Table 3.2 show that the anomer with an equatorial methyl group is hydrolysed approximately twice as fast as that with a methyl group axially disposed. This is characteristic of most pairs of glycosides. Note, however, that with glucopyranosides, the order is reversed above 132°C. Going from a methyl group to an ethyl group shows no considerable effect, but the rate of hydrolysis of phenyl glycosides is clearly higher (examples 1 and 4). A comparison of examples 3 and 5 shows that furanosides are extremely labile. When there is no hydroxyl group at position 2, hydrolysis is also considerably accelerated (compare first examples 1 and 8, then 7 and 9). A well-known property of simple acetals can be extended to the chemistry of glycosides: the hydrolysis of the diethyl acetal $CH_2OH–CH(OEt)_2$ is 300 times slower than that of $CH_3–CH(OEt)_2$. The effect is also shown, but to a lesser extent, when there is a hydroxyl missing at a position furthest from the acetal function; compare examples 6 and 10. Compound **3.28** is a glycoside of fucose which is an important sugar in living cells. It is necessary to keep in mind that in the chemistry of fucosylated oligosaccharides the fucosides are particularly labile.

Pyranosides with an amide function, such as **3.29** and its D-*galacto* isomer, are very important in living cells. At the same time as acid hydrolysis of the acetal function of **3.29** at 80°C in 1 N HCl ($k = 7.25 \times 10^{-3}$ min^{-1}), simultaneous hydrolysis of the amide ($k = 2.31 \times 10^{-3}$ min^{-1}) takes place which, for a notable fraction of the starting product, leads to the accumulation of the amino glycoside **3.30**, protonated in the medium. The positive charge on nitrogen conflicts with the protonation of the acetal oxygen giving a dication to the point where hydrolysis of **3.30** is very slow. Compare examples 1, 11, and 12 of Table 3.2. The oligosaccharides of cell walls contain *N*-acetylglucosamine or *N*-acetylgalactosamine units attached to the chain by glycosidic bonds and the determination of the exact separation conditions of these units by acid hydrolysis is especially important. Example 15 shows the fragility of ketosides. Among the natural ketose sugars, one which is of special interest is sialic acid, whose glycosides are hydrolyzed under very mild conditions. Glycosides of sialic acid combine the properties of ketone glycosides and those of sugars deoxygenated in the immediate neighbourhood of the acetal function. Sialic acid, in the form of sialoside, is located externally in the oligosaccharides of cell walls, as outlined in structure **3.32**, and is separated quantitatively by heating for 1 h in 0.1 M HCl at 80°C. Sometimes sialic acid takes on an acetylated form such as **3.33**. The analytical problem is to hydrolyse selectively without deacetylation, and 0.01 M HCl or formic acid is used for 1 h at 60°C, but the hydrolysis of the glycosidic function is not complete under these conditions (Schauer 1982).

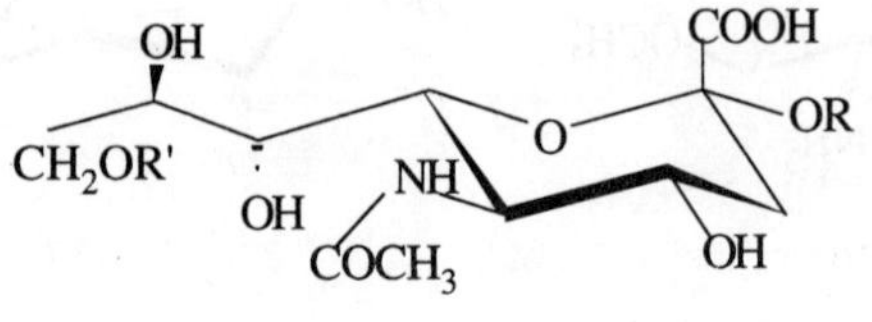

OH COOH O OR CH_2OR' NH OH $COCH_3$ OH

3.32 R'=H

3.33 R'=Ac

Examples 13 and 14 show that glycosides of tertiary alcohols are very labile. It is thought that their hydrolysis is carried out by a different mechanism from the one in Fig. 3.2 involving, as for the halides of these alcohols, a carbenium ion such as Me_3C^+.

Kinetic studies allow the activation parameters to be calculated, using equation (3.10).

$$\Delta G^{\ddagger} = \Delta H^{\ddagger} - T\Delta S^{\ddagger} \quad (3.10)$$

For 24 pyranosides at 60°C, the numerical equation (3.11a) is observed; other published values are of the same order.

$$+4.1 \leq \Delta S^{\ddagger}/\text{cal mol}^{-1}\ \text{deg}^{-1} \leq 23.0 \quad (3.11a)$$

average: 13.7

On the contrary, a negative value was observed for a limited number of furanosides (BeMiller 1967).

$$-11.1 \leq \Delta S^{\ddagger} \leq -8.3 \quad (3.11b)$$

These numbers can be interpreted as indicating a different hydrolysis mechanism (Bochkov and Zaikov 1979).

3.5.2 Enzymic hydrolysis and transfer

The glycosyl-hydrolase enzymes, more simply known as 'glycosidases', catalyse the hydrolysis of glycosidic bonds under conditions close to neutrality. Table 3.3 gives the list of commercial glycosidases which are moderately enough priced for their use in preparative chemistry. The efficiency of this catalysis is incredibly increased with respect to that of an acid. Direct comparison with alkyl glycosides is not possible because non-enzymic hydrolysis is not visible in conditions close to neutrality, but the reaction rate can be measured at pH 5 with the β-D-glucopyranoside of *p*-nitrophenol **3.34**. In water at 25°C (pH 5.0) the very weak pseudo-monomolecular constant is $52 \times 10^{-12}\ s^{-1}$. In the presence of the β-D-glucosidase of bitter almonds, under the same conditions, a k value equal to $78\ s^{-1}$ is measured, that is, around 10^{12} times higher (Legler 1990).

Table 3.3 Properties of a few commercial glycoside hydrolases

Enzyme	Source	Type
α-D-Glucosidase	yeast	p (a–a)
β-D-Glucosidase	almonds	p (e–e)
α-D-Galactosidase	green coffee beans	p (a–a)
β-D-Galactosidase	*Escherichia coli*	p (e–e)
α-D-Mannosidase	Jack beans	p (a–a)
α-D-Fucosidase	bovine kidney	p (a–a)
α-D-Neuraminidase	*Arthrobacter ureofaciens,*	p (e–e)
(Sialidase)	*Clostridium perfringens*	p (e–e)
N-Acetyl-β-D-gluocosaminidase	bovine kidney	–
β-D-Fructosidase (Invertase)	yeast	f (r)

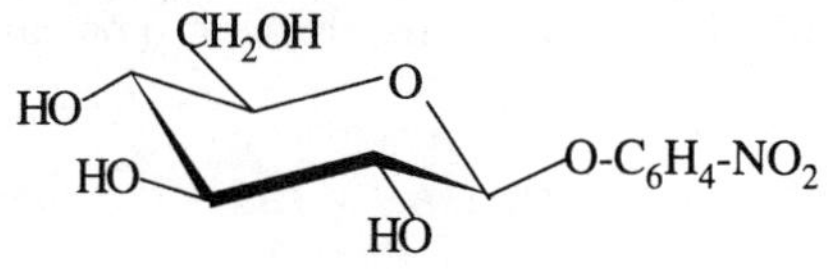

3.34

Rates were also compared under conditions in which they both obey a second-order law in relation to substrate and catalyst concentrations. It was found that the enzyme was 10^{14} times more efficient than hydronium ions. Such an increase in rate corresponds to a drop in the activation energy close to 18 kcal mol^{-1}.

Distinction is made between the '*exo*-glycosidases' and '*endo*-glycosidases'. The former hydrolyse terminal glycosidic bonds in the oligosaccharide chains (see Chapter 9) and the latter hydrolyse internal bonds. Within the framework of this section essentially devoted to preparative chemistry, we shall only deal with the *exo*-glycosidases because they also catalyse the hydrolysis of alkyl and aryl glycosides. They are quite numerous. Every glycosidase is specific towards the configuration of the sugar involved in the glycosidic bond by its anomeric oxygen. It is relatively indifferent to the nature of the organic group linked to this oxygen. This observation allows the classification by groups: α-D-glucosidases which hydrolyse all the α-D-glucosides, β-D-galactosidases which hydrolyse all the β-D-galactosides, etc. Within each group, there are small variations in property according to their source. If we now look closely at the configuration, we see that there are four different cases (*a,b,c,d*) for the pyranosides, and two (*e,f*) for the furanosides (Fig. 3.4). The hemiacetal is represented as it is immediately after its separation from the active site. Hydrolyses proceed globally with configurational retention (*a,b,e*) or inversion (*c,d,f*). The symbols of classification for *a,b,c,d,e* and *f* are thus p(e-e), p(a-a), p(e-a), p(a-e), f(r), and f(i), respectively. All the glycosidases in Table 3.3 function with retention of configuration.

Because of the speed of mutarotation, it is difficult to know with certainty the anomeric configuration of the hemiacetal leaving the enzyme. The answer can be found via another reaction of glycosidases which occurs with retention; they catalyse the glycosyl transfer of one aglycon to another according to equation (3.12) where G–OR and G–OR′ are two glycosides of the G–OH sugar.

$$\text{G–OR} + \text{R}'\text{–OH} \rightleftharpoons \text{G–OR}' + \text{R–OH} \qquad (3.12)$$

Hydrolysis thus appears as the special case of transfer to water, H–OH. There is no longer any ambiguity as to the anomeric configuration of the glycoside produced, incapable of mutarotation. We find that it is the same as that of the starting glycoside. This conclusion is moreover essential according to the reversibility of reaction (3.12) by applying the microreversibility principle. If the anomeric configuration is not the same in both cases, the enzyme should be able to hydrolyse the two opposing α- and β-configurations by the same path.

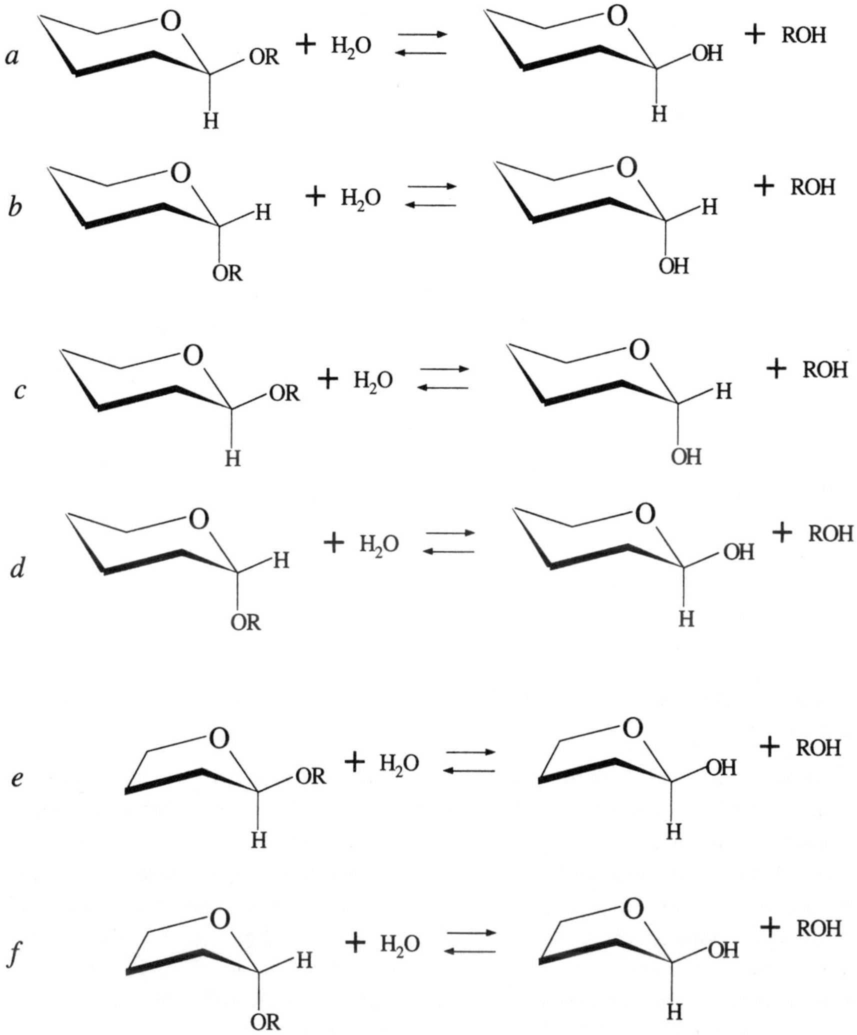

Fig. 3.4 Possible enzymic hydrolysis routes of a glycoside.

Reaction (3.12) suggests synthetic applications of glycosidases other than hydrolysis. Thus, we observe reaction (3.13) in aqueous solution in the presence of an α-D-galactosidase.

This is how a disaccharide, **3.35**, is rapidly constructed, characteristic of the human blood group B. Although the yield is mediocre, other known purely chemical methods involve several steps and in the end, are hardly more efficient. In glycosidase protocols, yields are low but the cost for raw materials is negligible with respect to the value of oligosaccharides. The disadvantage does not lie in the low yields, but rather in the necessity to separate the products from an

(3.13)

HO CH2OH O HO HO OAr + HO CH2OH O HO HO OCH3 →

HO CH2OH O HO HO O HO CH2OH O HO OCH3

3.35

excessive amount of very close and polar substances, a problem which has not yet been resolved economically on a large scale.

The mechanism of enzymic hydrolysis is still a controversial subject (Sinnott 1990; Legler 1990). There is one point of agreement in that retention is the result of two consecutive inversions, one at the time of the attachment to the enzyme, and the other when the glycosyl is transferred to the acceptor.

3.5.3 Glycosides as protected sugars under neutral or alkaline conditions

Glycosides are only hydrolysed in an alkaline medium under extreme conditions, entirely foreign to current synthetic practices. For every reaction which is carried out in a neutral or alkaline medium, glycosides have cyclic polyol properties. Hemiacetal sugars cannot tolerate alkaline conditions because of the aldol nature of the carbonyl tautomer. To effect transformations at the level of hydroxyl groups involving alkaline conditions, it is better to first transform the sugar into a glycoside. For example, the benzylated fucose **3.37** is a required intermediate in the activation of fucose to introduce the α-fucopyranosyl unit into oligosaccharides. This is prepared by the following three steps: glycosidation of fucose to give methyl glycoside **3.28**, benzylation of hydroxyls with benzyl chloride and sodium hydride in *N,N*-dimethylformamide to give **3.36**, and finally acid hydrolysis to give **3.37**.

The route via methyl glycoside as the protected form and the final return to the hemiacetal by acid hydrolysis is only possible if the final product is stable under the necessary acidic conditions. Thus, if the substituents on the intermediate **3.36** were acetyl rather than benzyl, they would disappear partially during deprotection. Most sugars themselves are very stable in an acidic medium. However, fructose is partially transformed into levulinic acid ($CH_3COCH_2CH_2COOH$) in an acidic medium at 100°C. As to deoxyribose, very mild acidic conditions suffice for its quantitative transformation into levulinic acid. The most frequent case is when acid removal of the protecting substituent

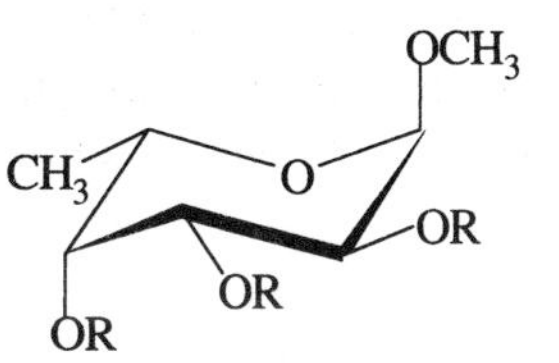

3.28 R=H
3.36 R=CH_2Ph(Bn)

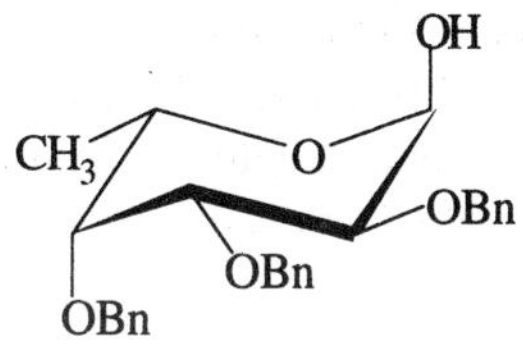

3.37

risks hydrolysing other desired glycosidic bonds. In this case, protection will be carried out by preparing a glycoside with benzylic alcohol, and removal of the protecting group under neutral conditions by hydrogenolysis over palladium on charcoal (Fig. 3.5).

3.6 Glycosylamines and nucleosides

3.6.1 General

These compounds will be dealt with here because of their structural analogy with oxygenated glycosides, although their role is quite different. These intermediates are not often used in synthesis. Universally widespread natural structures are found in this family. Certain transformations described below are good models of well-established biosynthetic pathways.

3.6.2 Glycosylamines (Paulsen and Pflugthaupt 1980)

Preparation

Treatment of aldohexoses or aldopentoses with liquid ammonia in an alcohol solution replaces the anomeric hydroxyl group with NH_2 and gives an anomeric mixture of glycosylamine pyranosides. The equatorial derivative predominates. Its direct crystallization is often observed from the reactive medium. Thus glucose gives β-D-glucosylamine **3.38**. The primary and secondary aliphatic amines lead to substituted glycosylamines **3.39** and **3.40**. With the less nucleophilic arylamines, it may be necessary to use a mild acid catalyst such as an ammonium salt in order to obtain glycosamines **3.41**, generally quite crystalline. The ketoses treated as described above give alkyl ketosylamines, which are very

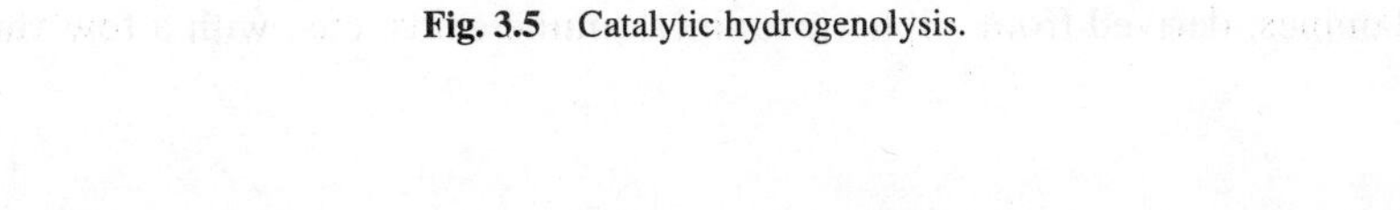

Fig. 3.5 Catalytic hydrogenolysis.

labile in an acidic medium. We see that the carbonyl functions of free sugars do not give Schiff bases. These can be obtained with aldehydo-sugars whereby the hemiacetal is prevented from forming by blocking the alcohol functions (hence the name *aldehydo*).

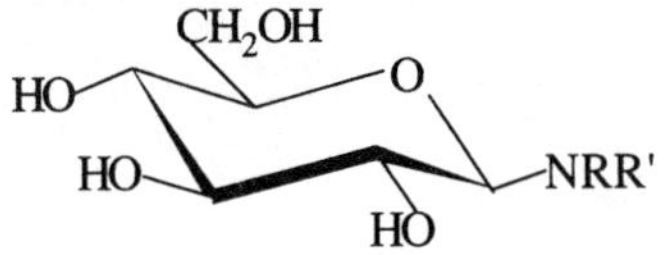

3.38 R=R'=H **3.40** R,R' = alkyl

3.39 R=H, R'= alkyl **3.41** R=H, R'= aryl

As in the syntheses of oxygenated glycosides,it may be necessary to start from an activated sugar derivative as, for example, the peracetylated halogenose **3.7**. Likewise, the formation of a 1,2-*trans* derivative is observed when there is a participating group at position 2 of the sugar. By way of example, we will describe in detail the preparation of compound **3.42**, a structure that is found at the anchor point of the oligosaccharide chain on the polypeptide backbone of glycoproteins (Garg and Jeanloz 1985; Augé *et al.* 1989) (see Chapter 13). The amide function of nitrogen would probably not be nucleophilic enough to consider doing a direct condensation on the free sugar. Starting from the peracetylated halogenose **3.43** derived from *N*-acetylglucosamine, substitution by tetrabutylammonium azide gives the equatorial azide **3.44**, reduced to amine **3.45** by catalytic hydrogenation on platinum. Amine **3.45** is condensed with the free carboxyl group of the partially protected aspartic acid **3.46** by means of *N,N'*-dicyclohexylcarbodiimide. The synthesis is completed by alkaline hydrolysis of the ester functions and liberation of the α-amino function of the aspartic acid part by catalytic hydrogenation.

Properties of glycosylamines

First there is mutarotation in solution. Figure 3.6 illustrates one possible mechanism. Protonation of the ring oxygen is followed by ring opening to give an intermediate in which the anomeric carbon has lost its chirality. The same intermediate can evolve into the formation of the hydrolysis product which is very fast with glycosylamines, except in extreme conditions where the pH is greater than 9 or less than 1.5. In a very acidic medium, protonation of nitrogen introduces a positive charge which conflicts with that of oxygen. For example, β- D-glucosylpiperidine is stable for 17 h at 0°C in 2 N HCl. The hydrolysis rate is at a maximum when the pH is around 5. Arylglycosylamines are more stable.

Upon heating in the presence of traces of acid, *N*-glycosylamines rearrange to arylamino derivatives of ketoses. For example, glucosylamine **3.41** gives the amino fructose **3.47**. In fact, this reaction takes place with the most varied glycosylamines, derived from aliphatic amines, amino acids, etc., with a few varia-

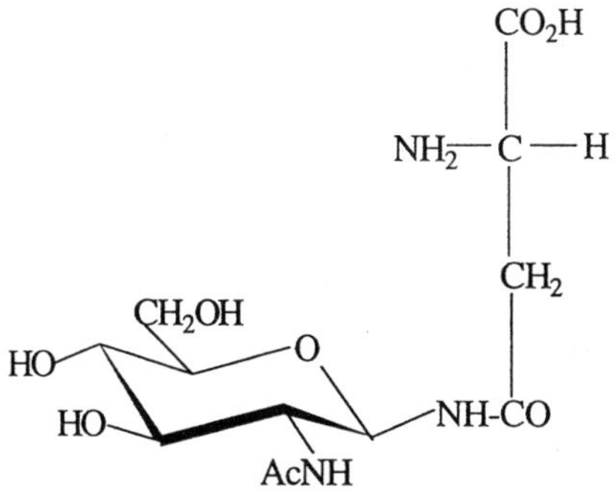

3.42

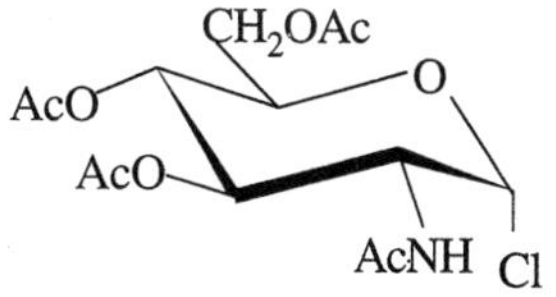

3.43

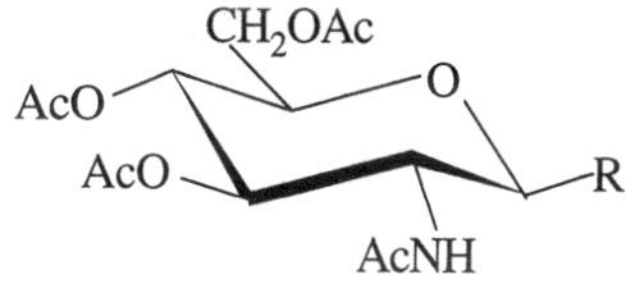

3.44 R=N_3

3.45 R=NH_2

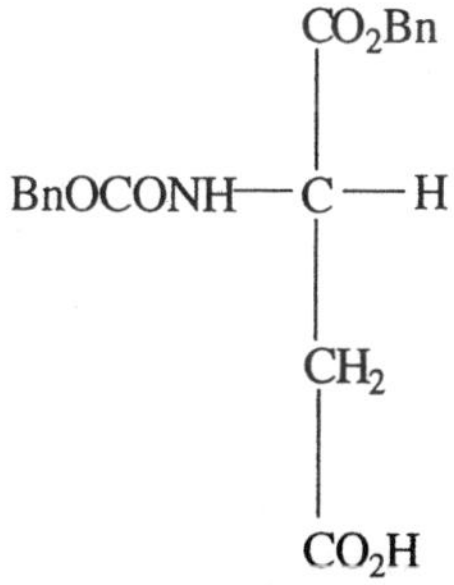

3.46

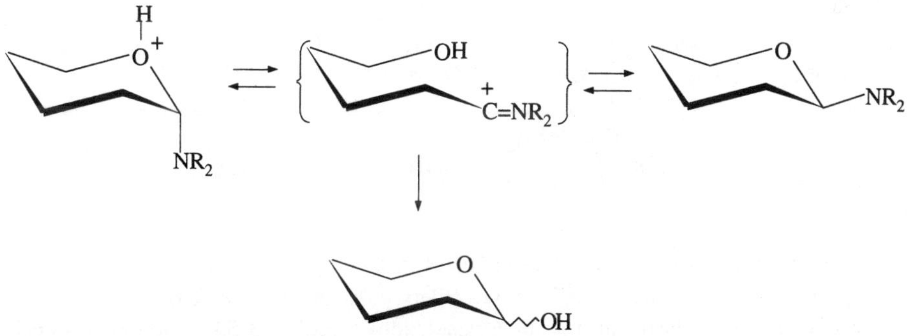

Fig. 3.6 Proposed mechanism for the hydrolysis of a glycosylamine.

tions in the procedures. This is known as the Amadori rearrangement. The biosynthesis of indole in enterobacteria involves such a rearrangement in which each step is catalysed enzymically. An activated phosphate derivative of ribose **3.48** condenses with anthranilic acid **3.49** to give *N*-aryl ribosylamine **3.50**. Rearrangement of **3.50** leads to **3.51** having a free carbonyl group. Electrophilic attack on the *ipso* carbon of the carboxyl group, followed by dehydration, gives the indole ring linked to phosphate triol **3.52**. The latter is detached thereafter by retro-aldol cleavage.

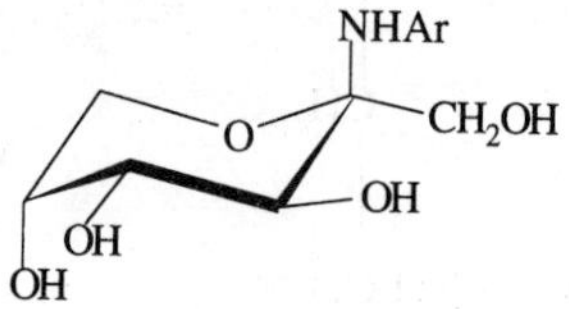

NHAr
O
CH_2OH
OH
OH
OH

3.47

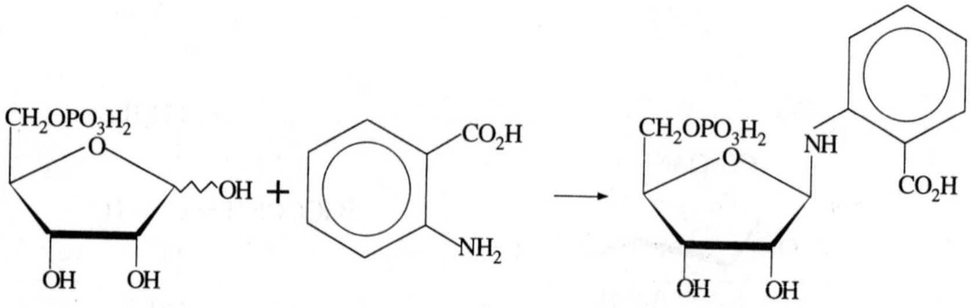

$CH_2OPO_3H_2$
O
OH +
CO_2H
NH_2
OH OH

3.48 **3.49**

$CH_2OPO_3H_2$
O
NH
CO_2H
OH OH

3.50

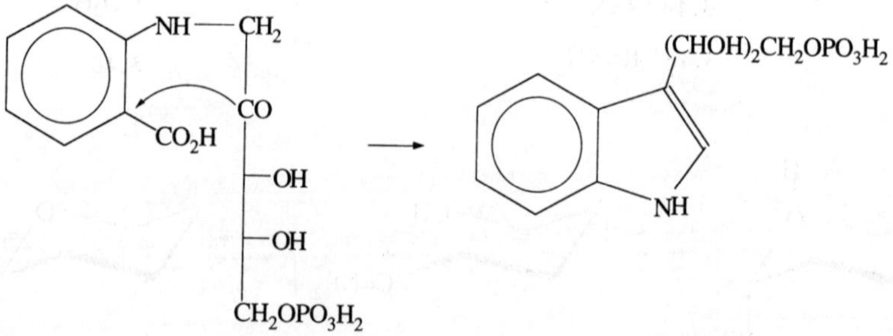

NH — CH_2
CO
CO_2H
OH
OH
$CH_2OPO_3H_2$

3.51

$(CHOH)_2CH_2OPO_3H_2$
NH

3.52

Another remarkable rearrangement in this series is that of Heyns. Ketosylamines are transformed into 2-amino-2-deoxyaldoses *in a stereospecific fashion.* Fructose in alcoholic ammonia gives fructosylamine **3.53** which is spontaneously transformed into 2-amino-2-deoxy-D- glucose **3.55**. The same spontaneous reaction is observed with other fructosylamines substituted at nitrogen, as in the example where **3.54** gives **3.56**. Sometimes it is necessary to use an acid catalyst. It is by this rearrangement that cells manufacture the amino sugar 2-amino-2-deoxy-D-glucose, fundamental to their existence, starting from fructosyl phosphate. The donor of the NH_2 group is the amino acid glutamine. We have already pointed out this biosynthesis, which explains why *N*-acetyl glucosamine belongs to the D-series of sugars (Chapter 1, Section 1.2).

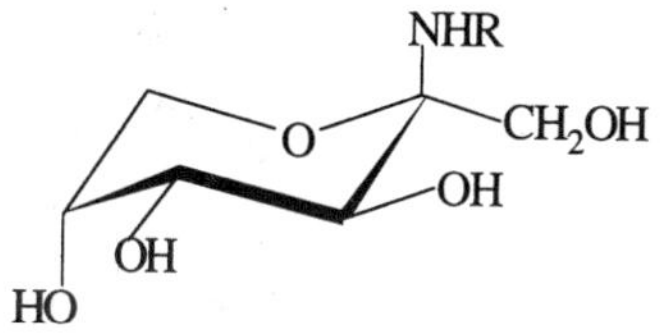

3.53 R=H
3.54 R=$CHMe_2$

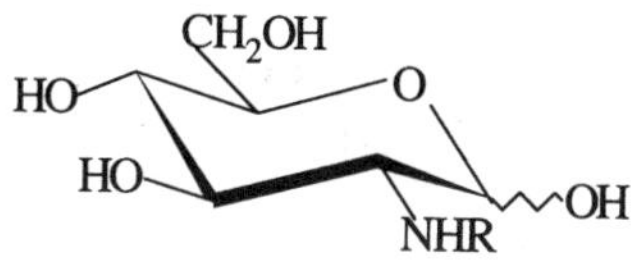

3.55 R=H
3.56 R=$CHMe_2$

3.6.3 Nucleosides (Secrist 1988)

In these aminoglycosides, nitrogen is part of a heterocycle and the sugar is D-ribose or its deoxygenated derivative at position 2, 'deoxyribose' (whose correct name is 2-deoxy-D-*erythro*-pentose), both of which are furanoses. Ribosides, linked by phosphodiester functions between the alcohol functions at positions 3 and 5, form ribonucleic acid (RNA) and deoxyribosides form deoxyribonucleic acid (DNA) in the same manner. The progressive elucidation of the genetic role of these two molecules is without a doubt the most important discovery in the last half of this century, and the author feels that any educated person ought to be aware of formulas **3.57**, **3.58**, **3.59** and **3.60**, on the one hand, and **3.61**, **3.62**, **3.63** and **3.64** on the other, whatever his or her educational background. The chemistry of DNA and RNA does not fit into the framework of this book, and only the aminoglycoside aspect will be treated. Hence there are no grounds for comparing the length of this paragraph with the importance of these molecules.

In nucleic acids, especially in transfer ribonucleic acid (tRNA), there are, in addition, about 50 modified nucleosides, generally rather close to the basic types. Finally, in nature there are a good number of nucleosides bearing diverse heterocyclic bases, and varying sugars, often possessing more or less important therapeutic properties. Since the products of this last category are not universal constituents of the living cell, they are also outside the domain of this work.

Fundamental nucleosides are easily prepared by chemical or enzymic hydrolysis starting from abundant sources such as yeast RNA and the DNA of the soft roe of fish, both commercially available. The syntheses that we are going to present are only of interest in their extension to the preparation of modified nucleosides, or isotopically labelled compounds.

Adenosine is obtained by condensation of the activated ribofuranosyl chloride **3.65** with the chloromercurial derivative of 6-benzamidopurine **3.66**, followed by debenzoylation (reaction (3.14)).

Pyrimidine nucleosides are often prepared by the Vorbrüggen method (Niedballa and Vorbrüggen 1970). The silylated derivative of the base is condensed with the peracetate of the sugar at room temperature in the presence of $SnCl_4$. Reaction (3.15) gives the preparation of homocytidine **3.67** by this method (David and de Sennyey 1979). Vorbrüggen has recently introduced other catalysts. The derivative β-1,2-*trans*, isolated in excellent yield, is again

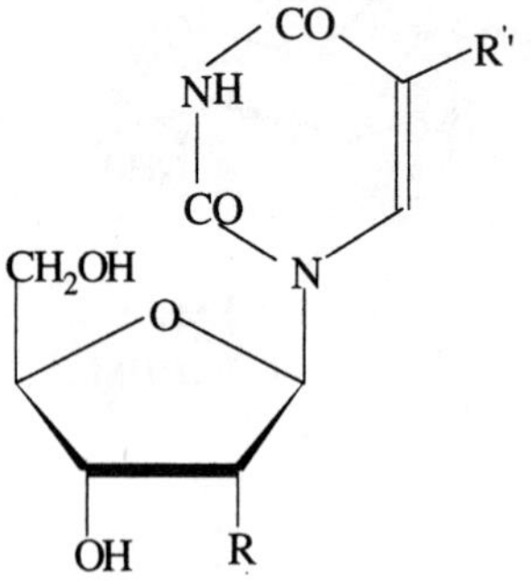

3.57 R=OH, R'=H

3.58 R=H, R'=CH_3

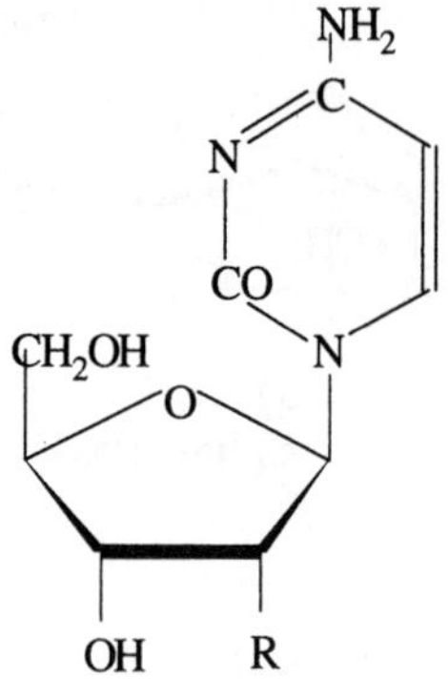

3.59 R=OH

3.60 R=H

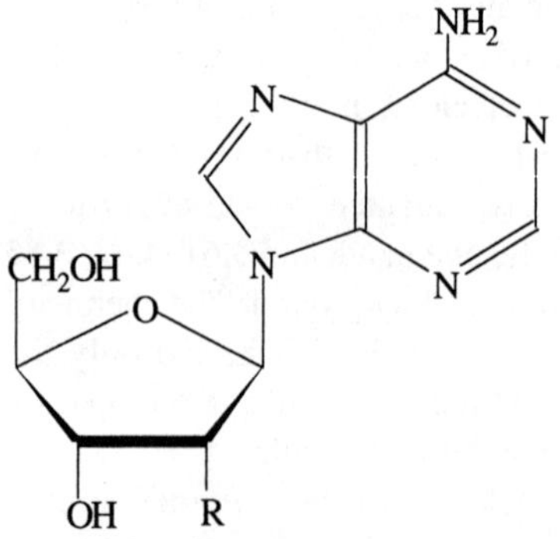

3.61 R=OH

3.62 R=H

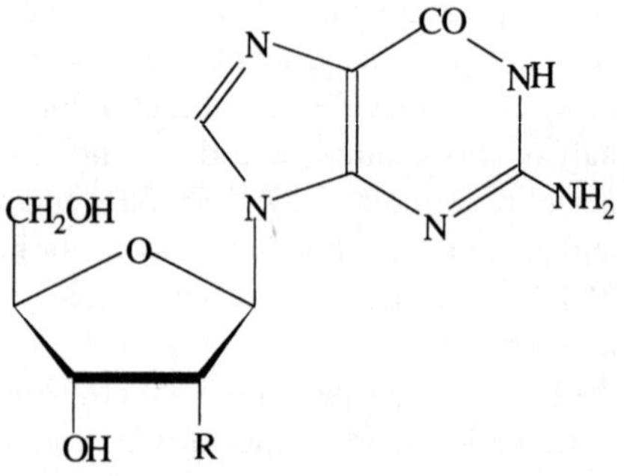

3.63 R=OH

3.64 R=H

(3.14) **3.65** + **3.66** → → **3.61**

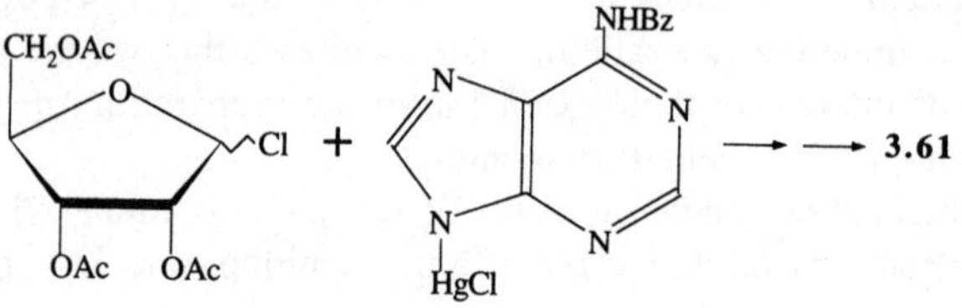

the result of the presence of a participating group at C-2. Moreover, when these methods are applied to the preparation of deoxyribonucleosides, a mixture of α- and β-nucleosides are generally obtained.

(3.15)

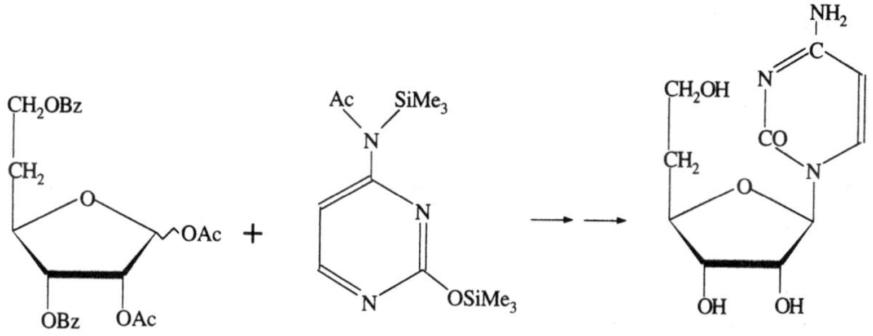

3.67

For other compounds, it was found to be advantageous to build the heterocycle on the nitrogen of the ribosylamine. Treatment of ribose by ammonia gives ribopyranosylamine **3.68** which undergoes ring contraction on isopropylidenation (Section 5.2); **3.69** is obtained. Condensation of **3.69** with the methyl cyanoacetate derivative **3.70** gives β-ribosylimidazole **3.71**. Deacetalation and decarboxylation of **3.71** leads to nucleoside **3.72**, of which the phosphate is the natural precursor for nucleosides of adenine (Cusack *et al.* 1973) and for the pyrimidine of thiamine (Estramareix and David 1990).

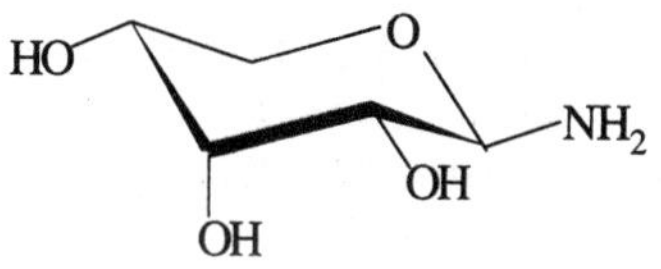

3.68

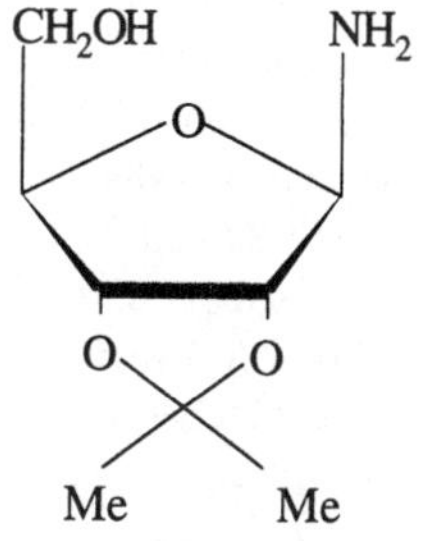

3.69

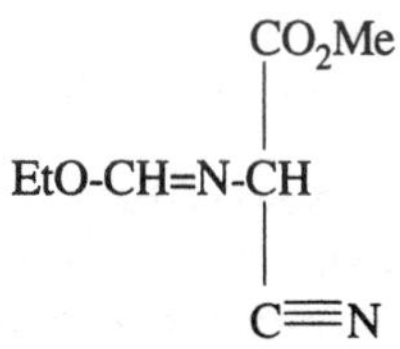

3.70

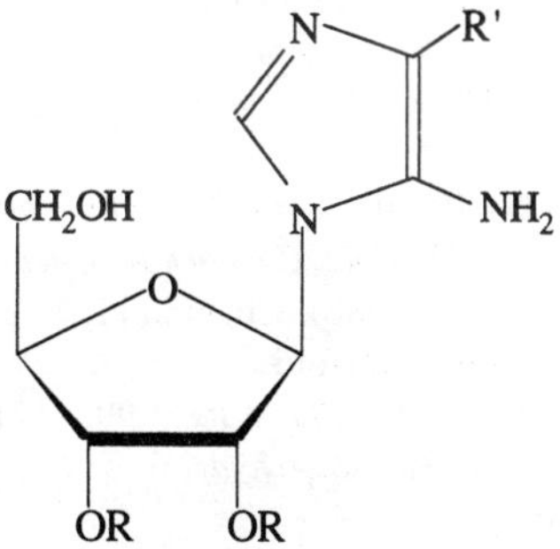

3.71 R,R=CMe_2 R'=CO_2Me

3.72 R=R'=H

References

Angyal, S. J. and Dawes, K. (1968), *Aust. J. Chem.*, **21**, 2747–2760.

Angyal, S., Evans, M. E., and Beveridge, R. J. (1980), *Methods Carbohydr. Chem.*, **8**, 233–235.

Augé, C., Gautheron, C., and Pora, H. (1989), *Carbohydr. Res.*, **193**, 288–293.

Augestad, I. and Berner, E. (1954), *Acta Chem. Scand.*, **8**, 251–256.

BeMiller, J. N. (1967), *Adv. Carbohydr. Chem.*, **22**, 25–108.

Bethell, G. S. and Ferrier, R. J. (1973), *Carbohydr. Res.*, **31**, 69–80.

Bochkov, A. F. and Zaikov, G. E. (1979), *The chemistry of the O-glycosidic bond*, Pergamon Press, Oxford.

Bollenback, G. N. (1963), *Methods Carbohydr. Chem.*, **2**, 326–328.

Brimer, K. and Vasella, A. (1989), *Helv. Chim. Acta*, **72**, 1371–1382.

Coleman, G. H. (1963), *Methods Carbohydr. Chem.*, **2**, 397–399.

Cusack, N. J., Hildick, B. J., Robinson, D. H., Rugg, P. W., and Shaw, G. (1973), *J. Chem. Soc., Perkin Trans I*, 1720–1731.

David, S. and de Sennyey, G. (1979), *Carbohydr. Res.*, **77**, 79–97.

Edward, J. T. (1955), *Chem. Ind. London*, 1102.

Estramareix, B. and David, S. (1990), *Biochim. Biophys. Acta*, **1035**, 154–160.

Ferrier, R. J. (1988), The synthesis and reaction of monosaccharide derivatives. In *Carbohydrate chemistry* (ed. J. F. Kennedy), Clarendon Press, Oxford.

Ferrier, R. J. and Collins, P. M. (1972), Chapter 3. In *Monosaccharide chemistry*, Penguin, Harmondsworth.

Garg, H. G. and Jeanloz, R. W. (1985), *Adv. Carbohydr. Chem. Biochem.*, **43**, 135–201.

Green, J. W. (1966), *Adv. Carbohydr. Chem.*, **21**, 95–142.

Halcomb, R. H. and Danishefsky, S. J. (1989), *J. Am. Chem. Soc.*, **111**, 6661–6666.

Hannessian, S. and Banoub, J. (1980), *Methods Carbohydr. Chem.*, **8**, 243–254.

Isbell, H. S. and Frush, H. L. (1940), *J. Res. Nat. Bur. Stand.*, **24**, 125–151.

Legler, G. (1990), *Adv. Carbohydr. Chem. Biochem.*, **48**, 319–384.

Lemieux, R. U. and Huber, G. (1956), *J. Am. Chem. Soc.*, **78**, 4117–4119.

Lubineau, A. and Fischer, J. C. (1991), *Synth. Commun.*, **21**, 815–818.

Mowery, D. F. (1975), *Carbohydr. Res.*, **43**, 233–238.

Niedballa, U. and Vorbrüggen H. (1970), *Ang. Chem., Int. Ed. Engl.*, **9**, 461–462.

Overend, W. G. (1972), Glycosides. In *The carbohydrate*, Vol. 1A. (2nd edn.) (ed. W. Pigman and D. Horton), Academic Press, New York.

Paulsen, H. and Pflugthaupt, K. W. (1980), *The carbohydrates, chemistry and biochemistry*, Vol. 1B. (2nd edn.). (ed. W. Pigman and D. Horton), p. 881, Academic Press, New York.

Ritchie, C. D. (1990), *Physical organic chemistry*, Marcel Dekker, New York.

Schauer, R. (1982), *Adv. Carbohydr. Chem. Biochem.*, **40**, 131–234.

Secrist, III, J. A. (1988), Chapter 11. In *Carbohydrate chemistry* (ed. J. F. Kennedy), Clarendon Press, Oxford.

Sinnott, M. L. (1990), *Chem. Rev.*, **90**, 1171–1202.

Szejtli, J. (1976), *Säuerhydrolyse Glykosidischer Bindungen*, Akadémiai Kladó, Budapest.

4 Nomenclature

4.1 Introduction

Until now the topic of carbohydrate nomenclature has been deferred because of its rather tedious nature. But we hope the reader's interest was sufficiently stimulated during the first three chapters for the reader to feel brave enough to tackle the present one. Only the nomenclature rules applicable to the work described in this book will be given. Knowing the correct terms will allow the reader to undertake reading of the primary literature and enable him or her to extrapolate these terms in unexpected situations due to a minimal sense of the general picture. These rules are extremely practical. The corresponding names are those utilized (in minute detail) in the tables of chemical compounds of the *Chemical Substances Index* of *Chemical Abstracts*. They are also the names used in the experimental sections of papers and theses, as paragraph headings describing procedures.

4.2 Nomenclature of aldoses

4.2.1 Common names of sugars and configurational symbols

The basic names of aldoses having a chain with 3–10 carbons or more are called triose, tetrose, pentose, hexose, heptose, octose, nonose, decose, etc. The chain is numbered starting from the carbon bearing the aldehyde function, real or potential. To these names are added a symbol which describes the configuration of the hydroxyl groups. These symbols are derived from common names of sugars. Table 4.1 gives the configuration and common name of sugars of the D-series, from C-3 to C-6. Sugars are drawn according to the Fischer projection. To find the systematic name of a sugar derivative, we must always come back to this projection.

This table should be completed by a table of the L-series, the members of which are enantiomers of those in Table 1. Remember that the D-series is defined by the orientation to the right of the secondary alcohol hydroxyl group with the highest-numbered atom. Sometimes the series descriptor is omitted for glucose, galactose, mannose, and ribose. In this case we must remember that they belong to the D-series. The recommended common names in Table 4.1 do not come from the systematic nomenclature.

The sugars in Table 4.1 are distinguished by the number and relative position of the chiral centres which are vicinal secondary alcohol functions. In a general manner, their names (modified) serve as a basis for the description of the

Table 4.1 Configuration of pentoses and hexoses of the D-series

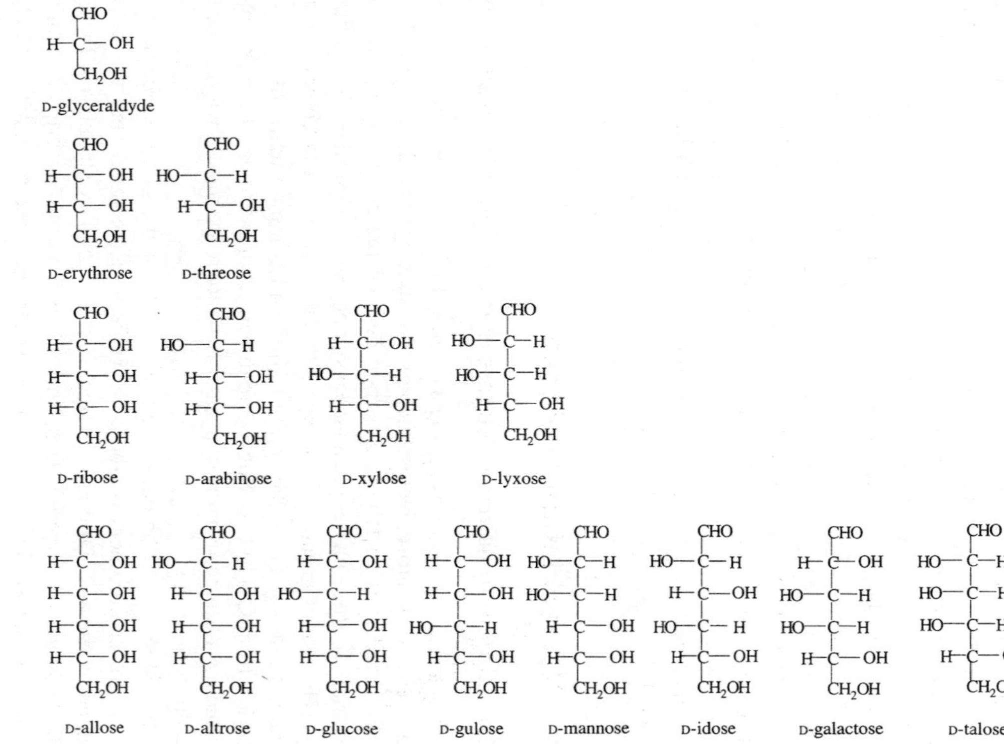

configuration of the chiral centres on a chain, *even if they are not adjacent*. The configuration of D-glyceraldehyde is called D-*glycero*. For the other configurations these symbols were coined by suppressing the suffix 'se'. To avoid any confusion, they are written in italics and without a capital letter, as in D-*erythro*, D-*ribo*, L-*galacto*, etc. Some examples of systematic names are D-*ribo*-pentose (ribose), D-*gluco*-hexose (glucose). The 'deoxyribose' **4.1** and its isomers **4.2** and **4.3** are the three possible D-*erythro*-pentoses.

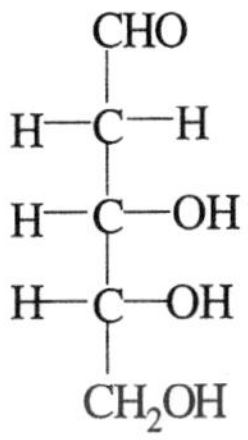

CHO
H—C—H
H—C—OH
H—C—OH
CH_2OH

4.1

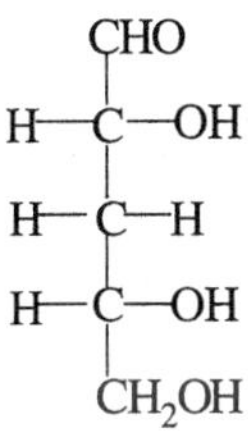

CHO
H—C—OH
H—C—H
H—C—OH
CH_2OH

4.2

CHO
H—C—OH
H—C—OH
H—C—H
CH_2OH

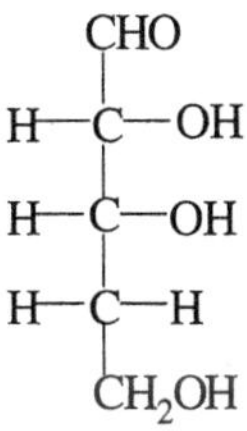

4.3

For heptoses and higher sugars, there may be more than four chiral centres on the chain. It was not considered desirable to create new names to designate, for example, all the possible configurations of a system having five secondary alcohol functions. These configurations are described by an association of symbols. Starting with the closest chiral centre of the aldehyde function, the chiral centres are brought together in groups of four, until there are only three, two, or one remaining which form the last group on the side of the non-reducing end. A configurational symbol can be attributed to each of these groups. In the systematic name, they are enumerated in the order starting from the non-reducing end: D-*glycero*-D-*gluco*-heptose **4.4** or L-*ribo*-D-*manno*-nonose **4.5**.

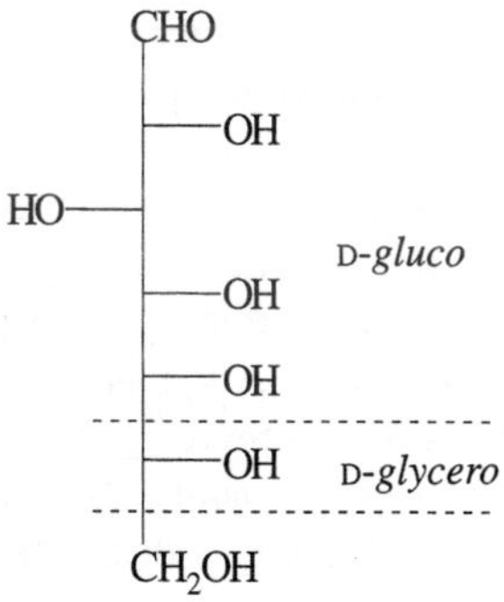

CHO
—OH
HO—
—OH
—OH
D-*gluco*
—OH
D-*glycero*
CH_2OH

4.4

CHO
HO—
HO—
—OH
—OH
D-*manno*
HO—
HO—
HO—
L-*ribo*
CH_2OH

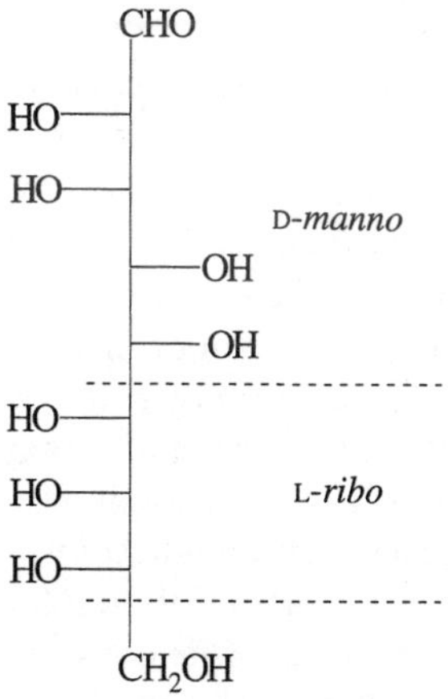

4.5

4.2.2 Deoxygenated sugars

The replacement of an alcohol hydroxyl group by a hydrogen is indicated by the prefix ‘deoxy’, preceded by a number which gives the deoxygenated position. The complete systematic names can now be given for the three sugars **4.1**, **4.2**, and **4.3**, i.e. 2-deoxy-D-*erythro*-pentose, 3-deoxy-D-*erythro*-pentose and 4-deoxy-D-*erythro*-pentose, respectively. Care must be taken with this nomenclature not to introduce a term which suggests the presence of more chiral centres than there are in reality. This is why the common name ‘deoxyribose’ is a misnomer because this sugar has only two asymmetric carbons whereas the root ‘ribo’ suggests the presence of a ribose configuration with three asymmetric carbons. Briefly, a chiral centre should not be introduced at one end of a systematic name only to erase it at the other end.

Deoxygenation of the primary alcohol function does not eliminate a chiral centre. For example, the most abundant natural enantiomer of fucose, L-fucose **4.6**, is systematically called 6-deoxy-L-*galacto*-hexose.

```
        CHO
         |
  HO—C—H
         |
     H—C—OH
         |
     H—C—OH
         |
  HO—C—H
         |
        CH3
```

4.6

4.2.3 Sugars substituted by NRR′, F, Cl, Br, I, N_3, *S*-alkyl, and *S*-phenyl

First the non-oxygenated group is replaced by a hydroxyl group, which gives the basic configuration. Thus the configuration of the amino derivative **4.7** is that of D-*galacto*-hexose **4.8**. The relationship between **4.8** and **4.7** is indicated by two prefixes of which one is deoxy, introduced in the preceding paragraph, and the other specifies the nature of the substituent. The most common descriptors are amino, acetamido, fluoro, chloro, bromo, iodo, azido, thioalkyl, and thiophenyl. The location is given by a number which precedes each prefix as, for example, compound **4.7** is 2-amino-2-deoxy-D-*galacto*-hexose. The names galactosamine, glucosamine, and mannosamine are the names for 2-amino-2-deoxy-D-*galacto*-D-*gluco*-, and-D-*manno*-hexoses.

The same principle allows **4.9** to be designated as 3-azido-3-deoxy-D-*allo*-hexose, **4.10** as 4-deoxy-4-iodo-D-*talo*-hexose, and **4.11** as 5-deoxy-5-thiomethyl-D-*ribo*-pentose. Prefixes are placed in alphabetical order, before the configurational descriptors.

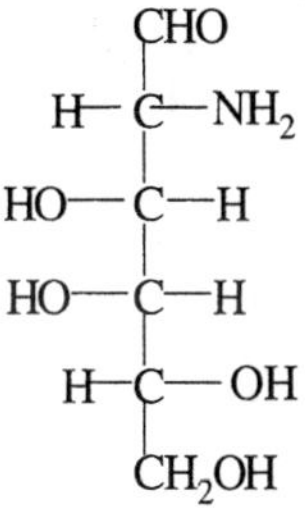

4.7 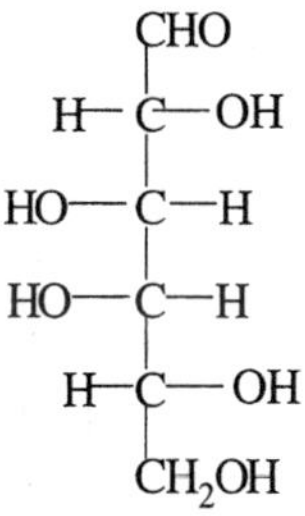 4.8

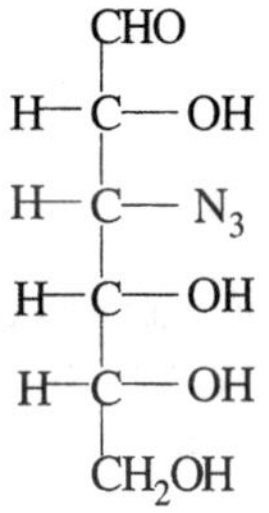

4.9 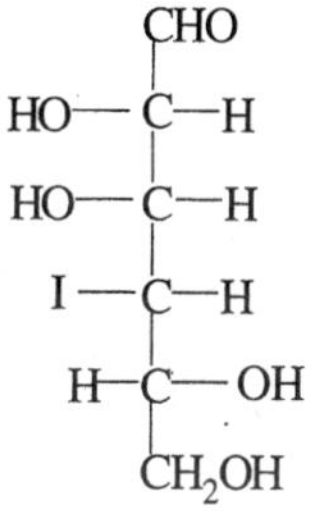 4.10 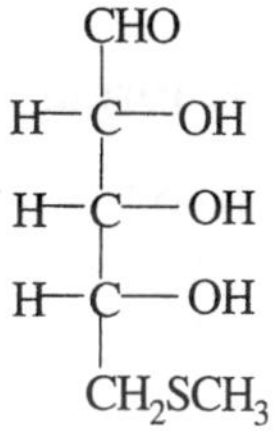 4.11

4.2.4 Derivatives substituted on the oxygen atom

Substitution on the oxygen atom is indicated by a capital *O*- (in italics followed by a hyphen), then the name of the substituent, preceded by a number indicating the position of the oxygen atom substituted on the chain. Identical substituents are not repeated. Some examples are 3-*O*-benzyl -D-*galacto*-hexose **4.12**, 3-*O*-benzyl-6-*O*-methyl-D-*galacto*-hexose **4.13**, and 3,6-di-*O*-benzyl-D-*galacto*-hexose **4.14**. For these derivatives from sugars as shown in Table 4.1 (or from enantiomeric sugars), these names are simplified to 3-*O*-benzyl-D-galactose, 3,6-di-*O*-benzyl-D-galactose, etc.

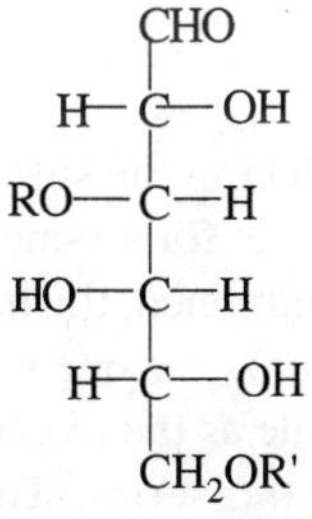

4.12 R=Bn, R'=H

4.13 R=Bn, R'=CH_3

4.14 R=R'=Bn

Substitution on the carbon atom is indicated in the same way with an italicized capital *C* as in 4-*C*-methyl-D-*gulo*-hexose **4.15**.

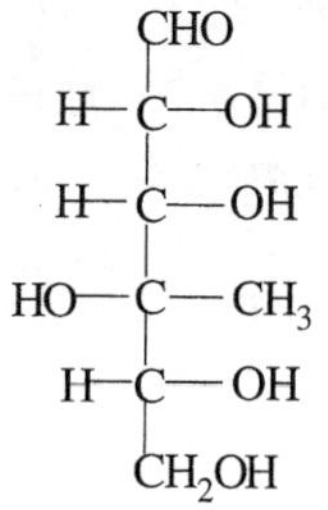

$$\begin{array}{rcl} & CHO & \\ H- & C & -OH \\ H- & C & -OH \\ HO- & C & -CH_3 \\ H- & C & -OH \\ & CH_2OH & \end{array}$$

4.15

4.2.5 Acyclic forms

Because of the unusual properties of acyclic forms, certain authors prefer to underline this situation. Compound **4.16** would be called 2,3,4,5,6-penta-*O*-acetyl-*aldehydo*-D-*gluco*-hexose or, more briefly, *aldehydo*-D-glucose-pentaacetate.

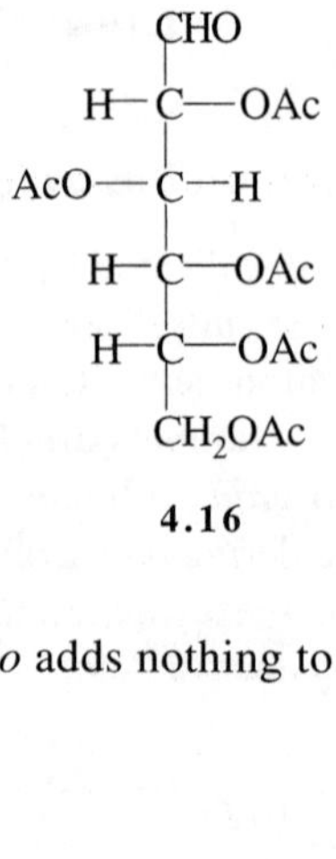

$$\begin{array}{rcl} & CHO & \\ H- & C & -OAc \\ AcO- & C & -H \\ H- & C & -OAc \\ H- & C & -OAc \\ & CH_2OAc & \end{array}$$

4.16

Note that the word *aldehydo* adds nothing to the first name but is indispensable to the second.

4.2.6 Cyclic forms

The size of a ring is indicated by replacing the suffix 'se' of the sugar name by 'furanose' for five-membered, 'pyranose' for six-membered, and 'septanose' for seven-membered rings. As already mentioned, the hemiacetal hydroxyl group is referred to as anomeric. The α-anomer is the one which, in Fischer projections such as **4.17** to **4.19**, is on the same side as the oxygen linked to the asymmetric carbon whose configuration defines the series. The symbol of the anomeric configuration, α or β, is placed before the symbol of the series as shown in the following examples: β-D-galactopyranose **4.17**, α-D-glucopyranose **4.18**, L-*glycero*-β-D-*gluco*-heptopyranose **4.19**.

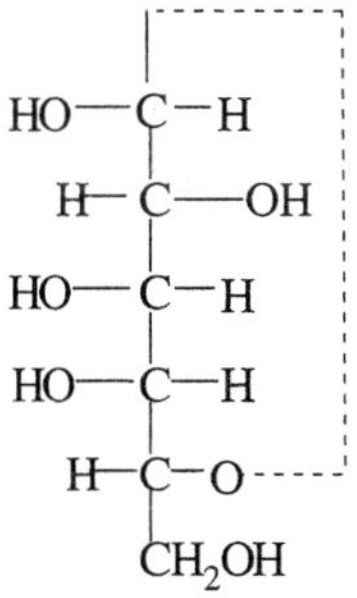

4.17

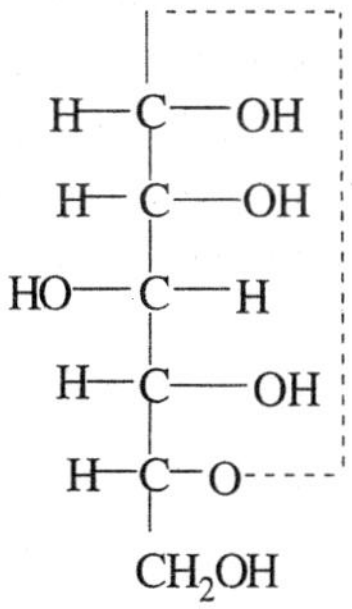

4.18

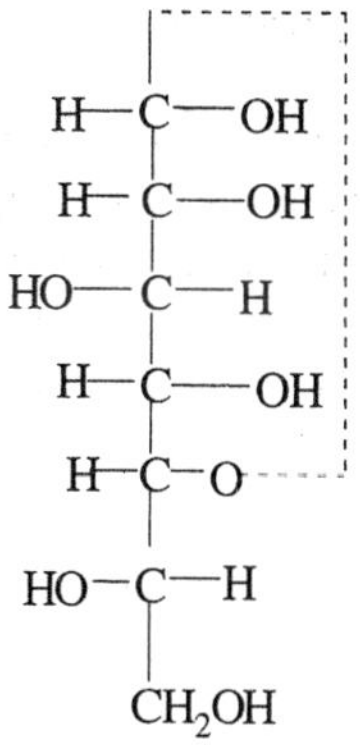

4.19

The names for glycosides (see Chapter 3) are formed by replacing the suffix 'se' of the cyclic form by 'ide'. The name of the group that substitutes the hemiacetal hydrogen is placed in front as a separate word, e.g. methyl α-D-gulofuranoside **4.20**.

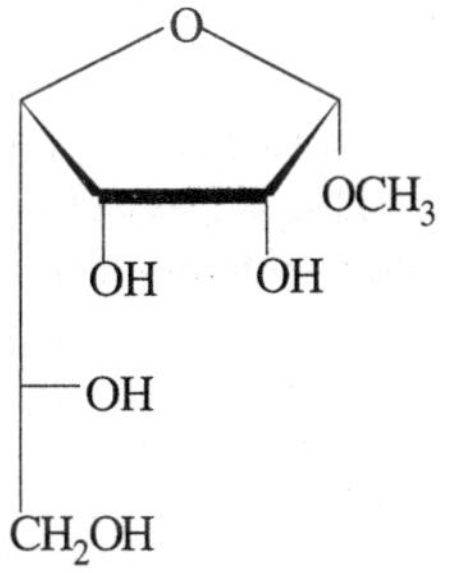

4.20

The radical formed by detachment from the anomeric hydroxyl group is indicated by replacing the final letter 'e' by 'yl' as in bromide **4.21** or dipotassium phosphate **4.22** of tetra-*O*-acetyl-α-D-glucopyranosyl. In the case of a replacement by an amino group, the suffix 'amine' is added, as in 2-acetamido-2-deoxy-β-D-glucopyranosylamine **4.23**.

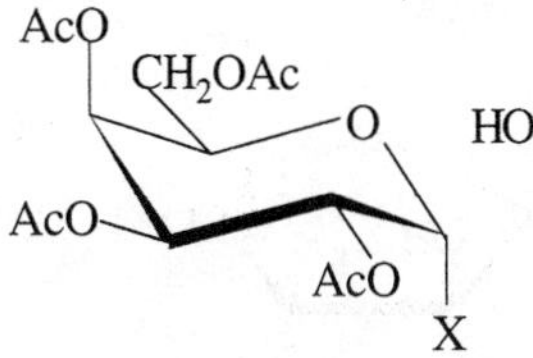

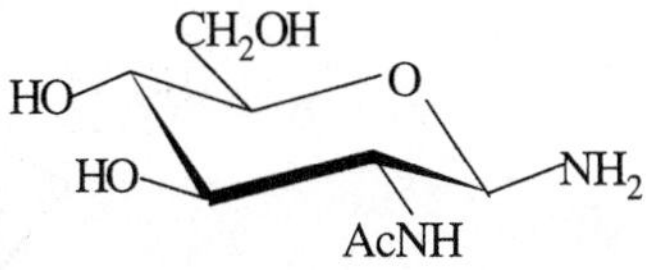

4.21 R=Br

4.22 $R=OPO_3K_2$

4.23

4.2.7 Alditols

The alditols are derived from the parent sugar name by replacing 'ose' by 'itol'. The same alditol can come from several parent sugars. In this way D-arabinitol **4.24**, named after a derivative of D-arabinose, is also obtained by the reduction of D-lyxose and could be called D-lyxitol. Diverse rules, which we will not give here, make it possible to select a unique name.

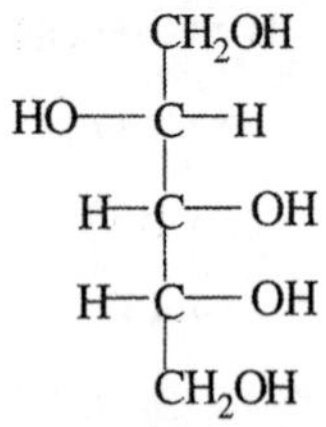

CH_2OH
HO—C—H
H—C—OH
H—C—OH
CH_2OH

4.24

4.2.8 Aldonic acids

The carboxylic acids, resulting from the oxidation of the aldehyde function, are named by replacing 'ose' by 'onic'. Salts, esters, halides, amides, and nitriles are designated following the usual rules of organic nomenclature. Some examples are: sodium D-gluconate **4.25**, methyl tetra-*O*-acetyl-L-arabinonate **4.26**, D-glucono-1, 4-lactone **4.27**, and D-glucononitrile **4.28**.

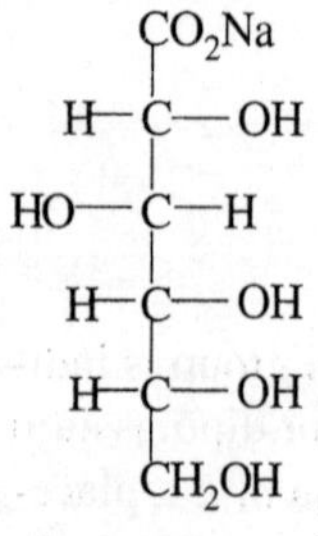

CO_2Na
H—C—OH
HO—C—H
H—C—OH
H—C—OH
CH_2OH

4.25

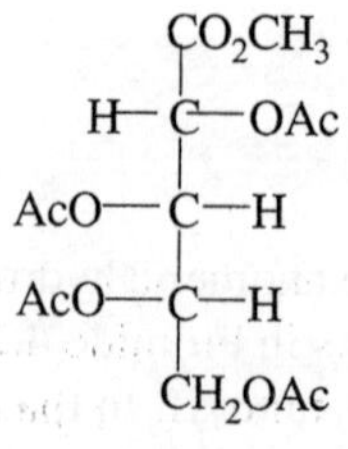

CO_2CH_3
H—C—OAc
AcO—C—H
AcO—C—H
CH_2OAc

4.26

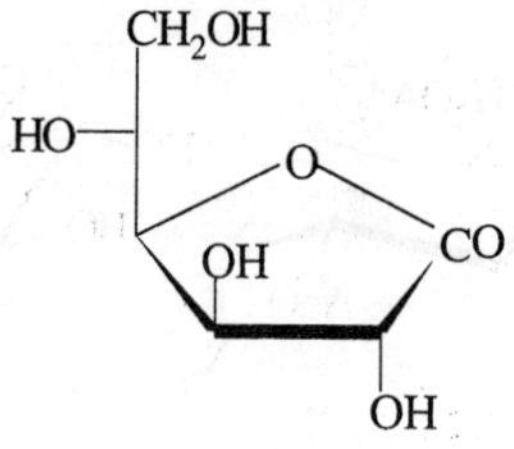

CH_2OH
HO—
O
OH
CO
OH

4.27

CN
H—C—OH
HO—C—H
H—C—OH
H—C—OH
CH_2OH

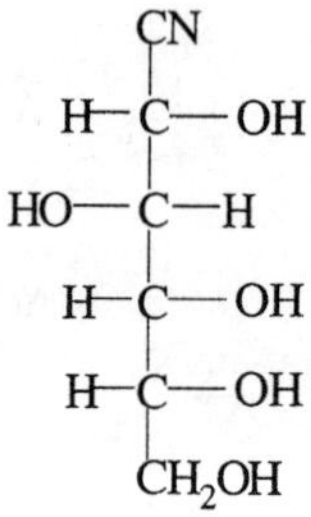

4.28

4.2.9 Uronic acids

Monocarboxylic acids resulting from the oxidation of the primary hydroxyl group of aldoses are referred to as uronic acids. They are designated by replacing the suffix 'ose' of the common or systematic name by 'uronic', and the suffix 'oside' of a glycoside by 'osiduronic'. Two examples are α-D-galacturonic acid **4.29** and methyl α-D -galactopyranuronate **4.30**.

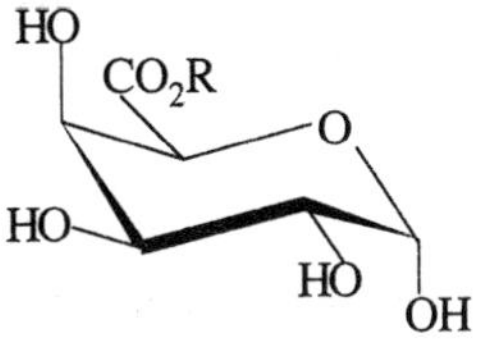

HO CO$_2$R O HO HO OH

4.29 R=H
4.30 R=CH_3

4.2.10 Cyclic acetals

The cyclic acetals are written by extending the rules applicable to alcohol derivatives with, as a prefix, the name of the divalent radical as from general nomenclature. Two such examples are methyl 4,6-*O*-benzylidene-α-D-glucopyranoside **4.31** and 1,2:5,6-di-*O*-isopropylidene-α-D-glucofuranoside **4.32**.

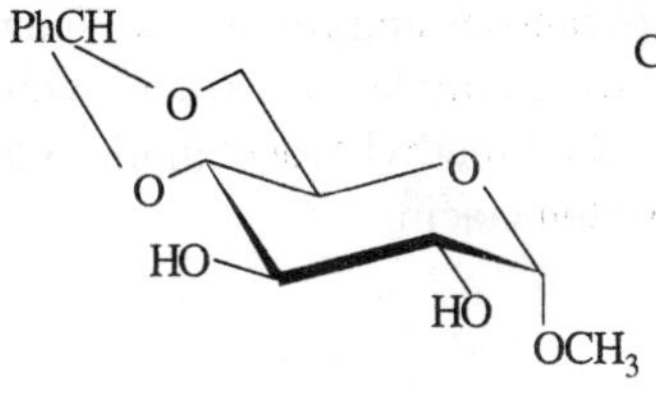

PhCH O O O HO HO OCH$_3$

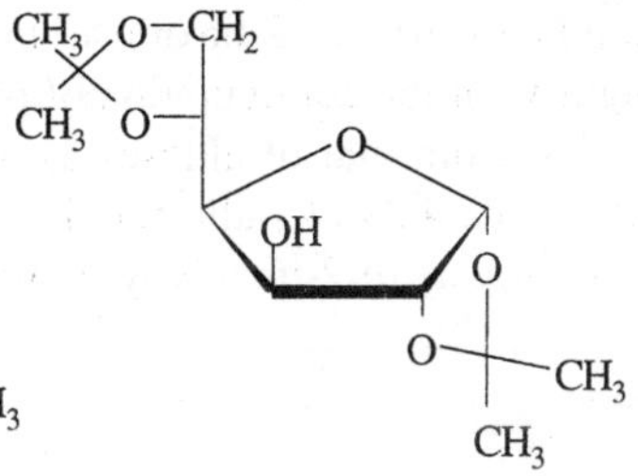

4.31

CH$_3$ O–CH$_2$ CH$_3$ O O OH O O CH$_3$ CH$_3$

4.32

4.2.11 Acetals and thioacetals

The words acetal and thioacetal are written after the name of the parent sugar as in D-glucose dimethyl acetal **4.33** and D-glucose dimethyl dithioacetal **4.34**.

4.2.12 Intramolecular anhydrides

The name of the sugar takes the prefix 'anhydro' followed by a hyphen. This prefix is preceded by two numbers which indicate the numbers of the bridged carbons as in 1,6-anhydro-β-D-glucopyranose **4.35**.

Oxiranes derived from sugars enter this category. Note that **4.36**, methyl 2,3-anhydro-4, 6-*O*-benzylidene-α-D-allopyranoside, is named as an *allo* derivative.

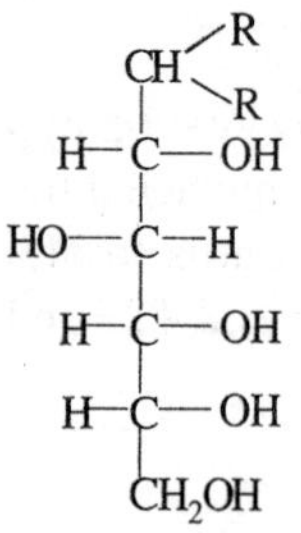

4.33 R=OCH_3
4.34 R=SCH_3

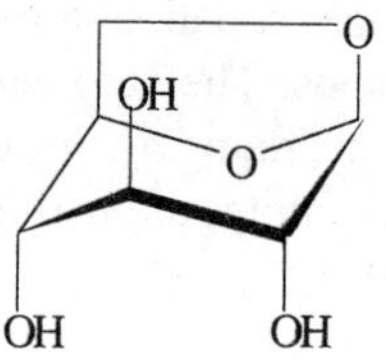

4.35

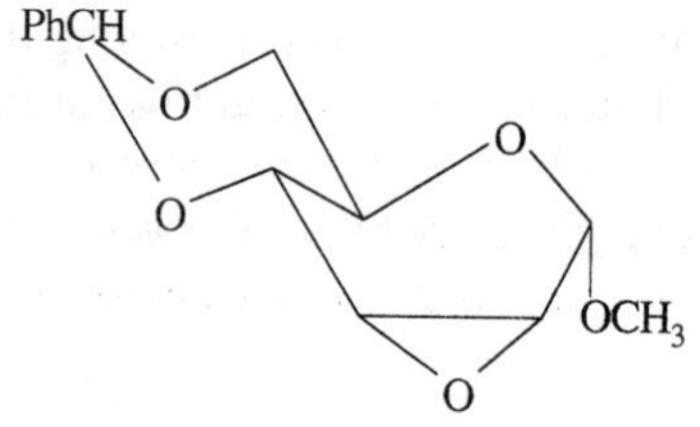

4.36

In the parent D-allose, each of the hydroxyl groups carried by carbons 2 and 3 has the same orientation as the oxygen bridge of **4.36** relative to the pyranose ring.

4.3 Ketoses

The ketoses are classified as 2-ketoses, 3-ketoses, etc., following the carbonyl position on the chain. The 2 of 2-ketoses, a common natural structure, can be removed. The suffix 'ose' is replaced by 'ulose' in the parent name. Likewise, fructose is a 2-hexulose, or more simply, a hexulose. For the complete name, it is preceded by the configuration descriptor. The systematic name of fructose **4.37** would be D-*arabino*-2-hexulose. In the case of 2-ketoses, there is no possible ambiguity for the configuration of pyranoses and furanoses. The nomenclature is copied from that of aldoses as, for example, β-D-fructopyranose **4.38**. Methyl glycoside **4.39** of sialic acid is called methyl 5-acetamido-3,5-dideoxy-D-*glycero*-α-D-*galacto*-2-nonulopyranosidonic acid.

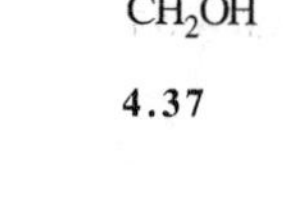

4.37

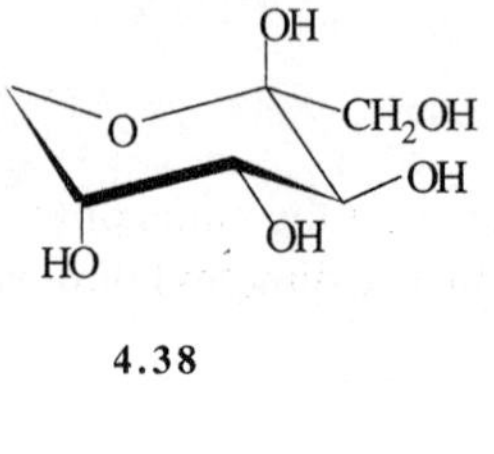

4.38

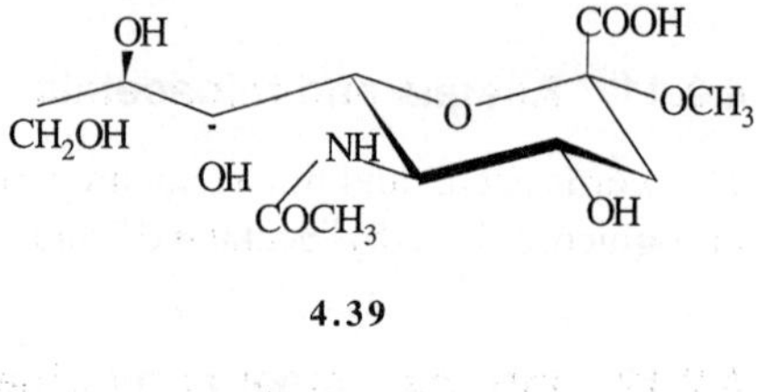

4.39

5 Reactions of hydroxyl groups

5.1 Functional derivatives

5.1.1 The importance of protecting groups in carbohydrate chemistry

In the majority of sugars, all carbons are functional. When a particular function is to be modified using traditional organic chemistry methods, the other functions need to be 'protected'. This means they are first transformed into derivatives that are inert under the desired reaction conditions, then reconverted into the starting function during a subsequent step. This last phase is called the 'deprotection' step. It is often useful to protect diverse functions of the same molecule by derivatizations of a different nature in order to effect selective deprotections. We have already encountered the most general type of protection of the hemiacetal hydroxyl group, that is the glycosidation reaction (see Chapter 3). In this chapter we will not talk only about protection reactions. Certain functional derivatives play an analytical role and it is quite useful to know how to synthesize phosphates and sulfates, widely distributed natural products. The sulfonates will be treated in Section 6.1 as an introduction to reactions with inversion of configuration.

5.1.2 Ethers

Ethers are commonly prepared by treating alkoxides with alkyl halides or sulfates. Because these basic conditions would destroy free sugars, they must be first transformed into glycosides. Methyl ethers are employed only in analytical applications and structure determinations. In this utilization, their great stability is advantageous. Their preparation and applications will be dealt with in Section 9.3.3.

Benzyl ethers are widely used and an example has already been seen in Section 3.5.3 in the description of the synthesis of methyl tri-*O*-benzyl-α-L-fucopyranoside **3.36**. Most often benzylation is carried out in an alkaline medium. Oxolane can be employed as solvent, but the reaction is much faster in *N,N*-dimethylformamide (DMF), a better solvent for only partially protected sugars. As base, KOH has been used but now the preferred method is to make alkoxides with a very slight excess of NaH. The alkylating reagent is benzyl bromide. Benzyl ethers are stable under moderate acidic conditions and in an alkaline medium. In a general fashion, benzyl ethers are cleaved by catalytic hydrogenation according to equation (5.1).

$$\text{R–O–CH}_2\text{Ph} + \text{H}_2 \xrightarrow{\text{Pd/C}} \text{R–OH} + \text{PhCH}_3 \tag{5.1}$$

Deprotection takes place by catalytic hydrogenation in the presence of palladium under a hydrogen pressure of a few atmospheres. This reforms the hydroxyl groups and toluene is easily separated. Hydrogenolysis by transfer does not require any special equipment. An alcohol solution is heated at reflux with the protected sugar in the presence of palladium over charcoal with, as hydrogen donor, cyclohexene or cyclohexadiene which aromatizes in the reaction.

p-Methoxybenzyl ethers are prepared in the same way as benzyl ethers. They are cleaved by oxidation with 2,3-dicholoro-5,6-dicyano-1,4-benzoquinone (DDQ) or cerium ammonium nitrate, $(NH_4)_2Ce\ (NO_2)_6$ (reaction 5.2). The driving force in this reaction is the oxidation at the benzylic position of the protecting group, greatly facilitated by a methoxy group in the *para* position.

(5.2) CH_2OR … OMe → R-OH + CHO … OMe (reagent: Cl, Cl, CN, CN, O, O)

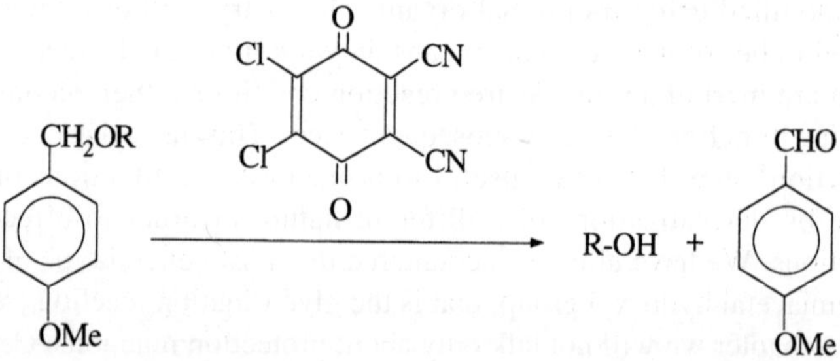

Chlorotriphenylmethane (Ph_3CCl) reacts under more moderate conditions on the sugar dissolved in pyridine. At room temperature, if treatment is not unduly prolonged, only the primary alcohol function is etherified. For example, ether **5.1** is obtained with methyl α-D-glucopyranoside. These ethers are hydrolysed under very mild acidic conditions, for example in aqueous acetic acid at 80°C. There is also the risk that hydrolysis will occur when they are purified by chromatography on silica gel; finally they are cleaved by hydrogenolysis under the same conditions as benzyl ethers.

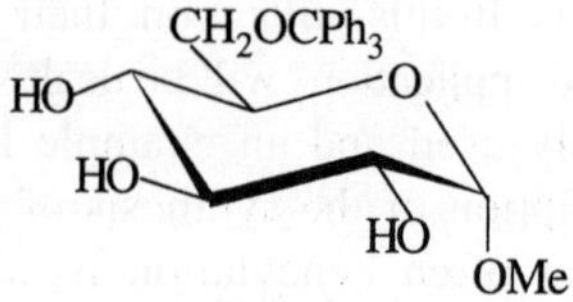

CH_2OCPh_3 O HO HO HO OMe

5.1

Allyl ethers (Manthorpe and Gigg 1980) are prepared in the same way as are benzyl ethers, but with allyl bromide. Deprotection is carried out in two steps: (a) isomerization to prop-1-enyl ether by heating in the presence of a strong base (Me_3COK) or catalytically, in the presence of a complex of a transition metal; (b) cleavage by treatment of prop-1-enyl ether with mercury salts reaction (5.3). Deprotection of a hydroxyl group transformed into an allyl ether is possible without affecting benzyl protecting groups in the same molecule. The reverse is not true as in the presence of palladium, an allyl ether is partially hydrogenolysed and partially reduced to propyl ether, a synthetic dead end.

$$\text{R–O–CH}_2\text{–CH=CH}_2 \xrightarrow{a} \text{R–O–CH=CH–CH}_3 \xrightarrow{b} \text{R–OH} \quad (5.3)$$

Phenyl ethers are obtained by treating diols with diacetoxytriphenylbismuth in boiling dichloromethane (David and Thieffry 1983). Reaction (5.4) is an example of this transformation, the mechanism of which is unknown.

(5.4)

HO, CH_2OH, O, BnO, OBn, OBn $\xrightarrow{Ph_3Bi(OAc)_2}$ PhO, CH_2OH, O, BnO, OBn, OBn

5.1.3 Silylated derivatives

Trimethylsilyl 'ethers' are too labile to assure protection of hydroxyl groups. Their synthesis and analytical applications have been described in Chapter 1. In synthetic work, the *t*-butyldimethysilyl ethers are used preferentially, obtained with *t*-butylchlorodimethylsilane (Me_3CSiMe_2Cl) by reacting in pyridine in the presence of imidazole. They are not perfectly stable, however. Greater stability can be achieved by converting alcohols to *t*-butyldiphenylsilyl derivatives obtained with *t*-butylchlorodiphenylsilane (Hanessian and Lavallée 1975; 1977). Etherification takes place at moderate temperature in DMF in the presence of imidazole. This last protection resists catalytic hydrogenolysis and hydrolytic conditions for acetals. All silyl ethers are hydrolysed in the presence of fluorides. These salts are the cause of slightly alkaline conditions which can lead the acetyl groups to migrate.

5.1.4 Esters and carbonates

Acetylation and benzoylation are carried out on glycosides by traditional techniques using $(CH_3CO)_2O$, CH_3COCl, and PhCOCl as reagents, and pyridine or triethylamine as base. What is novel here is the acylation of free sugars. Essentially peracetates of pyranoses are formed, and conditions have been found in which one or the other anomer is selectively produced. Thus, in the presence of $ZnCl_2$ or other acid catalysts, in anomerizing conditions, the most stable anomer is obtained having an axial acyl group such as penta-*O*-acetyl-α-D-glycopyranose. Unfortunately, this is not the most interesting anomer in synthetic work because of its low reactivity. Its formation is largely avoided by working in acetic anhydride at 100°C with sodium acetate as base, conditions which lead to the equatorial anomer **5.2**.

The acetylated glycosides are deprotected by alkaline methanolysis at room temperature with methanol containing about 1% of CH_3ONa. The acetyl groups, transfered onto the solvent, give the volatile methyl acetate. The anomeric oxygen cannot be deacetylated in this fashion because the hemiacetal function produced would rapidly decompose in the alkaline medium. Acetyl groups are stable in neutral or mildly acidic conditions, and were effectively used as protecting groups in certain syntheses of oligosaccharides. However, complete security is not assured because in partially acetylated products, migration can take place, probably via cyclic intermediates as depicted in reaction (5.5).

(5.5) HO OCOCH$_3$ → [O O OH] → CH$_3$CO-O OH

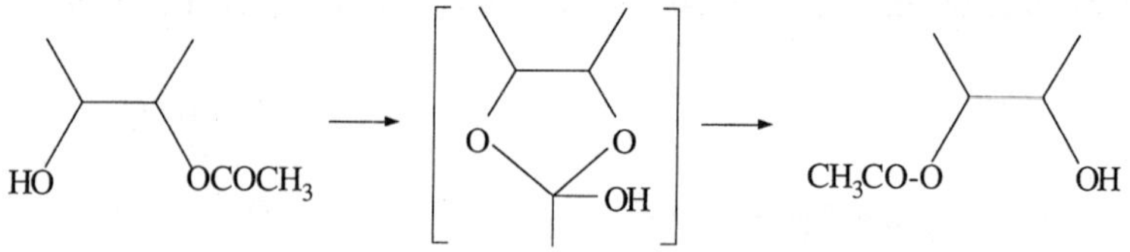

For the preparation of trichloroacetimidates used in glycoside synthesis (see Section 10.3.2), the anomeric position must be selectively deacetylated. This is possible starting from the β-anomer by treatment with hydrazine acetate (Excoffier *et al.* 1975). Thus, compound **5.3** can be prepared from **5.2** by this method.

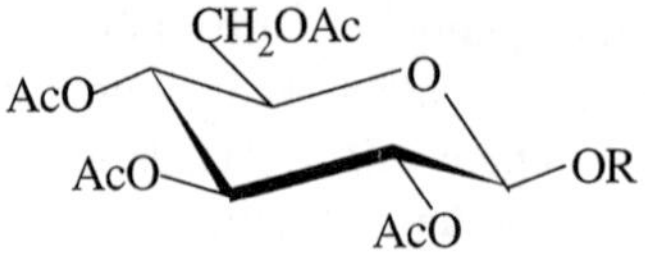

CH$_2$OAc AcO O AcO OR AcO

5.2 R=Ac
5.3 R=H

5.1.5 Selective etherification and acylation; organostannic derivatives and enzymic methods

We have already encountered selective derivatizations, for example with chlorotriphenylmethane and *t*-butylchlorodiphenylsilane which etherify selectively the primary alcohols. Pivaloyl chloride, Me_3CCOCl, also displays a preference for this position. Selective reactions can be observed with the mildly benzoylating reagent *N*-benzoylimidazole **5.4**. When it is a matter of differentiating between secondary positions, it can be advantageous to go through intermediates which are *formally* alkoxides of the dibutyltin cation, Bu_2Sn^{2+}.

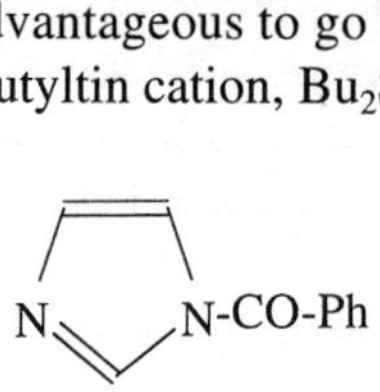

N N-CO-Ph

5.4

The reagent is an insoluble powder, the polymeric oxide $(Bu_2SnO)_n$. When the latter is heated with a diol in a benzene solution with water removal by a Dean–Stark apparatus, the powder is dissolved and a penta-coordinated complex of tin is formed, generally dimer **5.5** ($n = 0, 1$) as in reaction (5.6). The product can be transferred to a polar solvent where it becomes monomer **5.6**, with co-ordination to solvent. On this simpler structure, we can observe that, even if the glycol were symmetric, the two oxygens would be differentiated in the stannic derivative, one being apical and the other equatorial. This is equally true in each of the monomers associated in dimer **5.5**, but here there is a supplementary difference since one of the oxygens is tricoordinated, hence without nucleophilic power (David and Hanessian 1985). 'Stannylene', the name generally used for these products by the author as well, is not correct within the framework of organometallic nomenclature which, in this way, designates a divalent tin derivative, R–Sn–R.

(5.6) $HOCH_2-(CH_2)_n-CH_2OH$ + 2 Bu_2SnO ⟶

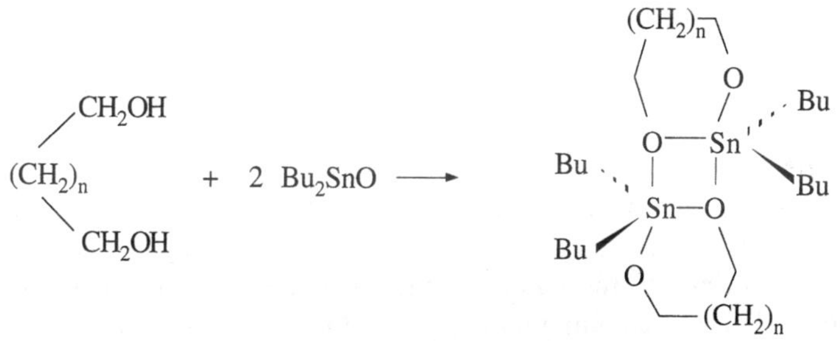

5.5

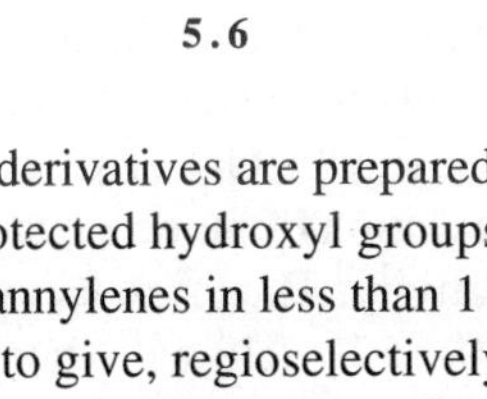

5.6

The stannylenes of sugar derivatives are prepared in the same way from derivatives in which two non-protected hydroxyl groups remain at appropriate sites. Acid chlorides react with stannylenes in less than 1 min at room temperature in a benzene or a DMF solution to give, regioselectively, the monoester. Benzyl and allyl halides lead to the corresponding ethers in warm DMF. Here again, the reaction is regioselective.

No reaction is observed between the benzyl or allyl halides and the stannylenes, even in warm benzene in which these derivatives are apparently polymers. However, these ethers can be prepared in good yields in refluxing benzene in the presence of catalytic quantities of tetraalkylammonium halides; and this is how the highest regioselectivity can be observed in difficult cases. Benzyl β-D-galactopyranoside **5.7**, transformed into a stannylene, then treated with benzyl bromide in the presence of tetramethylammonium bromide in boiling benzene, is benzylated exclusively at position 3 to give ether **5.8** in 67% yield. It is worth noting the preference for the oxygen at position 3 over the other three even though one is a primary alcohol function. An example of even greater selectivity is given in Section 10.1.

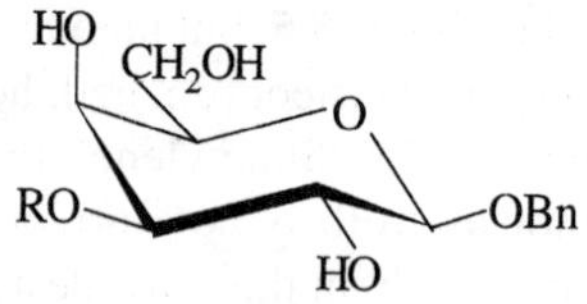

5.7 R=H
5.8 R=OBn

What could be the origin of this selectivity? On glycoside **5.7** there are at least three pairs of hydroxyls capable of forming a stannylene. We can suppose that all three are formed but are at equilibrium, and that the free enthalpies of all dimers or higher polymers possible are different enough for only one to exist practically at the end. This mechanism selects one hydroxyl pair, and diverse constraints due to the nature of the sugar molecule result in one of the two hydroxyl groups adopting preferentially the apical position.

The *lipases* are a family of enzymes of broad specificity which catalyse the hydrolysis of carboxylic acid esters. The known reversibility of enzymic reactions suggested that they could be used as esterification catalysts. These esterifications would, of course, be impossible in water, while, on the other hand, free sugars are only soluble in polar organic solvents in which most enzymes are inactive. Nevertheless, successful regioselective esterifications have been achieved in pyridine (Therisod and Klibanov 1986). It was found later that proteolytic enzymes are also active, the one which is most generally used being *subtilisine*, a commercial protease from *Bacillus subtilis*, both stable and active in numerous anhydrous organic solvents including pyridine and dimethylformamide. The acylating agents are esters of 2,2,2-trichloroethanol, acetone oxime and 'vinyl alcohol', $R\text{–}CO\text{–}O\text{–}CH_2CCl_3$, $R\text{–}CO\text{–}O\text{–}N{=}CMe_2$, and $R\text{–}CO\text{–}O\text{–}CH{=}CH_2$. In the acylation of monosaccharides, the primary hydroxyl group shows the greatest reactivity. Thus D-galactose in pyridine solution gave the 6-*O*-acyl derivatives in 70–85% yield with the oxime esters in the presence of the lipase from *Pseudomonas cepacia*. When the primary hydroxyl group is protected or absent, selective acylation of the C-2, C-3, or C-4 hydroxyl group can be achieved depending on the choice of enzyme. Thus 6-*O*-trityl-D-glucose dissolved in oxolane was butyroylated at the C-3 position in quantitative yield with 3,3,3-trichloroethyl butyrate in the presence of the lipase from *Chromobacterium viscosum*. On the other hand, when the primary hydroxyl group of glucose was protected with a *t*-butyldiphenylsilyl group, acylation in dichloromethane in the presence of the *Candida cylindracea* lipase occurred exclusively at the C-2 position (75% yield).

Less surprising is de-*O*-acylation in the presence of lipases. This can also be selective in polyacylated derivatives.

A comprehensive review of enzymic esterification and de-esterification of carbohydrates was published recently (Bashir *et al.* 1995).

5.1.6 Phosphates

In the majority of cases, the metabolic conversion of a sugar in cells begins by its conversion to a phosphoric ester. Syntheses of sugar phosphates were first developed to give practical access to these convenient tools for research in biochemistry. More recently, the introduction of enzymic methods to preparative chemistry created the demand for greater quantities. It is precisely these enzymic methods which allow their preparation in large quantities. Thus, D-glucose 6-phosphate is obtained by phosphorylation of glucose by a triphosphate, adenosine triphosphate (ATP), more explicitly symbolized by A–O–PO(OH)–O–PO(OH)–O–PO_3H_2, which transfers its terminal phosphate in the presence of the enzyme hexokinase. The phosphorylating agent is transformed into adenosine diphosphate (ADP), or A–O–PO(OH)–O–PO_3H_2 (reaction 5.7).

(5.7)

CH_2OH HO HO O OH OH + A-O-PO(OH)-O-PO(OH)-O-PO_3H_2 ⟶

CH_2-OPO_3H_2 HO HO O OH OH + A-O-PO(OH)-O-PO_3H_2

Because of the very high price of ATP, reaction (5.7) must be coupled with a regenerating system, the transfer of phosphate to ADP starting from the enol phosphate of pyruvic acid (an easily accessible and inexpensive phosphate), catalysed by the enzyme pyruvate kinase (reaction (5.8). In the same flask are mixed glucose, phosphoenolpyruvate, hexokinase, pyruvate kinase, and a catalytic quantity of ATP (about 1% mol) and the system produces D-glucose 6-phosphate until the phosphoenolpyruvate runs out. The kinases are easily accessible and, if they are immobilized on an insoluble support (see Section 10.4.1), they are reusable a certain number of times. In this way glucose 6-phosphate can be easily prepared on a 250 g scale (Pollak *et al.* 1977).

(5.8) $$\text{A-O-}\underset{}{\overset{\text{OH}}{\overset{|}{\text{PO}}}}\text{-O-PO}_3\text{H}_2 + \text{CH}_2\text{=}\overset{\text{OPO}_3\text{H}_2}{\overset{|}{\text{C}}}\text{-COOH} \longrightarrow \text{A-O-}\overset{\text{OH}}{\overset{|}{\text{PO}}}\text{-O-}\overset{\text{OH}}{\overset{|}{\text{PO}}}\text{-O-PO}_3\text{H}_2 + \text{CH}_3\text{COCOOH}$$

Phosphorylation of the anomeric hydroxyl group gives a glycosyl phosphate. These compounds are prepared by non-enzymic methods. Brief heating of β-pentaacetate **5.2** at 50°C with anhydrous H_3PO_4 gives around 50% of the tetraacetylated β-D-glycopyranosyl phosphate which is transformed into the α-anomer by prolonged heating. These glycosyl phosphates are more labile in an acidic medium than ordinary phosphoric monoesters.

Generally, sugar phosphates are more acidic than phosphoric acid.

5.1.7 Hydrogensulfates

A certain number of recognition phenomena involve sugar sulfates. Synthetic examples are found in Chapter 17. Hydrogensulfates are prepared by treating an alcohol dissolved in DMF with sulfur trioxide in the presence of triethylamine (reaction 5.9).

(5.9) $$\text{R–OH} + \text{SO}_3 \rightarrow \text{R–O–SO}_3\text{H}$$

5.2 Acetals (Gelas 1981)

Acetals are constructed using a pair of oxygens from hydroxyl groups. In the acetalation of a sugar derivative which has more than two free hydroxyl groups, several acetals are, therefore, possible. Most of the time, an acid catalyst is used to accelerate the equilibrium between possible acetals, and the most stable one is finally isolated as the major product.

Aldehydes preferentially form 1,3-dioxanes. Treatment of glucose with benzaldehyde and methanol in the presence of zinc chloride readily gives methyl 4,6-*O*-benzylidene-α-D-glycopyranoside **5.9**. However, a 1,3-dioxolane ring can be obtained with aldehydes and two *cis vicinal* hydroxyl groups when there is no other possibility. If there is tautomeric equilibrium, the most favourable tautomer is selected. Glucose forms, with acetone and zinc chloride, the furanoid bis-acetal 1,2:5,6-di-*O*-isopropylidene-α-D-glucofuranose **5.10**, and galactose forms the pyranoid bis-acetal 1,2:3,4-di-*O*-isopropylidene-α-D-galactopyranose **5.11**. However, methyl α-D-galactopyranoside, which has only one *cis vicinal* hydroxyl pair, gives monoacetal **5.12**. These preferences are also observed in the acetalation of D-glucitol, represented by the zig-zag form, **5.13**; benzaldehyde leads to triacetal **5.14** and acetone to triacetal **5.15**. However, it is possible to prepare high-energy acetals using special techniques, such as **5.16** which is obtained from methyl α-D-glucopyranoside by treatment with the dimethyl acetal of cyclohexanone in DMF in the presence of *p*-toluenesulfonic acid.

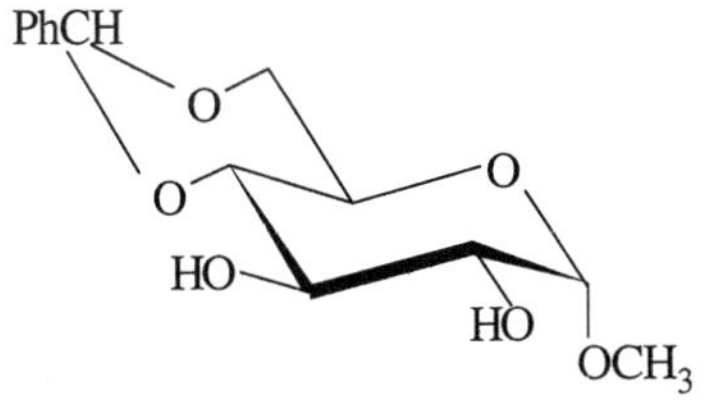

5.9

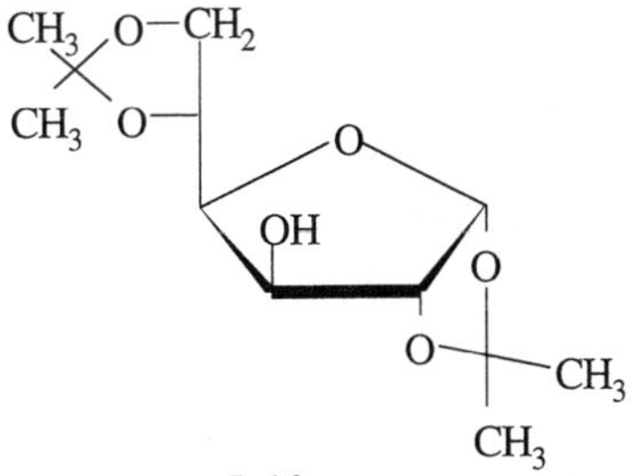

5.10

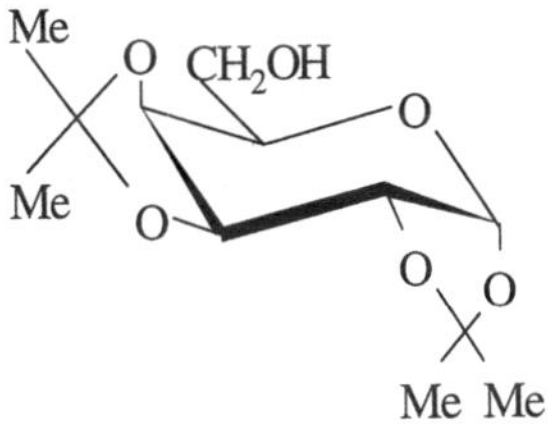

5.11

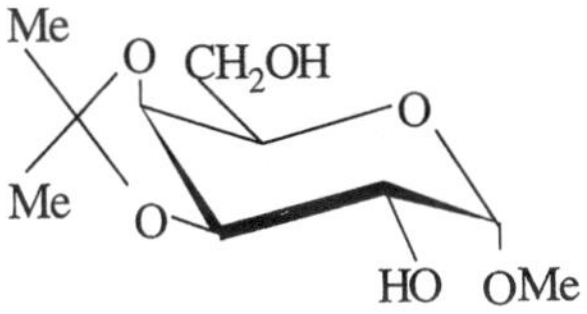

5.12

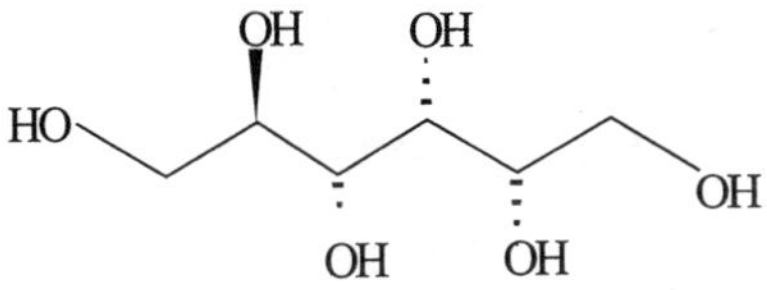

5.13

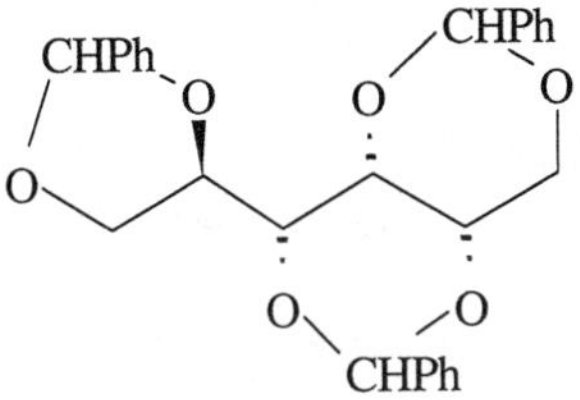

5.14

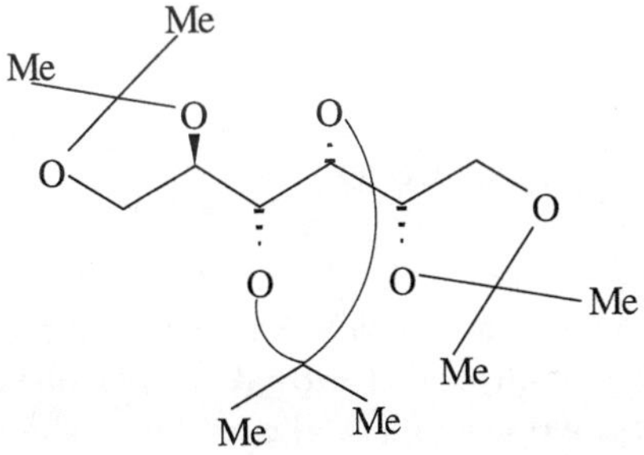

5.15

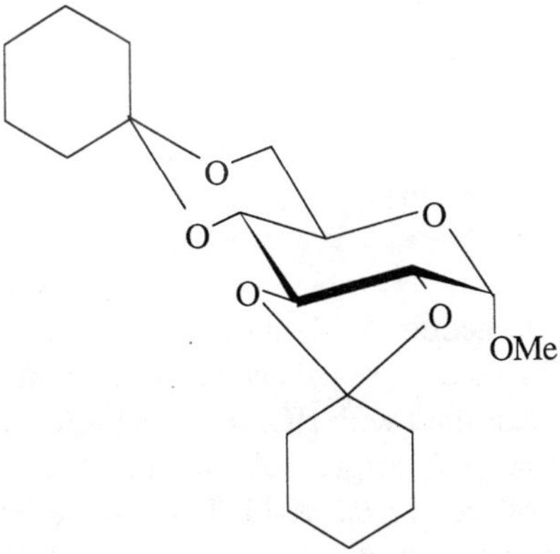

5.16

Benzylidene acetals, very fragile in an acidic medium, are hydrolysed by aqueous acetic acid or cleaved by hydrogenolysis. Selectivity can be observed in the hydrolysis of isopropylidene acetals. With compound **5.10**, only the ring spanning positions 5 and 6 is hydrolysed by aqueous acetic acid. Hydrolysis of the acetal function which involves the anomeric oxygen requires heating in mineral acid (0.1 M) for a few hours, comparable conditions to those of hydrolysis of a glycoside.

Acetals are used a great deal as protected derivatives. They are also the starting products of certain important reactions in synthetic work. Benzylidene acetals are converted into bromobenzoates by treatment with *N*-bromosuccinimide in boiling CCl_4 in the presence of $BaCO_3$ (Hanessian 1966). The reaction is especially interesting in the case of 4,6-*O*-benzylidene acetals because it gives the primary bromide exclusively. Thus methyl 4-*O*-benzoyl -6-bromo-6-deoxy-α-D-galactoside is obtained in 90% yield (reaction (5.10). Reaction conditions are compatible with a great variety of functions. The reaction mechanism appears to be the replacement of the benzylic hydrogen by a bromine atom, followed by ionization and bromide attack of the benzoxonium cation (reaction 5.11).

(5.10)

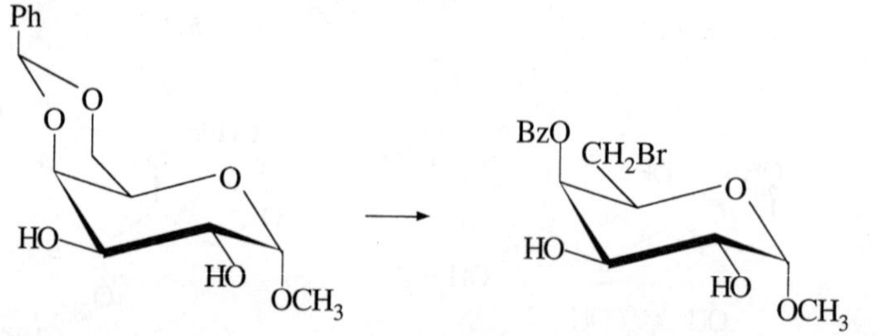

Ph O O O HO HO OCH_3 → BzO CH_2Br O HO HO OCH_3

(5.11)

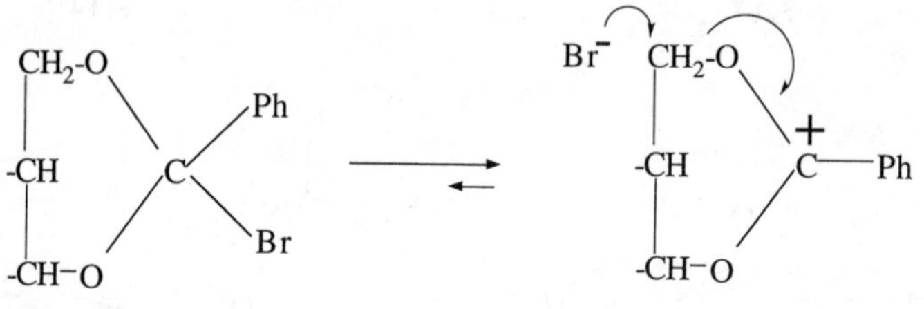

CH_2-O -CH -CH–O C Ph Br ⇄ Br^- CH_2-O -CH -CH–O C^+—Ph

Reaction (5.12) shows an example of a reductive opening of a benzylidene. The mixture of the acetal and sodium cyanoborohydride ($NaBH_3CN$) in oxolane is acidified with HCl in ether until no more gas is emitted. The reaction, which appears to be general (Garegg *et al.* 1982), is completed in 5 min at room temperature in 87% yield. This can be interpreted as resulting from regioselective protonation at O-4 of the sugar, followed by displacement of oxonium by a hydride.

(5.12)

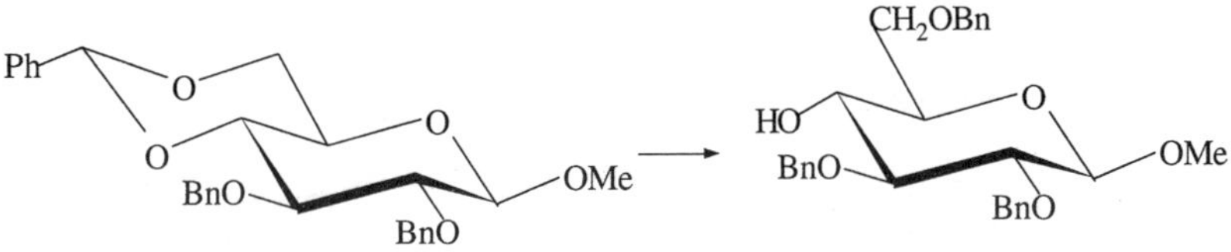

Treatment of dibenzylidene mannoside **5.17** (reaction 5.13) by butyllithium gives a keto deoxy sugar (Klemer and Rodemeyer 1974; Horton and Weckerlé 1975). We can suppose that the strong base extracts the H-3 proton which induces the elimination of the *vicinal* O-2, then the departure of PhCHO to give the enolate.

(5.13)

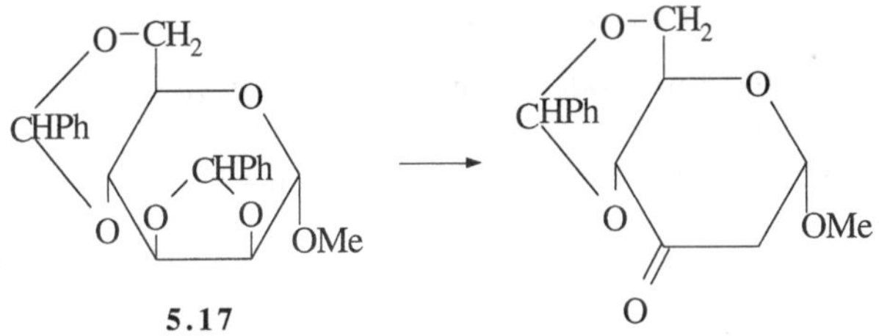

5.17

A new family of ketals has been reported more recently. For instance, the reaction of the bis-vinyl ether **5.18** with methyl α-D-galactopyranoside in acidic conditions gives the 2,3-dispiroketal-protected derivative **5.19** in good yield (Ley *et al.* 1992). Noteworthy is the preference of the reagent for spanning a *trans*-diequatorial vicinal diol. A similar reaction with the bis-dimethylacetal of 1,2-cyclohexanedione **5.20** gave the protected thiomannoside **5.21** (Ley *et al.* 1994). Such protections introduce rigidity in the pyranose ring, and make more difficult the generation of a carbenium ion on C-1. For this reason they have applications in oligosaccharide synthesis (Section 10.2) (Grice 1995).

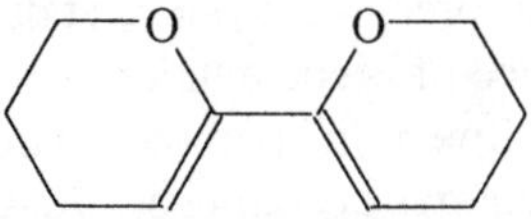

5.18

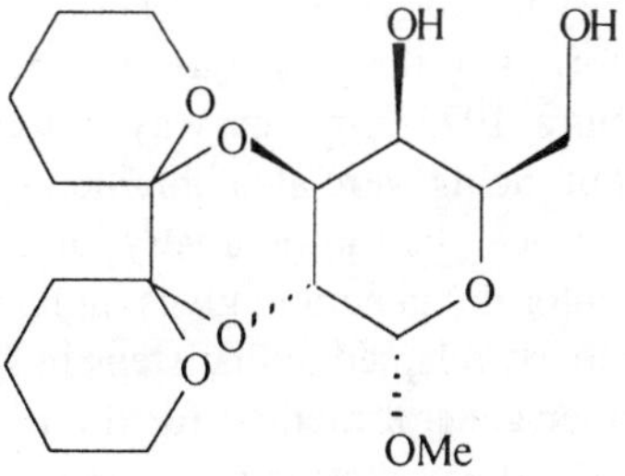

5.19

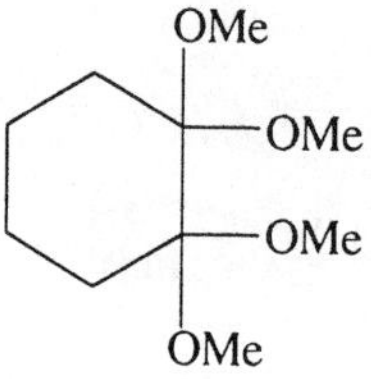

OMe OMe OMe OMe

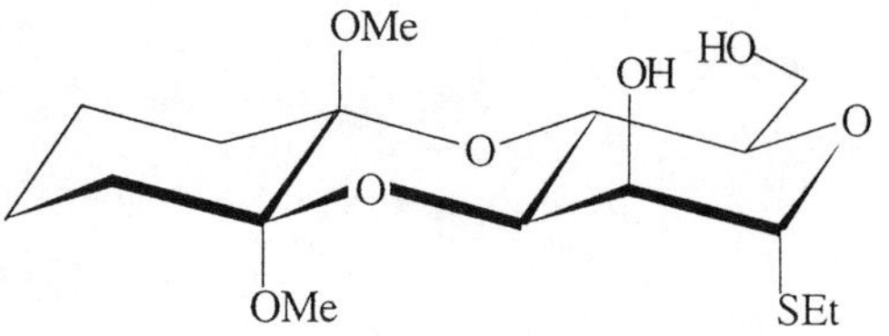

5.20

OMe OH HO O O O OMe SEt

5.21

5.3 Oxidations to aldehydes or ketones

5.3.1 Isolated hydroxyl groups

The most utilized methods are the Pfitzner and Moffat reaction and its subsequent variations (Jones and Moffatt 1972). The oxidizing agent is dimethyl sulfoxide (DMSO). The generally accepted intermediate is structure **5.22** which leads to compound **5.23** by an elimination reaction (reaction 5.14).

(5.14) $$RR'CH\text{-}O\text{-}\overset{+}{S}(CH_3)_2 \xrightarrow{-H^+} RR'CH\text{-}O\text{-}\overset{+}{S}(CH_3)\overset{-}{C}H_2 \longrightarrow RR'CO + Me_2S$$

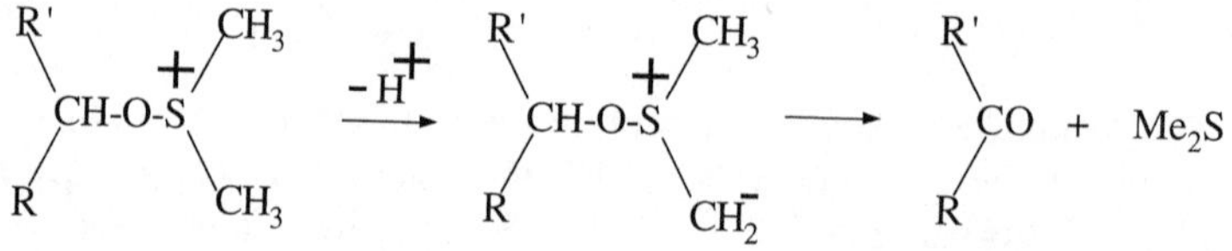

5.22 **5.23**

The variations are distinguished by the reagents used to prepare the sulfoxonium derivative of alcohol **5.22**. For the oxidation of the primary hydroxyl group of sugars to an aldehyde, the dicyclohexylcarbodiimide (C_6H_{11}–N=C=N–C_6H_{11}, 3 molar eq.) system is recommended, associated with a mild acid such as dichloroacetic acid (0.5 eq.). Aldehyde **5.24** is thus obtained (2.5 h at room temperature, isolated in 80% yield) from the ribofuranoside. For the oxidation of the secondary hydroxyl groups to ketones, phosphoric anhydride can simply be the promotor, as in the preparation of **5.25**, obtained in 92% yield (Onodera and Kashimura 1972). In this way ketone **5.26** was prepared from acetal **5.10**. Excellent yields were also obtained in the presence of acetic anhydride. These oxidations can be run on a very large scale. The Swern method (activation by oxalyl chloride) generally gives high yields, but requires operating at –70°C and seems poorly adapted to first steps in synthetic work.

Another general method for the preparation of ketones was also applied to sugars, i.e. the oxidation of secondary alcohols by ruthenium tetroxide, RuO_4. Only catalytic quantities of this very costly compound are used which is regen-

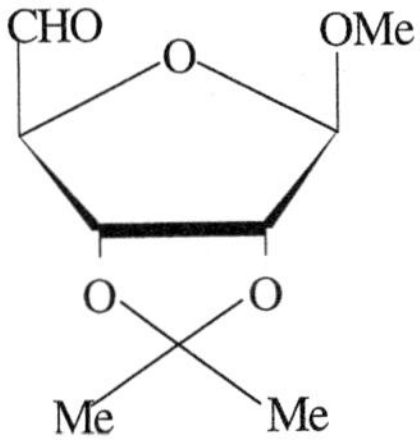

5.24

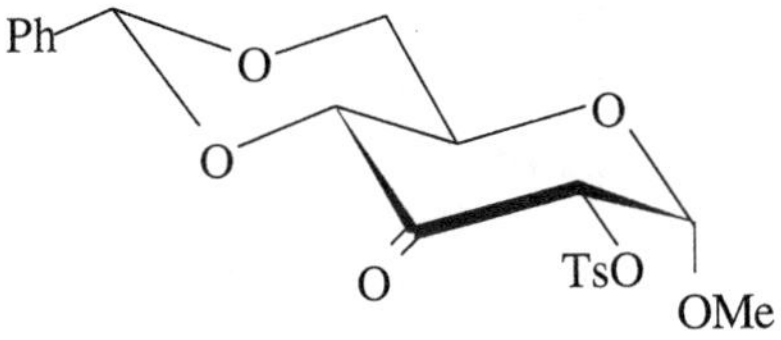

5.25

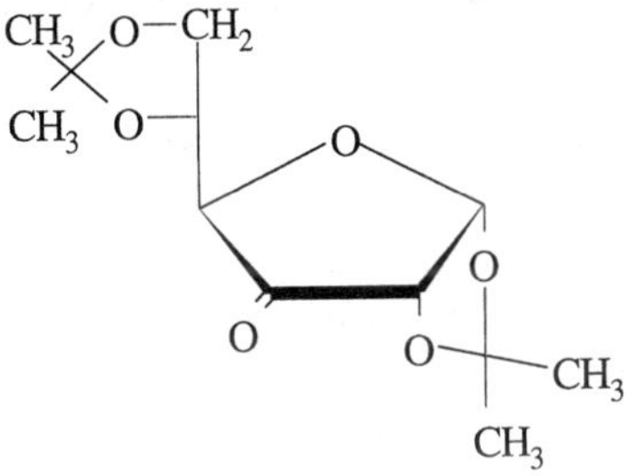

5.26

erated by an oxidant, KIO_4 in aqueous solution for example, buffered with K_2CO_3 or an alkaline hypochlorite. (Baker *et al.* 1972).

5.3.2 Oxidation of diols to hydroxy ketones

The treatment of stannylenes (see Section 5.1.5) dissolved in benzene with a bromine solution in the same solvent in the presence of molecular sieves or Bu_3SnOMe as base, gives, at room temperature, the hydroxyl ketone at the rate of titration (David and Thieffrey 1979). The reaction is nearly uniformly regiospecific, that is to say that it gives only one of the two possible hydroxyl ketones (reaction 5.15). The same selective activation by stannylation, already observed in benzylation, allylation, and acylation, is found here. Thus, compound **5.28** is prepared from diol **5.27** in 75% yield. The replacement of bromine by *N*-bromosuccinimide without any other reagent has recently been recommended (Kong and Grindley 1993).

(5.15) R-CH-O / ( )$_n$ / R'-CH-O $\rangle SnBu_2 + Br_2 \longrightarrow$ R-CO / ( )$_n$ / R'-CHOH + $SnBu_2Br_2$

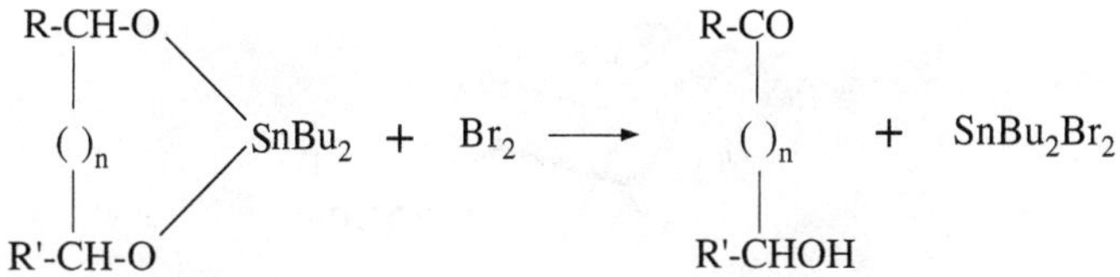

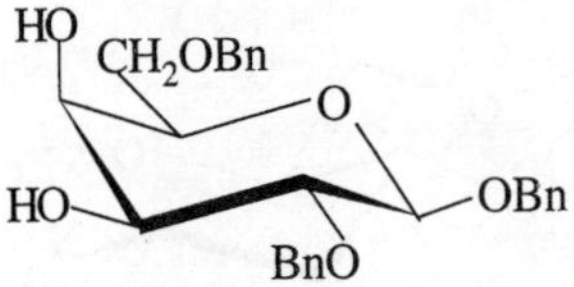

HO CH$_2$OBn O HO OBn BnO

5.27

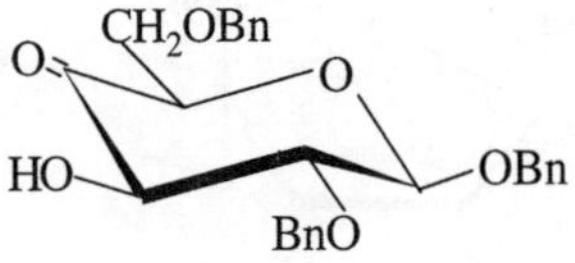

CH$_2$OBn O O HO OBn BnO

5.28

5.3.3 Synthetic usefulness of sugar aldehydes or ketones

Reduction by $NaBH_4$ or $LiAlH_4$ gives a mixture of epimers. In the most favourable cases, the compound with inversion of configuration relative to the starting product can be separated in acceptable yields. Thus, 1,2:5,6-di-*O*-isopropylidene-α-D-allose is prepared by oxidation of bis-acetal **5.10** to ketone **5.26** followed by reduction with $NaBH_4$. The latter is stereospecific and gives the α-D-*allo* configuration, the epimer at position 3 of the starting α-D-*gluco* configuration. This method of inversion of configuration, apart from exceptional cases, has lost its value for S_N2 substitution of sulfonates ever since the discovery of polar aprotic solvents and very good leaving groups. Naturally, labelled hydrides allow the specific introduction of deuterium or tritium on the carbon bearing the hydroxyl group. These sugar aldehydes or ketones can easily be hydrated. By prolonged incubation in $[^{18}O]H_2O$, hydration–dehydration equilibrium can introduce the ^{18}O isotope into the carbonyl group, which, after reduction, provides a labelled hydroxyl group. Finally, the major routes to the preparation of branched-chain sugars pass through the carbonylated derivatives (see Section 7.6).

5.3.4 Catalytic oxidation over platinum

This reaction is particularly efficient and selective with non-protected glycopyranosides; the primary hydroxyl group is converted into a carboxylic acid in good yield. Uronic acids are obtained. Thus, derivative **5.29** of glucosamine is obtained in 75% yield. In practice, the sugar is added to a vigorously shaken aqueous suspension of Adams platinum, in which oxygen is bubbled. Oxidation of secondary hydroxyl groups can take place with a certain selectivity when a primary alcohol is absent.

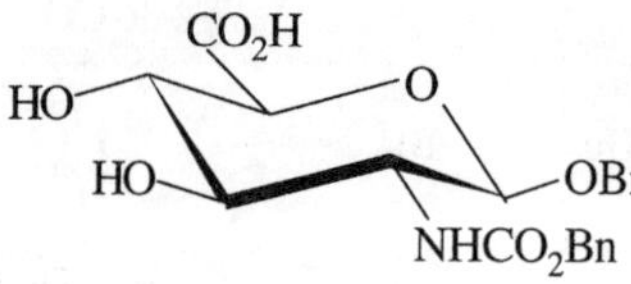

CO$_2$H HO O HO OBn NHCO$_2$Bn

5.29

5.4 Alkaline periodates and lead tetraacetate

We may recall the well-known reactions of these two oxidants on *vicinal* glycols, reactions (5.16) and (5.17).

$$\begin{array}{c} \text{R-CHOH} \\ | \\ \text{R'-CHOH} \end{array} + IO_4^- \longrightarrow \begin{array}{c} \text{R-CHO} \\ + \\ \text{R'-CHO} \end{array} + IO_3^- \quad (5.16)$$

$$\begin{array}{c} \text{R-CHOH} \\ | \\ \text{R'-CHOH} \end{array} + Pb(OAc)_4 \longrightarrow \begin{array}{c} \text{R-CHO} \\ + \\ \text{R'-CHO} \end{array} + Pb(OAc)_2 + 2\ AcOH \quad (5.17)$$

Alkaline periodates and lead tetracetate are complementary, periodate being used in aqueous solution and lead tetraacetate in organic solvents. One supposes that these reactions occur via cyclic intermediates, for example by addition of hydroxyl groups to the I–O bonds of IO_4^- giving **5.30**, and by dehydration of **5.30** to **5.31**, the former an octahedral and the latter a trigonal bipyramid. We will point out that because of the length of the I–O bond, the O–I–O angle in the five-membered ring has a very low value close to 75° and this no doubt affects the reaction. Be that as it may, dehydration as shown with compound **5.31** seems fundamental. In a very basic medium, **5.30** loses a proton, dehydration is no longer possible, and the reaction slows down. Likewise, tridentate complexes such as **5.32**, which cannot be dehydrated, are stable and visible in NMR studies (Perlin and von Rudlof 1965). The reader will recognize that **5.32** is a complex of 1,2-*O*-isopropylidene-α-D-glucofuranose, a partially hydrolysed product from **5.10**. In fact, periodate does not cleave *vicinal* glycols in a rigid *trans*-diaxial position. However, it is worthwhile to note that with lead tetraacetate, it is possible to observe the oxidation of *vicinal* glycols incapable of participating in five-membered rings.

5.30 **5.31** **5.32**

Besides the glycols, periodate and lead tetracetate cleave other *vicinal* polyoxygenated compounds that are encountered in sugar chemistry such as 2-hydroxyaldehydes, *vicinal* diketones, α-keto and α-hydroxy acids, and amino alcohols. On paper, scission of α-hydroxyaldehydes by periodates corresponds to the general mechanism of reaction (5.17) in supposing that the aldehyde is hydrated (R = OH), which gives on the right of the equation HO-CHO, that is to say formic acid. The reader may verify that with this mechanism, glucitol **5.33** consumes five molecules of periodate, supplies two molecules of formaldehyde coming from the terminal carbons, and four molecules of formic acid from the central carbons. The reaction is stoicheometric. Oxidation of α-D-glucopyranose begins by cleavage of the C-1–C-2 bond and the formation of pentose **5.34** with a formyl group at position 4. This continues and, if the formate hydrolyses in the medium, five molecules of periodate are consumed, five molecules of formic acid, and one molecule of formaldehyde coming from C-6 are formed. Methyl α-D-glucopyranoside is cleaved between C-2 and C-3, and C-3 and C-4 to afford one molecule of formic acid and 'dialdehyde' **5.35** (in fact hydrated and cyclized).

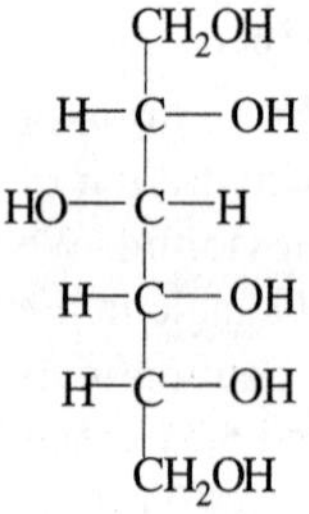

5.33 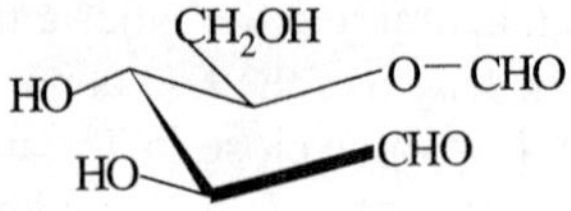 **5.34** 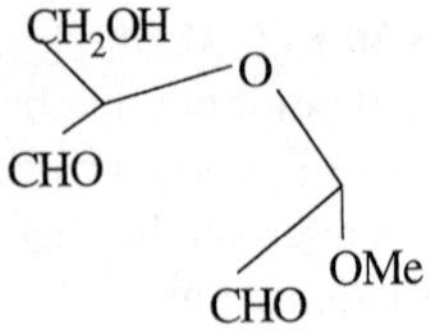 **5.35**

The reagents used in the scission of glycols have played an important role in the development of carbohydrate chemistry. We will only cite examples of recent applications. The stoicheometric nature of these reactions allow them to be used in quantitative determinations. The possibility of separating carbons, C-6 for example, the only one giving CH_2O from glucose, or else C-3, giving formic acid in the oxidation of methyl α-D-glucopyranoside, etc., makes it possible to locate an isotopic carbon in metabolic studies. The most important contemporary utilization is in syntheses by degradation. D-Erythrose is commonly prepared from D-glucose by acetalation with acetaldehyde, catalysed by H_2SO_4, periodic oxidation, and acid hydrolysis (reaction 5.18). Oxidation of 'diacetone mannitol' **5.36** by lead tetraacetate gives two molecules of isopropylidene glyceraldehyde **5.37**, an important raw material in chiral synthesis.

(5.18)

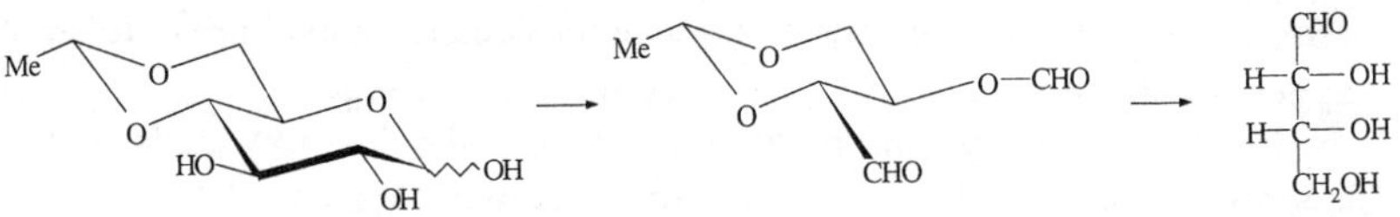

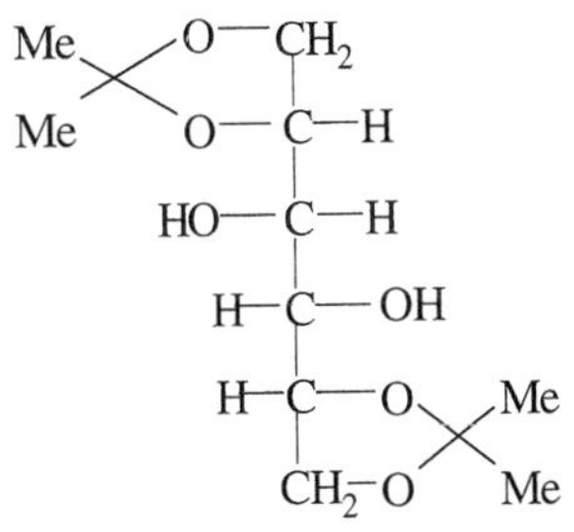

5.36 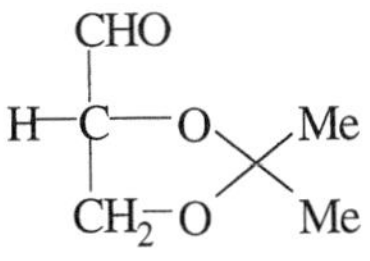 **5.37**

5.5 Deoxygenation

We have already seen two examples of deoxygenated sugars generally present in cells, i.e. deoxyribose and fucose. There are a good many others, particularly in bacterial compounds, and because of their being rare, they are often prepared synthetically. Also abundant sugars, which are very useful in chiral synthesis, are much too functionalized to be used as they are, providing another use for deoxygenation reactions.

Primary tosylates are reduced by $LiAlH_4$. The diacetal of D-fucose **5.39**, an enantiomer of L-fucose, is thus prepared by reduction of tosylate **5.38** derived from the diacetal of D-galactose **5.11**. A hydroxyl group can also be replaced by a halogen (C1, Br, I) at any position of a protected sugar. The halide is easily reduced by a radical mechanism with tributylstannane, Bu_3SnH (reaction 5.19).

(5.19) $$R\text{-}Cl + Bu_3SnH \rightarrow R\text{-}H + Bu_3SnCl$$

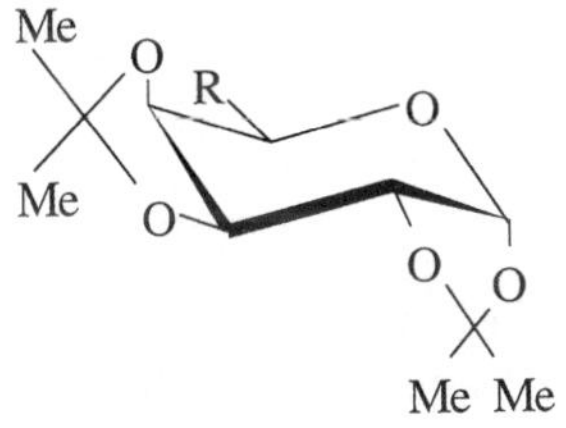

5.38 $R{=}CH_2OTs$
5.39 $R{=}CH_3$

In this reaction, tributylstannane could undoubtedly be replaced by hypophosphorous acid (H_3PO_2) or a triethylammonium salt, as in the case of the deoxygenation reaction described at the end of this chapter. The most direct method to deoxygenate an isolated hydroxyl group is by the formation of dithiocarbonate (reaction 5.20).

(5.20) $R\text{–}OH \xrightarrow{NaH} R\text{–}O^- \xrightarrow{CS_2} R\text{–}O\text{–}CS\text{–}S^- \xrightarrow{ICH_3} R\text{–}O\text{–}CS\text{–}SCH_3$

The reduction by tributylstannane is a chain reaction which must be initiated by azaisobutyronitrile. The overall result is given by reaction (5.21). The tin product decomposes rapidly and an excellent yield of the reduced product is obtained.

(5.21) $R\text{–}O\text{–}CS\text{–}SCH_3 + Bu_3SnH \rightarrow RH + (Bu_3Sn\text{–}S\text{–}CO\text{–}S\text{–}CH_3)$

The use of Bu_3SnH has its drawbacks as it is toxic and expensive; 291 g of the reagent are needed to obtain 1 g of hydrogen, and removal of the excess reagent and by-products of the reaction is difficult. As an alternate hydrogen donnor, hypophosphorous acid (H_3PO_2) was recently proposed, or its triethylammonium salts when acidity is to be avoided, and these give results at least as good as with stannanes in radical reactions. Thus the deoxygenated diacetal of glucofuranose at position 3, **5.40** is obtained in 91% yield by reduction of the methyl dithiocarbonate of 'diacetone glucose' **5.10** (Barton *et al.* 1993).

5.40

References

Baker, C. D., Horton, D., and Tindall, Jr., C. G. (1972), *Carbohydr. Res.*, **24**, 192–197.

Barton, D. H. R., Jang, D. O., and Jaszberenyi, J. Cs. (1993), *J. Org. Chem.*, **58**, 6838–6842.

Bashir, N. B., Phythian, S. J., Reason, A. J., and Roberts S. M. (1995), *J. Chem. Soc., Perkin Trans 1*, 2203–2222.

David, S. and Hanessian, S. (1985), *Tetrahedron*, **41**, 643–663.

David, S. and Thieffrey, A. (1979), *J. Chem. Soc., Perkin Trans I*, 1568–1573.

David, S. and Thieffry, A. (1983), *J. Org. Chem.*, **48**, 441–447.

Excoffier, G., Gagnaire, D., and Utille, J. P. (1975), *Carbohydr. Res.*, **39**, 368–373.

Garegg, P. J., Hultberg, H., and Wallin, S. (1982), *Carbohydr. Res.*, **108**, 97–101.

Gelas, J. (1981), *Adv. Carbohydr. Chem.*, **39**, 71–156.

Grice, P., Ley, S. V., Pietruszka, J., Priepke, H. W. M., and Walther, E. P. E. (1995), *Synlett*, 791.

Hanessian, S. (1966), *Carbohydr. Res.*, **2**, 86–88.

Hanessian, S. and Lavallée, P. (1975), *Can. J. Chem.*, **53**, 2975–2977; (1977) **55**, 562–565.
Horton, D. and Weckerlé, W. (1975), *Carbohydr. Res.*, **44**, 227–240.
Jones, G. H. and Moffatt, J. G. (1972), *Methods Carbohydr. Chem.*, **6**, 315–322.
Klemer, A. and Rodemeyer, G. (1974), *Chem. Ber.*, **107**, 2612–2614.
Kong, X. and Grindley, T. B. (1993), *J. Carbohydr. Chem.*, **12**, 557–571.
Ley, S. V., Leslie, R., Tiffin, P. D., and Woods, M. (1992), *Tetrahedron Lett.*, **33**, 4767–4770.
Ley, S. V., Priepke, H. W. M., and Warriner, S. L. (1994), *Angew. Chem., Int. Ed. Eng.*, **33**, 2290–2292.
Manthorpe, P. A. and Gigg, R. (1980), *Methods Carbohydr. Chem.*, **8**, 305–311.
Onodera, K. and Kashimura, N. (1972), *Methods Carbohydr. Chem.*, **6**: 331–336.
Perlin, A. S. and von Rudlof, E. (1965), *Can. J. Chem.*, **43**, 2071–2077.
Pollak, A., Baughn, R. L., and Whitesides, G. M. (1977), *J. Am. Chem. Soc.*, **99**, 2366–2367.
Therisod, M. and Klibanov, A. M. (1986), *J. Am. Chem. Soc.*, **108**, 5638–5640.

6 Reactions of carbonyl groups and hemiacetals

6.1 Introduction

When protection of hydroxyl groups prevents the transformation of a carbonyl into a hemiacetal by cyclization, the reactivity of a sugar cannot be distinguished from that of an aldehyde or an ordinary ketone, linked to very electronegative groups possibly with an oxygen atom at the β-position. Thus, we can observe the formation of hydrates stable in water and β-elimination, if the stereochemistry allows it, in an alkaline medium. As for the usual hemiacetal sugar, all the reactions of the carbonyl compounds are possible due to the rate at which tautomeric equilibrium is established. In fact, it is rather the possible instability of a reagent in the solvent used which limits its application. We have already seen a typical hemiacetal reaction, glycosidation. Another example is the oxidation by halogens, with which we will begin.

6.2 Oxidation by halogens (Green 1980)

The oxidants Cl_2, Br_2, and I_2 are used in different ways. If a sugar in a non-buffered aqueous solution is treated with chlorine or bromine, oxidation leads to the acidification of the medium owing to the formation of HCl or HBr. It seems that this slows down the reaction and the best yields are obtained, nearly quantitatively, by adding an insoluble salt such as $BaCO_3$ which acts as a buffer. When oxidation is by bromine, the pH is close to 5.4. It is the cyclic form which is oxidized. D-Glucopyranose (α or β) gives D-glucono-1,5-lactone (reaction 6.1). Industrially, the reaction is done in an electrolytic cell; bromine is regenerated by anodic oxidation of bromide, which is added in only a catalytic amount. The β-anomer reacts much faster than the α-anomer. Oxidation is essentially due to molecular bromine. It is supposed that hypobromite is formed which then undergoes E_2 elimination as outlined by the partial formula **6.1**. Oxidation by bromine is a preparative route giving access to gluconic and galactonic acids.

(6.1) [glucopyranose: CH_2OH, HO, HO, HO, O, OH] + Br_2 ⟶ [gluconolactone: CH_2OH, HO, HO, HO, O, CO] + 2 HBr

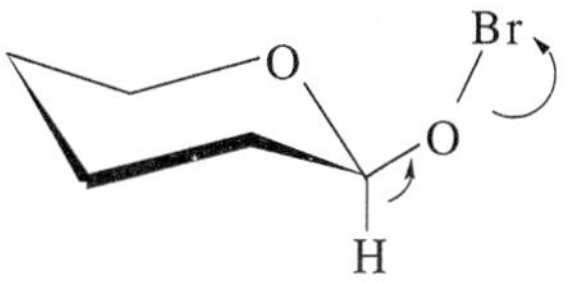

6.1

Iodine is not reactive in an acidic medium. In a basic medium, it is converted into hypoiodite, a powerful oxidant, and quantitatively oxidizes aldoses into aldonic acids according to reaction (6.2), which is of analytical as well as preparative value.

$$RCHO + I_2 + 3NaOH \rightarrow RCO_2Na + 2NaI + 2H_2O \tag{6.2}$$

6.3 Nucleophilic reagents

6.3.1 Sodium borohydride

This reagent, stable in water in slightly alkaline conditions, is ideal for reducing sugars. The aldehyde function is reduced to a primary alcohol and the ketone function to a mixture of secondary alcohol epimers. The simplest treatment after reduction consists in eliminating the sodium on a cation exchanger column, then reaction with boric acid in the form of volatile methyl borate by co-evaporation with methanol. The alditol is recovered by evaporation of the solvent to dryness or lyophilization.

6.3.2 Thiols (Wander and Horton 1976)

Dithioacetals, obtained by reaction of thiols with sugars, constitute the largest class of acyclic sugar derivatives. Diethyl dithioacetal **6.2** is prepared by treating glucose dissolved in 11 M HCl with ethanethiol (EtSH) for 4 h at 0°C. The excess acid is removed with $BaCO_3$, $PbCO_3$, or an anion exchanger, and the product is obtained by evaporation of the solvent. This reaction is very general. Ethanethiol, benzylthiol, ethanedithiol, and 2,3-propanedithiol are commonly used. Thiophenol reacts very slowly. This reaction is observed with all aldoses according to the probable reversible mechanism (6.3).

$$>C{=}O + RSH \underset{}{\overset{+H^+}{\rightleftharpoons}} >C(\overset{+}{O}H_2)(SR) \underset{+\,RSH}{\overset{-\,H_2O}{\rightleftharpoons}} >C(\overset{+}{S}HR)(SR) \overset{-\,H^+}{\rightleftharpoons} >C(SR)(SR) \tag{6.3}$$

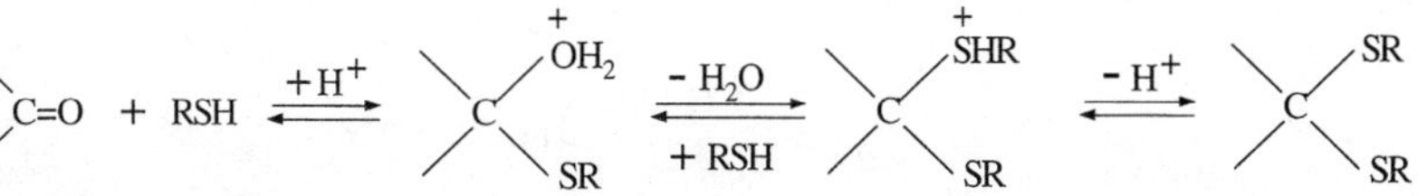

During the second step, we can observe that a second thiol molecule rather than a sugar hydroxyl displaces H_2O. This could be explained by the higher nucleophilicity of sulfur towards carbon. Ketoses are decomposed under these conditions.

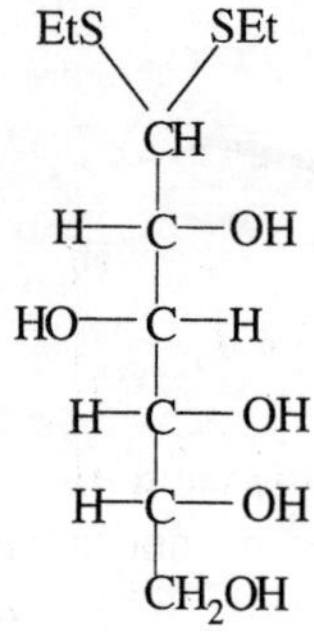

6.2

Preparation of fructose dithioacetals involves the *keto* peracetate **6.3** which is the normal peracetylation product. However, the carbonyl group of sialic acid is transformed into diethyl dithioacetal **6.4** in excellent yield at room temperature. The α-carboxyl and lactonization in this medium no doubt help in the conversion.

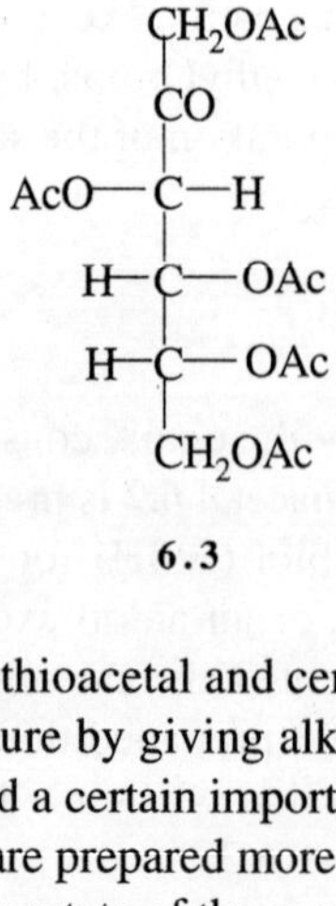

6.3 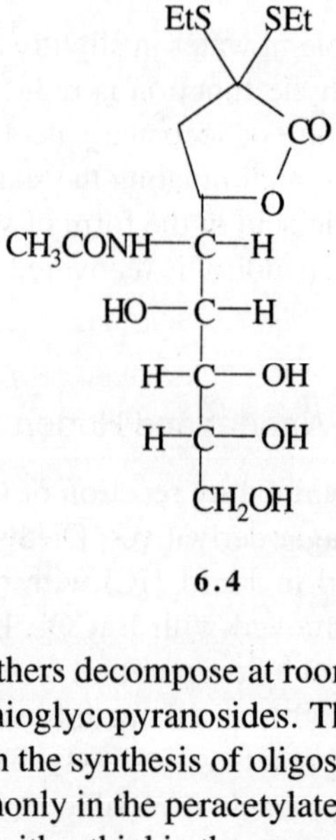 **6.4**

The glucose dithioacetal and certain others decompose at room temperature in the reaction mixture by giving alkyl 1-thioglycopyranosides. This class of compounds has gained a certain importance in the synthesis of oligosaccharides (reaction 10.4). They are prepared more commonly in the peracetylated form as **6.5**, by treating the β-peracetate of the pyranose with a thiol in the presence of $BF_3.OEt_2$.

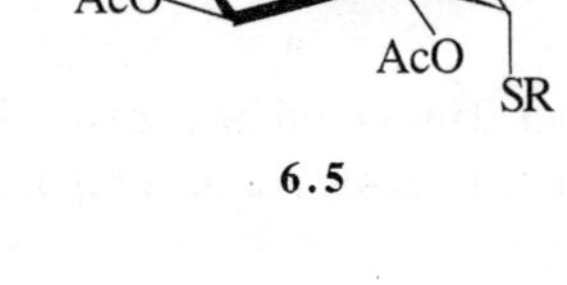

6.5

All the usual transformations of hydroxyl groups of sugars such as etherification, acylation, and acetalation can be carried out on dithioacetals. For example, the propylene dithioacetal of '2-deoxyglucose' gives the bis-isopropylidene acetal **6.6**. The rules of preference for acetalation positions are similar to those of alditols. Acylated dithioacetals can be deacylated by alkaline methanolysis, which indicates the stability of the dithioacetal function under such conditions. However, butyllithium removes the proton located between the two sulfur atoms, creating a negative charge at C-1 and, if there is an ether function at the α-positioned C-2, immediate elimination takes place to form the group, $-CH{=}C(SPh)_2$. On the other hand, the carbanion at C-1 is stable in dithioacetals of 2-deoxy sugars and it can be condensed with iodo derivatives. This is an easy method for the introduction of chiral chains, useful in total synthesis.

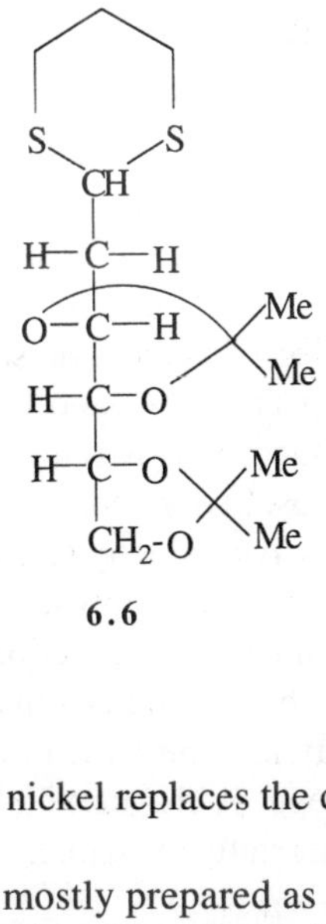

6.6

Desulfurization with Raney nickel replaces the dithioacetal group by a methyl group.

However, dithioacetals are mostly prepared as intermediates in the synthesis of acyclic derivatives. The carbonyl can indeed be regenerated under conditions which do not interfere with most of the common functional groups. Although regeneration with iodine is quantitative and bromine is usable in a preparative fashion, mercury salt-based systems seem to be favoured. If the cyclizable hydroxyl groups are protected, the acyclic sugar derivative can be prepared in this way. For example, peracetylation of **6.2** gives a pentaacetate, readily transformed into *aldehydo*-pentaacetate **6.7** by treatment with an excess of $HgCl_2$ in aqueous acetone. In a related manner, the *aldehydo*-bis-isopropylidene acetal **6.8** is prepared starting from the diethyl dithioacetal of D-arabinose. Solvolysis in methanol with a $HgCl_2$–HgO mixture leads to a dimethyl acetal such as **6.9** in the D-*galacto* series, deacetylated to **6.10** by $Ba(OMe)_2$.

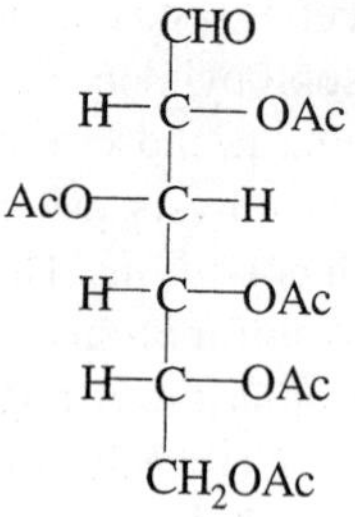

6.7

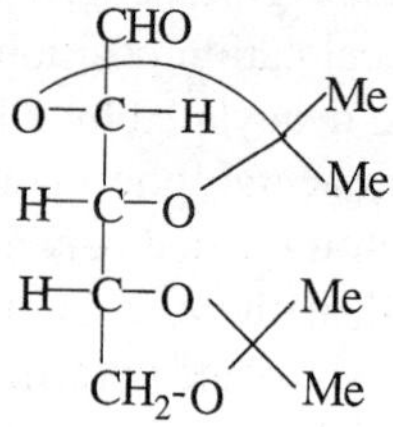

6.8

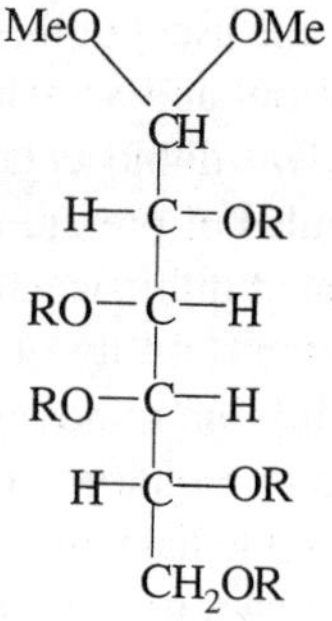

6.9 R=Ac
6.10 R=H

6.3.3 Cyanides of alkaline metals

We will take the example of the KCN reaction with D-arabinose in aqueous solution. Addition is carried out at room temperature to give a mixture of α-hydroxyl nitriles **6.11**. Heating in an alkaline medium hydrolyses these nitriles into carboxylates. The mixture of the carboxylic acids, gluconic **6.12** and mannonic **6.13**, is recovered with a cation exchanger, and the D-gluco epimer **6.12** is isolated by fractional crystallization of the barium salt. The regenerated acid is lactonized by prolonged heating at 100°C and reduction of lactone **6.14** by a sodium amalgam affords D-glucose. This sequence of reactions, known as the Kiliani–Fischer synthesis, is absolutely general. This is a method for ascending the sugar series starting from glyceraldehyde, as given in Table 4.1. It is currently used to prepare higher sugars (heptoses from hexoses) or rare sugars. Thus, by using the same sequence of reactions, the two rare compounds, L-glucose and L-mannose, can be obtained from a very abundant natural sugar, L-arabinose. This method may also be used in the preparation of specifically labelled sugars with an isotope of the carbon atom because [^{14}C]KCN and [^{13}C]KCN cyanides are commercially available.

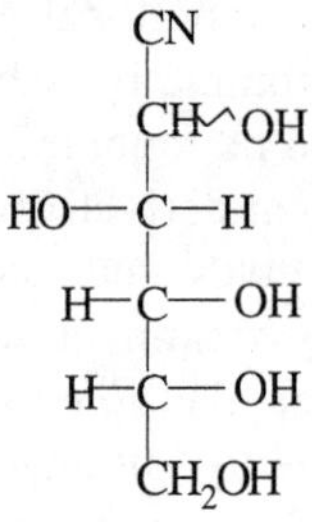

6.11

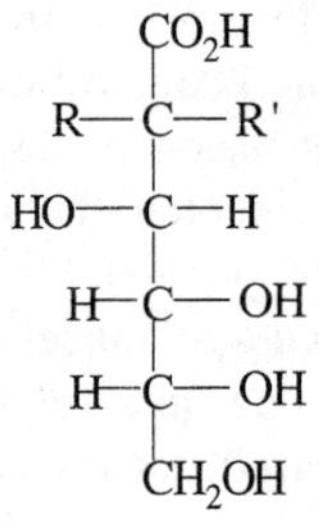

6.12 R=H, R'=OH
6.13 R=OH, R'=H

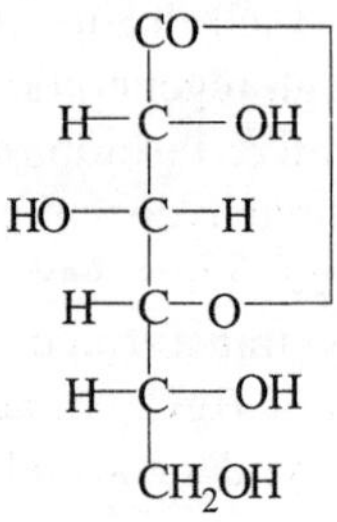

6.14

In one modification (Serianni *et al.* 1979), cyanide is added keeping the pH around 7.5, subsequently lowered to 4.2 at the end of the reaction. Nitriles are stable under these conditions. They are reduced to imines by hydrogenation with palladium over barium sulfate. These imines are rapidly hydrolysed into aldoses. This is the method recommended for the preparation of C-1-labelled sugars.

Here we will describe the Kuhn synthesis of amino sugars, even though the involvement of a glycosylamine is implied (see Section 3.6.1) it is conceptually similar (reaction 6.4). The addition of HCN to *N*-benzyl glycosylamine **6.15** gives the epimeric nitriles **6.16** and **6.17**. Each one gives, by catalytic hydrogenation on palladium in the presence of hydrochloric acid, the 2-amino-2-deoxy sugar with the same configuration (**6.18**, for example), the benzyl group being simultaneously eliminated in the form of toluene. The major success of this method is exemplified by the synthesis of lactosamine, as described in Section 10.1.

(6.4)

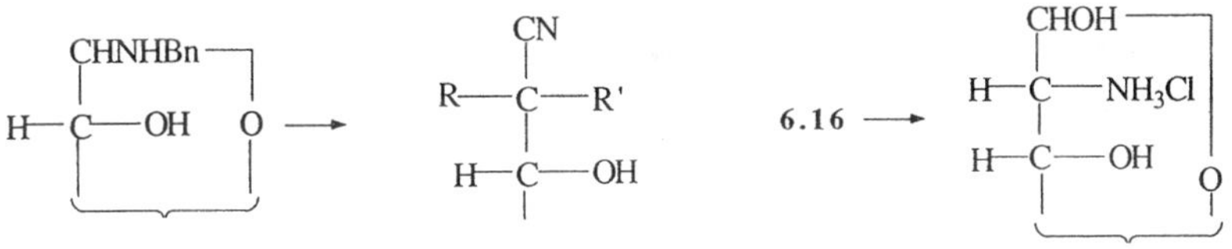

CHNHBn, H–C–OH, O → CN, R–C–R', H–C–OH; 6.16 → CHOH, H–C–NH_3Cl, H–C–OH, O

6.15

6.16 R=H, R'=NHBn
6.17 R=NHBn, R'=H

6.18

6.3.4 Wittig and organometallic reactions

As mentioned in the introduction, reactions involving a protected sugar in the *aldehydo* tautomeric form do not differ basically from those of common aldehydes. For example, we can observe additions of organomagnesium, organolithium reagents, etc. These reactions can be very useful, especially in total chiral synthesis, but this type of chemistry should be, in principle, familiar to the reader. Here we will only consider sugars that are totally free or capable of aldehyde hemiacetal tautomerism.

Under certain conditions, triphenylmethylene phosphorane ($CH_2 = PPh_3$) can be used with these substrates, as described in reaction (6.5) where methylenation is carried out in oxolane at –30°C with the addition of 1 eq. of NaH giving a 67% yield. The success of this reaction stems from the stability in an alkaline medium of the aldehyde which does not have an oxygen atom at the β-position, whereas deprotonation of the hydroxyl group at the α-position creates a negative charge, in conflict with a second deprotonation at CH (Nicolaou *et al.* 1988).

(6.5)

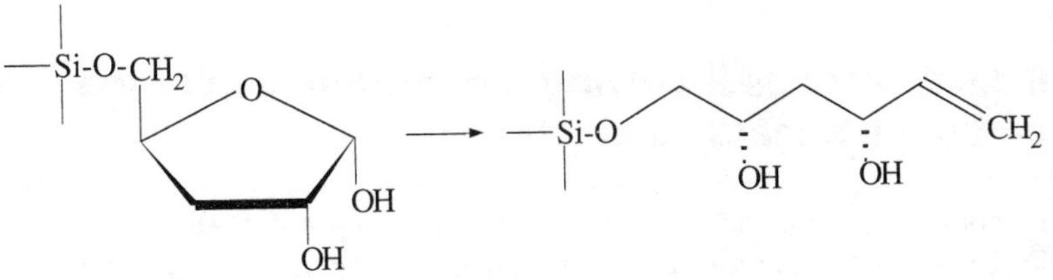

–Si-O-CH_2, O, OH, OH → –Si-O, OH, OH, CH_2

Other reactions have been described (Ohrui *et al.* 1975) with stabilized phosphoranes such as $Ph_3P = CHCO_2Et$. Ribofuranose **6.19** gives 54% of the *cis*-ethylenic isomer **6.20**, accompanied by 36% of the *trans* isomer. In the presence of NaOMe, cyclization of **6.20** by a Michael-type reaction gives pure 1,2-*trans*-oxolane **6.21**. Because of the superficial analogy with glycosides, compounds such as **6.21** are called *C*-glycosides. With the same phosphorane, the bis-acetal of mannofuranose **6.22** gives directly 1,2-*trans*-*C*-mannofuranoside **6.23**. The latter isomerizes into a 1,2-*cis* derivative in the presence of NaOMe, the probable mechanism being α-deprotonation of the ester function followed by elimination of the ring oxygen in the form of an alkoxide.

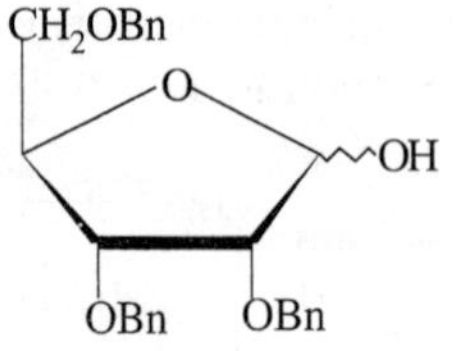

6.19 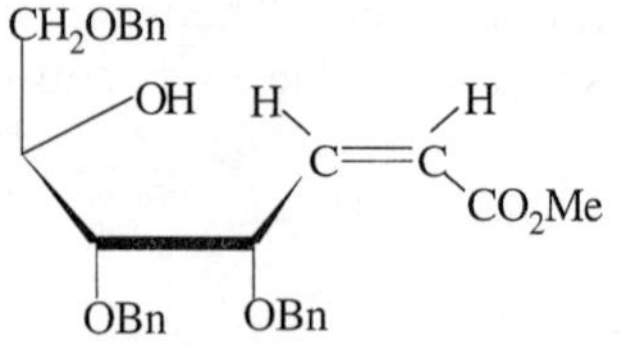 6.20 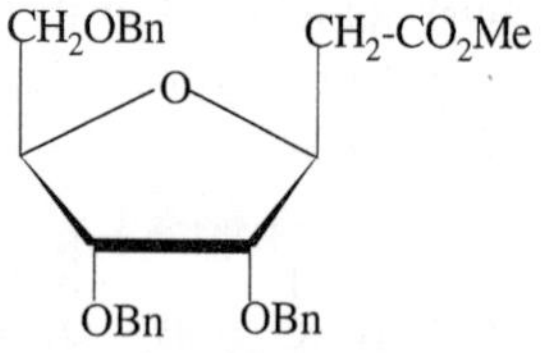 6.21

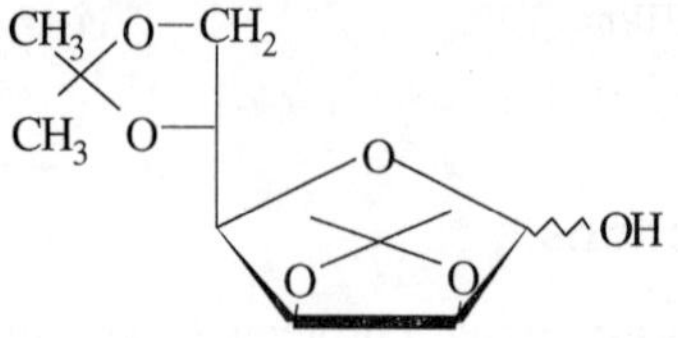

6.22 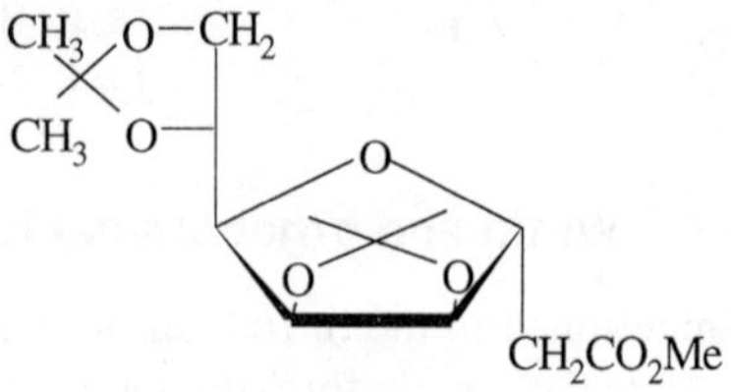 6.23

Organic derivatives of magnesium and lithium cannot be used in aqueous solvents and, in an anhydrous medium, the polyhydroxylated substrates bring about reagent waste. The discovery of a technique to produce and induce reaction of allylic tin derivatives in water (Luche and Damiano 1980) has allowed these organometallic reactions to be carried out in aqueous solution (Kim *et al.* 1993). An alcohol–water solvent was used in the presence of 2 eq. of tin powder and allyl bromide in an ultrasonic bath. Mannose, for example, gives a 90% yield of carbinol **6.24** with its epimer in a 6:1 ratio. Ozonization of the double bond gives the 2-deoxy sugar with two supplementary carbons. Tin can be replaced by indium.

6.4 Reactions involving α-deprotonation of the carbonyl; sugars as aldols

In a basic medium, equilibrium is anticipated between the sugar and the carbanion produced by capture of the α-proton. The first foreseeable result is the partial

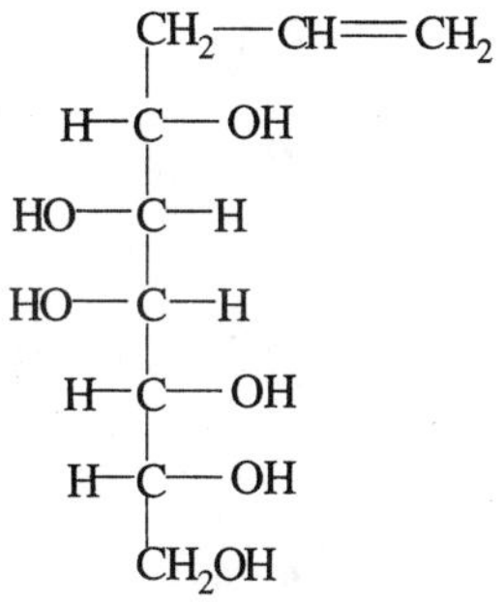

6.24

epimerization at this carbon. This reaction can be of preparative value if the starting sugar is abundant.

N-Acetylmannosamine **6.25** can be prepared by epimerization of *N*-acetylglucosamine (**1.7**) in aqueous solution at pH 11. After 3 days at room temperature the yield is 22%, but the starting material is very accessible.

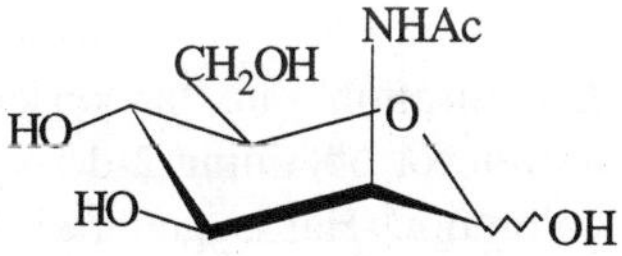

6.25

Although these epimerizations are general, they are only about as useful as the example given. In parentheses, we would like to compare them with the epimerization produced by molybdic acid (at 90°C in aqueous solution, pH 4.5) which does not involve a carbanion and is also general (Hayes *et al.* 1982). Glucose and mannose are at equilibrium, and the concentration ratio (2.5:1) is the one calculated from their free enthalphy difference. This is a typical reaction whose simplicity is misleading for there is simultaneous migration of the carbon atom at position 1 to position 2 (reaction 6.6).

(6.6) $[1-{}^{13}C]$D-glucose $\rightleftharpoons$ $[2-{}^{13}C]$D-mannose

Thus we have a very elegant method for labelling C-2 of a sugar, the introduction of the isotope at C-1 being easy with the reactions in Section 6.3.3. This is explained by the formation of a complex between dimolybdate $Mo_2O_5^-$ and the sugar as D-glucose in the *aldehydo*-form and reversible migration of C-3 to C-1 in the complex, according to the very simplified scheme **6.26**.

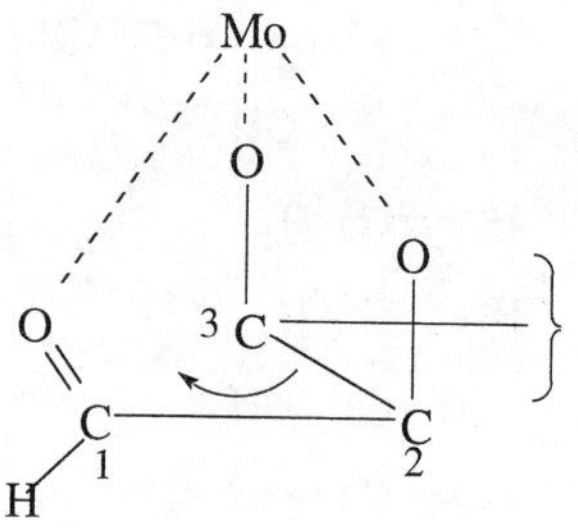

6.26

Let us now come back to the properties resulting from deprotonation of aldoses at C-2. The anion can react as a donor on an aldehyde to give an aldol condensation product. For example, 'diacetone mannose' gives the branched-chain sugar **6.27**. One of the limits of these reactions is that very often there is an etherified oxygen at C-3 which has a tendency to be eliminated if there are no particular steric constraints. This is what happens, for example, on alkaline treatment of certain glycoproteins (see Section 13.2.2). But elimination can be favoured, on the contrary, by introducing an appropriate leaving group at O-3 such as a methanesulfonyl group. Elimination in lukewarm aqueous solution of 3-*O*-methanesulfonyl-D-glucose by addition of aqueous potassium hydroxide is produced at an acidimetric titration rate, as observed in the presence of phthalein. The best preparation for obtaining 2-deoxy-D-ribose is achieved by stopping at the persistent coloration (Hardegger *et al.* 1957), which gives a 50% yield according to reaction (6.7).

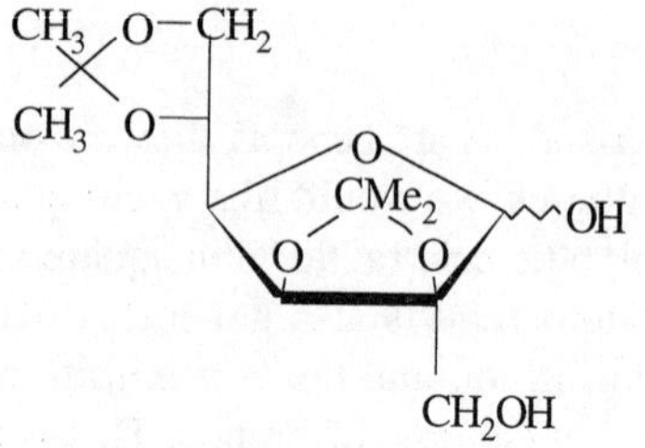

6.27

(6.7)

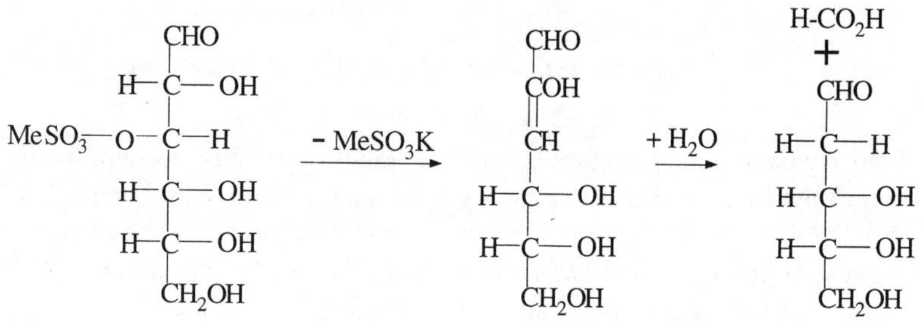

All the common aldoses possess a hydroxyl group in the β-position to the carbonyl. This structure is similar to that of aldols which are formed reversibly from aldehydes in an alkaline medium. Under these conditions, we could expect a retro-aldol cleavage between C-2 and C-3 to give a glycolic aldehyde molecule and an aldose with two carbons less and, conversely, the condensation of these two fragments to afford a higher aldose. In their non-enzymic form, these reactions are unknown or of little practical importance and, in living cells, schemes such as $C_4 + C_2 \rightleftharpoons C_6$, or else $C_3 + C_2 \rightleftharpoons C_5$ are not used in the major metabolic pathways. *Aldolase* enzymes exist, however, isolated from plant or microbial cells which catalyse the retro-aldol cleavage of a deoxyribose phosphate in acetaldehyde and a D-glyceraldehyde phosphate. Their role is purely catabolic in nature because cells produce the 2-deoxy-D-*ribo* structure by deoxygenation of the D-*ribo* structure. They could be useful in preparative chemistry since, to a certain extent, the donor and acceptor can be modified (Chen *et al.* 1992). But with the usual aldolases, the donor is a ketone.

Fructose-1,6-diphosphate aldolase of rabbit muscle has been studied very extensively and it is now commercially available. That of spinach leaves, obviously very accessible, was recently examined (Valentin and Bolte 1993). In the fundamental reaction of glycolysis (reaction 6.8), the donor is dihydroxyacetone phosphate. It can scarcely be varied, but there is more flexibility with the acceptor (David *et al.* 1991; Bednarski *et al.* 1989), and sometimes we can wander considerably from the subject of sugar chemistry. In any case, the *vicinal*-diol created at positions 3 and 4 (uloses numbering) has the D-*threo* configuration. Hence the condensation of the keto aldehyde **6.28** gives ketose **6.29** which, after isolation, is dephosphorylated enzymically in the presence of acid phosphatase.

(6.8) $CH_2OPO_3H_2$–CO–CH_2OH + CHO–H–C–OH–$CH_2OPO_3H_2$ $\rightleftharpoons$ $CH_2OPO_3H_2$, $CH_2OPO_3H_2$, O, HO, OH

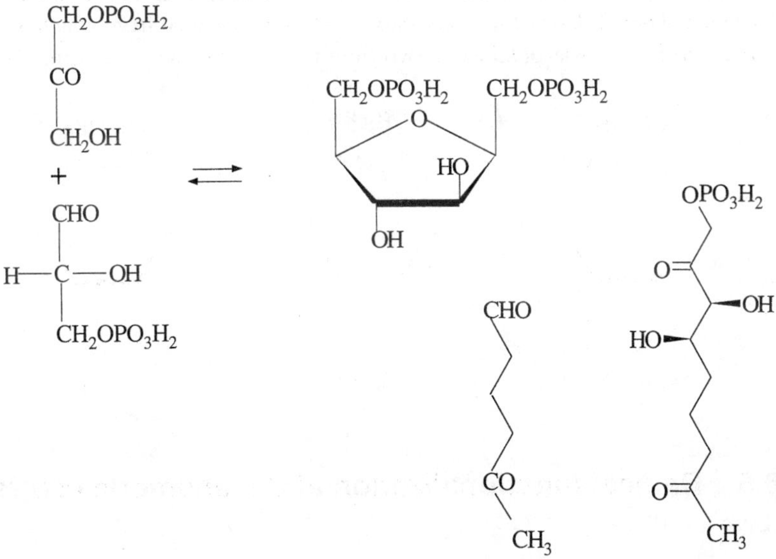

CHO, CO, CH_3

6.28

OPO_3H_2, O=, OH, HO, O=, CH_3

6.29

Acetalation with an acid catalyst provides an *anhydro*-type sugar system involving the more electrophilic carbonyl at position 8. Product **6.30**, obtained (Schultz *et al.* 1990) in a global yield of 48%, is transformed into a pheromone, (+)-*exo*-brevicomin by reduction of the group $COCH_2OH$ to CH_2CH_3. In most syntheses, enzymic condensation must be followed by phosphate removal, which is not a very serious disadvantage. There are far worse difficulties such as the impractical preparation of large quantities of dihydroxyacetone phosphate because of the great amount of solvents needed. The aldolization product is a ketose which is not easily isomerized to an aldehyde. Other aldolases function with pyruvic acid as donor, such as the 'sialylaldolase' whose synthetic use is described in Section 12.4, and the 'Kdo aldolase' usable in the synthesis of the plant and microbial sugar, 'Kdo' **6.31**.

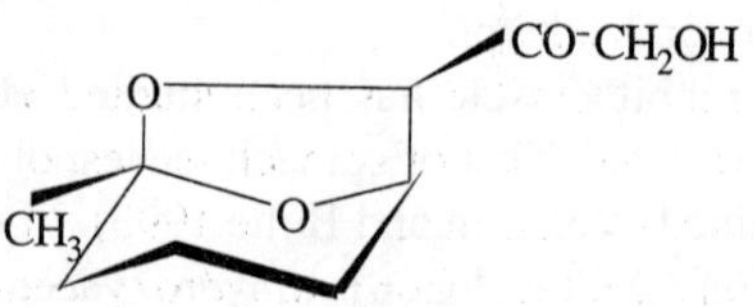

CO-CH_2OH O O CH_3

6.30

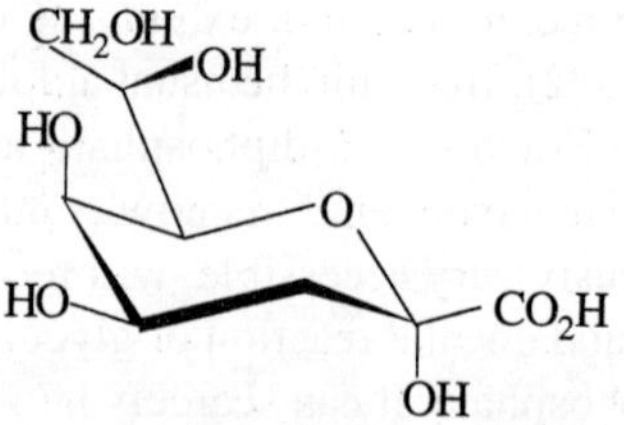

CH_2OH OH HO O HO CO_2H OH

6.31

Not unexpectedly, α-eliminations are also observed with sugar lactones (Section 6.2). Thus treatment of D-glucono-1,5-lactone **6.32** with an excess of benzoyl chloride and pyridine for 16 h at room temperature gives the crystalline elimination product **6.33** in quantitative yield. Catalytic hydrogenation then led to the benzoylated 3-deoxy-D-*arabino*-hexono-1,5-lactone **6.34** from which 3-deoxy-D-*arabino*-hexose was obtained in two steps (de Lederkremer and Varela 1994).

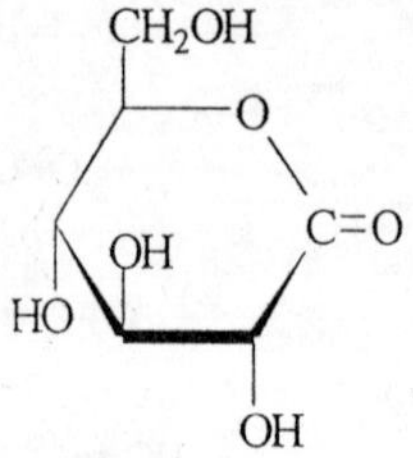

CH_2OH O C=O OH HO OH

6.32

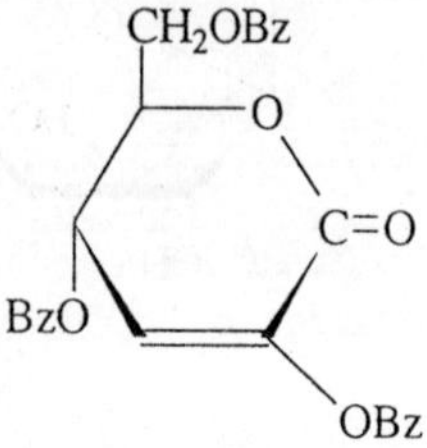

CH_2OBz O C=O BzO OBz

6.33

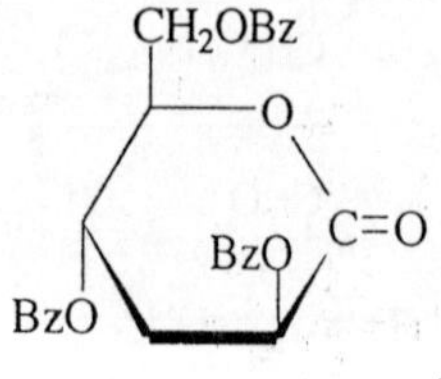

CH_2OBz O C=O BzO BzO

6.34

6.5 Radical functionalization at the anomeric centre

(Descotes 1992)

Irradiation of pyranoid derivatives produces a radical at the anomeric centre. In the presence of bromine or *N*-bromosuccinimide, the chloride of the peracetylated β-D-glucopyranoside gives the geminal dihalogenated sugar **6.35**, raw material for the synthesis of geminal analogues at C-1. Mannoside **6.36** undergoes a Norrish type II reaction to give oxolane **6.37**. In the presence of iodine and HgO, the glycol glucoside **6.38** gives the spiro-orthoester **6.39**, probably via the formation of a hypoiodite.

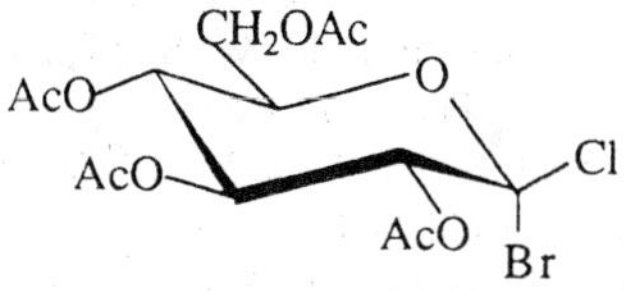

6.35

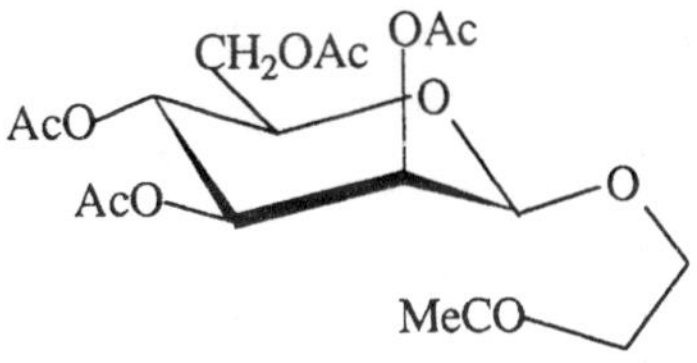

6.36

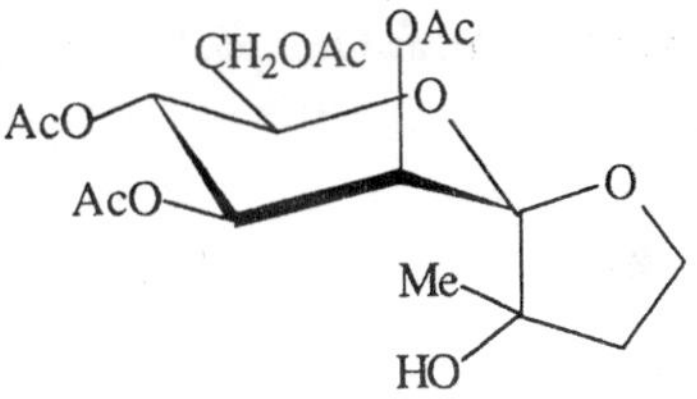

6.37

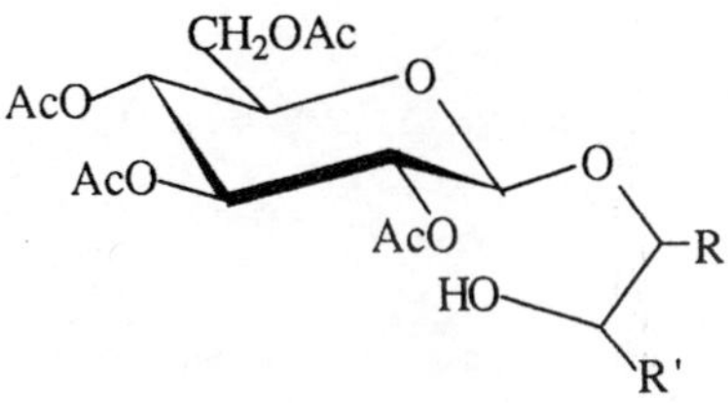

6.38

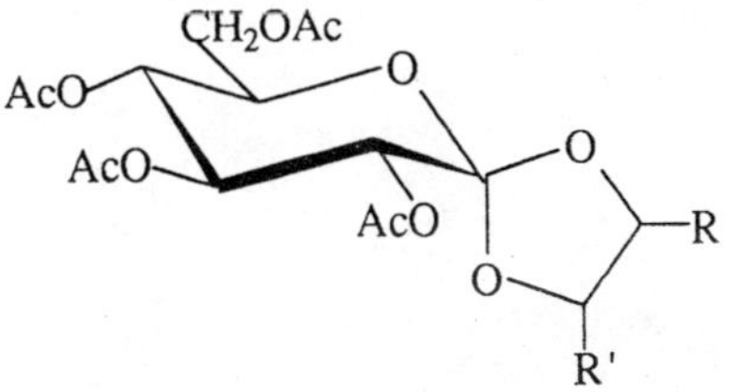

6.39

References

Bednarski, M. D., Simon, E. S., Bischofberger, M., Fessner, W. D., Kim, M. J., Lees, W., Saito, T., Waldmann, H., and Whitesides, G. (1989), *J. Am. Chem. Soc.*, **111**, 627–635.

Chen, L., Dumas, D. P., and Wong, C. H. (1992), *J. Am. Chem. Soc.*, **114**, 741–748.

David, S., Augé, C., and Gautheron, C. (1991), *Adv. Carbohydr. Chem. Biochem.*, **49**, 189–194.

de Lederkremer, R. M. and Varela, O. (1994), *Adv. Carbohydr. Chem. Biochem.*, **50**, 125–209.

Descotes, G. (1992), *Carbohydrates. Synthetic methods and applications in medicinal chemistry* (ed. H. Ogura *et al.*), p. 89, VCH, New York.

Green, J. W. (1980), Oxidative reactions and degradations. In *The carbohydrates, chemistry and biochemistry*, Vol. 1B, (2nd edn). (eds. W. Pigman and D. Horton), pp. 1106–1119, Academic Press, New York.

Hardegger, E., Schellenbaum, M., Huwyler, R., and Züst, A. (1957), *Helv. Chim. Acta*, **40**, 1815–1818.

Hayes, M. L., Pennings, N. J., Serianni, A. S., and Barker, R. (1982), *J. Am. Chem. Soc.*, **104**, 6764–6769.

Kim, E., Gordon, D. M., Schmid, W., and Whitesides, G. M. (1993), *J. Org. Chem.*, **58**, 5500–5507.

Luche, C. L. and Damiano, J. C. (1980), *J. Am. Chem. Soc.*, **102**, 7926–7927.

Nicolaou, K. C., Daines, R. A., Uenishi, J., Li, W. S., Paphatjis, D. P., and Chakraborty, T. K. (1988), *J. Am. Chem. Soc.*, **110**, 4672–4685.

Ohrui, H., Jones, G. H., Moffatt, J. G., Maddox, M. L., Christenson, A. T., and Byram, S. (1975), *J. Am. Chem. Soc.*, **97**, 4602–4613.

Schultz, M., Waldmann, H., Kunz, H., and Vogt, W. (1990), *Liebigs. Ann. Chem.*, 1019–1024.

Serianni, A. S., Munez, H. A., and Barker, E. (1979), *Carbohydr. Res.*, **72**, 71–78.

Valentin, M. L. and Bolte, J. (1993), *Tetrahedron Lett.*, **34**, 8103–8106.

Wander, J. D. and Horton, D. (1976), *Adv. Carbohydr. Chem. Biochem.*, **32**, 15–123.

7 Changes of configuration; unsaturated and branched-chain sugars

7.1 Displacement of hydroxyl groups

The most frequently used displacement method of a hydroxyl group begins with its conversion to sulfonate. Sulfonates of the secondary hydroxyl groups of sugars are not very reactive with respect to external nucleophiles. This is no doubt a result of the high functionalization of the molecule which combines steric hindrance with an unfavourable inductive effect. Substitution, practically impossible with former techniques, only became a current operation following two innovations. The first was the introduction of polar aprotic solvents; the favourite solvent is *N,N*-dimethylformamide since this is the easiest to eliminate at the end of the reaction, by extracting it with water from an ether solution. However, even in boiling DMF, some toluenesulfonates and methanesulfonates are inert. The second innovation was the introduction of two highly efficient leaving groups, trifluoromethanesulfonate (triflate) and imidazolylsulfonate (imidazylate). The efficiency of triflates, CF_3SO_2OR, is without a doubt linked to the stability of the anion CF_3SO_3 , itself a result of the high acidity of CF_3SO_3H. They are commonly prepared by reaction of the hydroxyl group with the anhydride $(CF_3SO_2)_2O$ in the presence of a base such as pyridine which can also act as solvent. The disadvantages of using triflates are that they are costly, unstable to humidity, and incompatible with the acetamido group, except under special care (Lubineau and Bienaymé 1991).

The reagent for the preparation of imidazylates is sulfuryl-bis-imidazole **7.1**, an indefinitely stable product, prepared by reaction of sulfuryl chloride with imidazole. It reacts with alkoxides to give imidazylate **7.2**. The nucleophilic substitution reaction (7.1) (Hanessian and Vatèle 1981) gives, as by-products, fragmentation products, SO_3 and imidazolate. The efficiency of this reaction can be explained as being a consequence of the fragmentation which makes irreversible the cleavage of the R–O bond due to the length of this bond in the S_N2 transition state. The disadvantage of the imidazylates is that alkoxide is needed in the most practical preparation, which for this reason involves highly basic conditions. The advantages are that they are stable at room temperature—even several days in water—and the cost of the reagent is negligible. These derivatives are therefore highly recommended for starting a synthetic sequence.

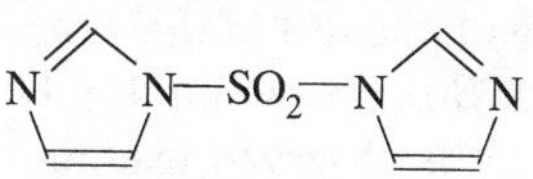

N N—SO_2—N N

7.1

(7.1) R-O-SO_2-N(imidazole)N + X^- ⟶ RX + SO_3 + N^-(imidazole)N

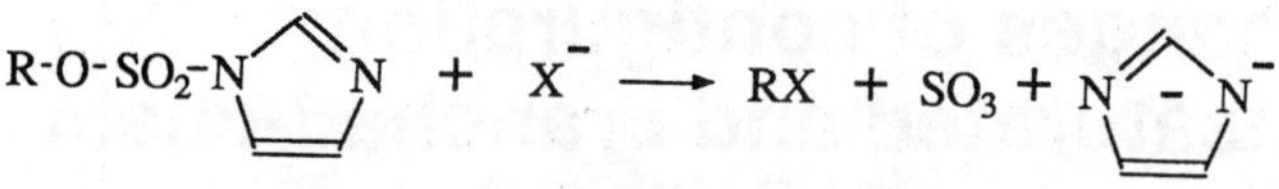

7.2

Depending on their location on a sugar, sulfonates have very different reactivities. This is apparent, for example, in the behaviour of tosylates. The tosylate of the primary alcohol function can be substituted without difficulty, even in solvents which are not polar or aprotic. Substitution at positions 3 and 4 are only possible in DMF solution. Substitution at position 2 is absolutely impossible. On the other hand, it is often observed without problems at position 2 starting from a triflate or an imidazylate. These reactivity differences clearly appear in the triple substitution reaction of the tris-triflate β-D-*galacto* **7.3** by benzoate (Alais and David 1990). The latter is prepared from the triol in 5 h at 0°C. It reacts quantitatively with tetrabutylammonium benzoate in toluene in 45 min at room temperature to give the D-gluco substitution product at C-4 and C-6. Heating for 1 h at 100°C then leads to the D-*manno* tribenzoate **7.4**.

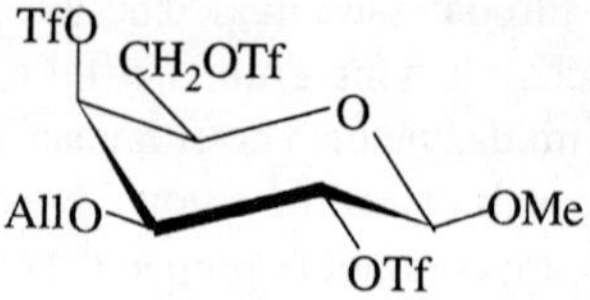

TfO, CH_2OTf, O, AllO, OMe, OTf — **7.3**

CH_2OBz, OBz, O, BzO, AllO, OMe — **7.4**

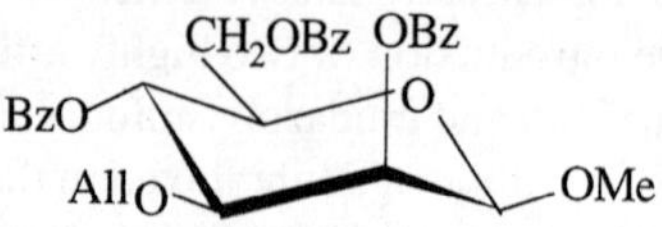

These general reactivity predictions must sometimes be revised because of steric constraints, even at the primary position. For example, α-D-*galacto* tosylate **7.5** is not very reactive, no doubt because of the bulky neighbouring group, isopropylidene. It is significant to compare the α-D-*allo* and α-D-*gluco* tosylates, **7.6** and **7.7**, respectively. Tosylate **7.6** is substituted by NaN_3 by heating for 4 h in DMF at 140°C. Under the same conditions, 15 days are required to obtain the substitution product of the α-D-*gluco* tosylate **7.7** because azide N_3^- must enter from *endo*-position between two five-membered rings (Baggett 1988). Within this context, one experiment (Hanessian and Vatèle 1981) brilliantly demonstrates the efficiency of new leaving groups whereby substitution of α-D-*gluco* imidazylate **7.8** in toluene for 5 h at 80°C gives α-D-*allo* azide **7.9** in 62% yield. As mentioned before, efforts to substitute 'diacetone glucose' at C-3 are thwarted because the nucleophile must come close to the *endo* face. A similar situation is observed when the nucleophile must be axially introduced, hindered by a 1,3-diaxial interaction with the rest of the molecule. Thus, when an attempt at substitution by benzoate of β-D-galacto imidazylate **7.10** is made, intramolecular attack by the ring oxygen is observed leading to the formation of the 2,5-*anhydro* derivative **7.11**.

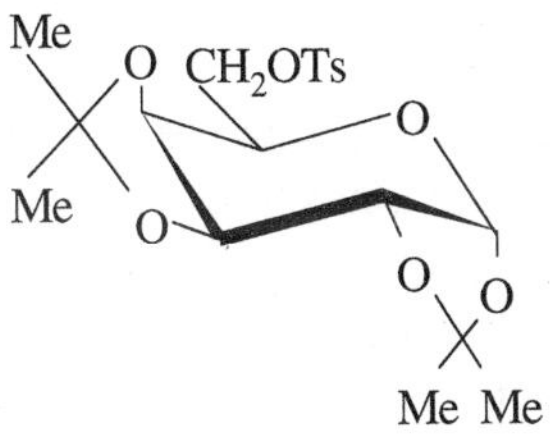

7.5 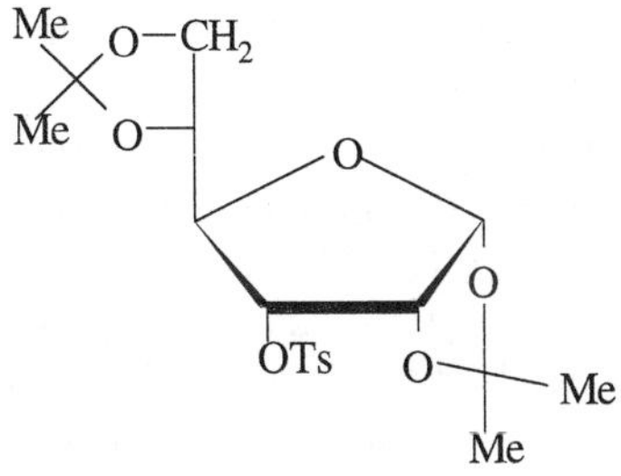 **7.6**

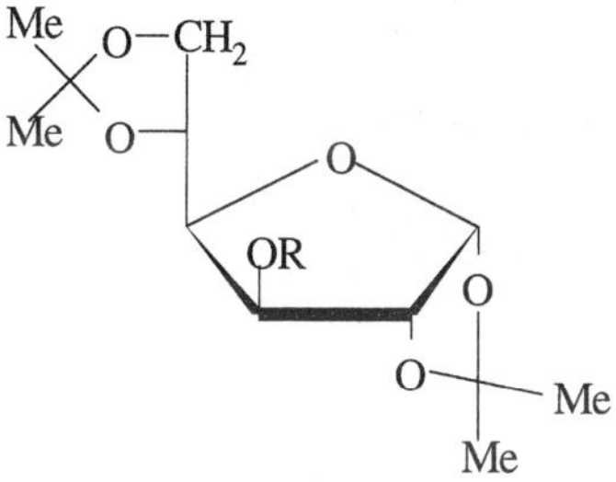

7.7 R=Ts
7.8 R=Ims 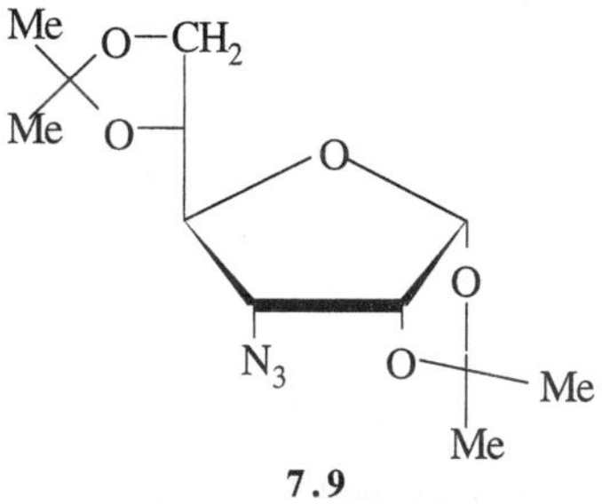
7.9

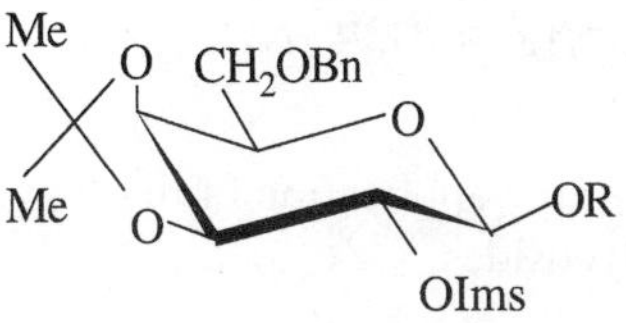

7.10 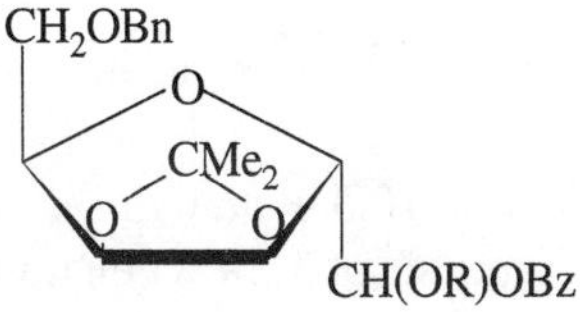 **7.11**

The most commonly used nucleophiles are salts of acetic and benzoic acids, which lead to acetates and benzoates of sugars with opposite configurations, the halides and thiolates, which allow the introduction of a halogen atom or an alkylthio radical, and finally the nitrogen nucleophiles, which will be discussed in more detail. Among the latter, NH_3 and NH_2NH_2 have been used and there may be cases in which their employment is preferable, but azide anion $(N{=}N{=}N)^-$ is presently the all-time favourite. It is not very bulky because of its cylindrical structure, in fact its conformational free enthalpy (see Section 2.5) is practically nonexistent, and it is only weakly basic. However, the risk of E_2 elimination is also present with sugars when a proton is antiparallel to the sulfonate on the neighbouring carbon atom. The organic azide is easily transformed into an amine, by reduction with $LiAlH_4$, or a thiol, or by catalytic hydrogenation. Its introduction is incompatible with allyl ether protection.

The following two examples show the usefulness of substitution reactions in the preparation of not very accessible sugars, important within the framework of this book. *N*-Acetylgalactosamine is very expensive while its epimer at position 4, *N*-acetylglucosamine, is available in large quantities. In order to produce a β-glycoside from *N*-acetylgalactosamine, it is advantageous to prepare the D-*gluco* analogue such as **7.12**, prepared for substitutions at positions 4 and 6, which leads to glycoside **7.13** after deprotection (Gross *et al.* 1967). *N*-Acetylmannosamine is also a very costly sugar whose chemistry has hardly been explored. To prepare these derivatives, it has been found useful (Augé *et al.* 1988) to use azide **7.15**, also obtained by substitution of imidazylate **7.14**.

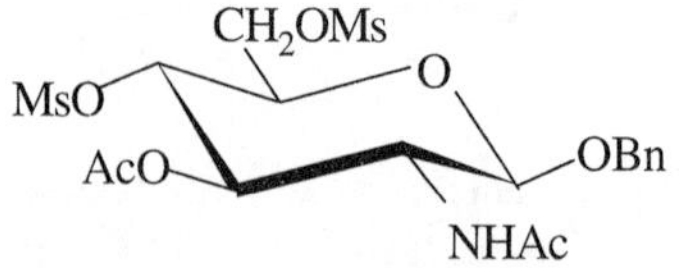

CH_2OMs, MsO, AcO, O, OBn, NHAc

7.12

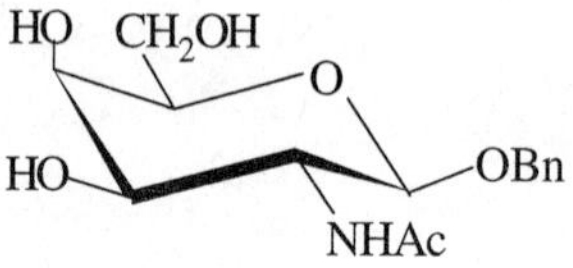

HO, CH_2OH, O, HO, OBn, NHAc

7.13

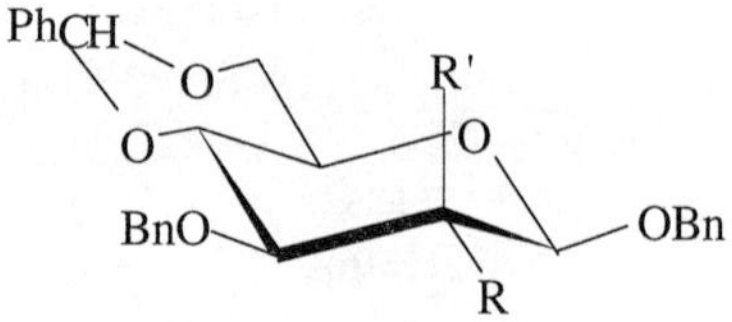

PhCH, O, O, R', O, BnO, OBn, R

7.14 R=O-Ims, R'=H
7.15 R=H, R'=N_3

In Section 10.3.4 we will see an example of the application of these methods to the synthesis of 1,2-*cis* equatorial–axial glycosides.

7.2 Epoxides (Baggett 1988; Williams 1970)

For ring formation to an epoxide to occur there must be two vicinal functions, sulfonate (or halide) and alkoxide, in antiparallel positions. This condition is not very restrictive in acyclic systems and an epoxide is easily constructed on the C-5–C-6 carbon atoms of a hexofuranose. In a six-membered ring, this involves *trans*-diaxial conformation **7.16** of the reactive functions. Reaction (7.2) shows an example which can appear quite unexpected as we can imagine that the epoxide obtained is a very rigid molecule due to its three rings. Moreover, this (internal) substitution reaction involves position 2 which is not very reactive. Nevertheless, it is complete in 10 min at room temperature in the presence of molar NaOMe. Closing of epoxides is also observed starting from *trans*-diequatorial substituents under mild conditions, or at the very most, a slightly prolonged heating at 100°C. We can suppose that the supplementary activation energy is due to the need for a less stable conformation where reactive functions approach a geometry as in disposition **7.16** Reaction (7.3) gives excellent access to epoxide **7.17**.

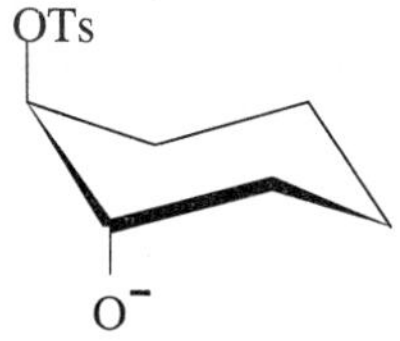

OTs

O⁻

7.16

(7.2)

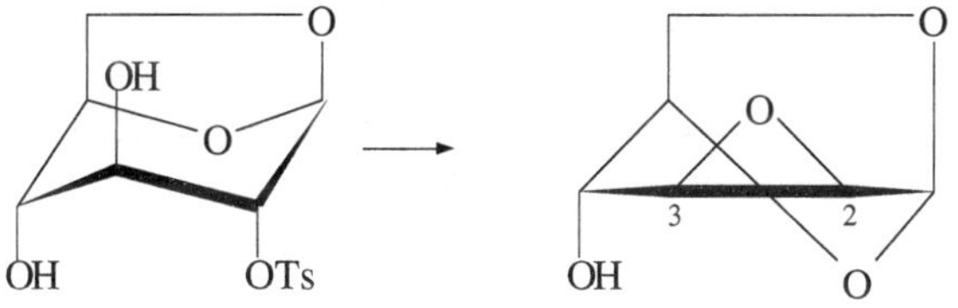

OH O O OH OTs → O O OH O 3 2

(7.3)

PhCH O O O HO TsO OMe → OCH$_2$ PhCH O O O OMe

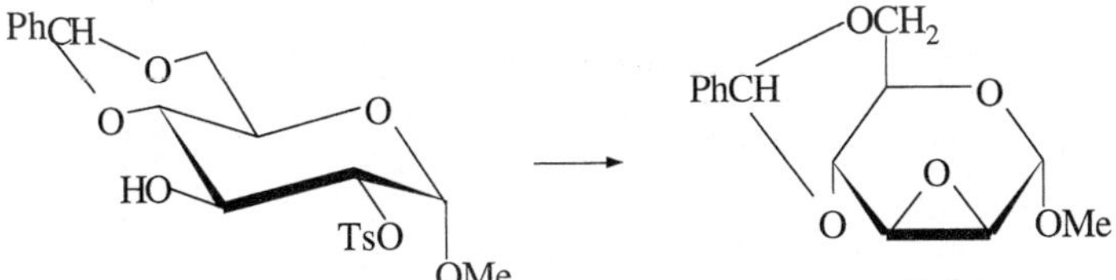

7.17

The vicinal hydroxyl group can be replaced by any function transformable to alkoxide under very basic conditions as, for example, acetate. More surprising is the obviously practical preparation starting from *trans*-ditosylates (reaction 7.4). This reaction occurs at 0°C, yet it involves heterolytic cleavage between the oxygen and sulfur atoms which is not at all simple to carry out by an intermolecular route. We can observe that the tosylate at C-3 is displaced leading to α-D-*allo* epoxide **7.18**. Here again we see the inertia of position 2 relative to the substitution.

(7.4)

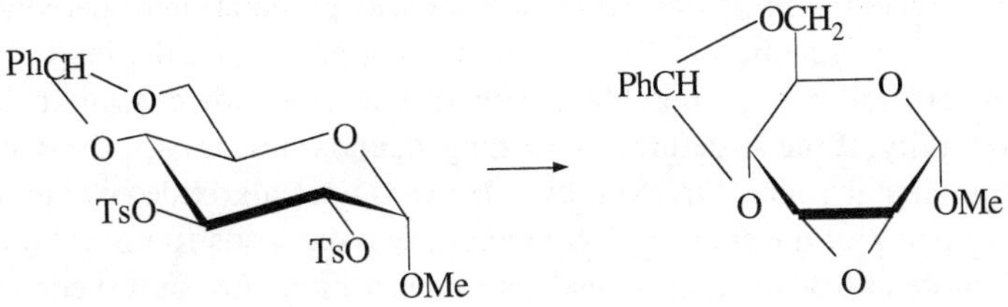

PhCH O O O TsO TsO OMe → OCH$_2$ PhCH O O OMe O

7.18

Closing the oxirane ring causes coplanarity of four atoms of the pyranose ring and half-chair conformation **7.19** (the summits of the hexagon are oxygen or carbon). Sugar epoxides show the same high reactivity as others, especially with nucleophiles. It is known that the openings of cyclohexane epoxides are '*trans*-diaxial' according to the Fürst–Plattner rule, formulated on rigid epoxides derived from steroids. If the 3, 4, 5, and 6 summits of the hexagon of structure **7.19** are maintained in a nearly rigid fashion, only one chair conformation is possible by opening of the oxirane ring, and the definition of the axial positions in the final product does not give rise to any ambiguity. In the case where a mobile chair is possible, one can argue that the conformation of this chair and, therefore, the definition of the axial positions, are not independent of the new substituents of the ring. In the case of pyranoses, let us say that the opening occurs in such a way as to give the diaxial product in the chair constructed on supposedly rigidly maintained atoms 3, 4, 5, and 6. If this conformation is unstable, the molecule adopts the alternative chair conformation and the two substituents introduced are found in the *trans*-diequatorial position. Examples of this situation are known, but very often in the end, substituents are found in the *trans*-diaxial position.

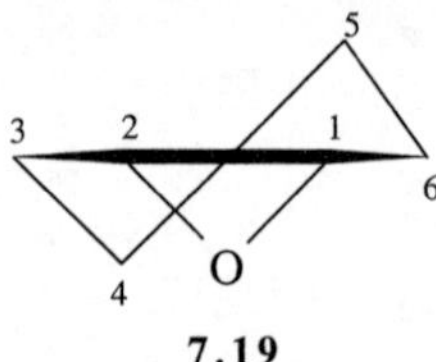

7.19

A good number of nucleophiles open sugar epoxides, thus these derivatives are synthetically very useful. The following reagents have been utilized: oxygen nucleophiles such as alkoxide, benzoate, phosphate, water in acidic medium, sulfur nucleophiles, such as thiolate, thiobenzoate, and thiocyanate, and nitrogen nucleophiles such as ammonia, amines, and azides in the presence of ammonium chloride. Thus we have typical diaxial openings such as in reaction (7.5) starting from the α-D-*manno* epoxide (X = OH, OCH_3, NH_2, N_3, SCH_3, SCN) and reaction (7.6) starting from the α-D-*allo* epoxide (X = OH, OCH_3, NH_2). Lithium aluminium hydride behaves as a nucleophile H^- (X = H) in the two preceding reactions, which allows easy preparation of deoxygenated sugars such as in reaction (7.7). Carbon nucleophiles open the bis-secondary epoxides, affording branched-chain sugars which are discussed in Section 7.6.3. Naturally, if the structure is so arranged, epoxides can also be opened by an intramolecular route. Opening by a *trans*-vicinal alkoxide is concomitant with the closing of another ring. A vicinal alkoxide leads to migration of the epoxide, more or less reversibly, such as isomerization (7.8) observed in a basic medium.

(7.5) **7.17** ⟶

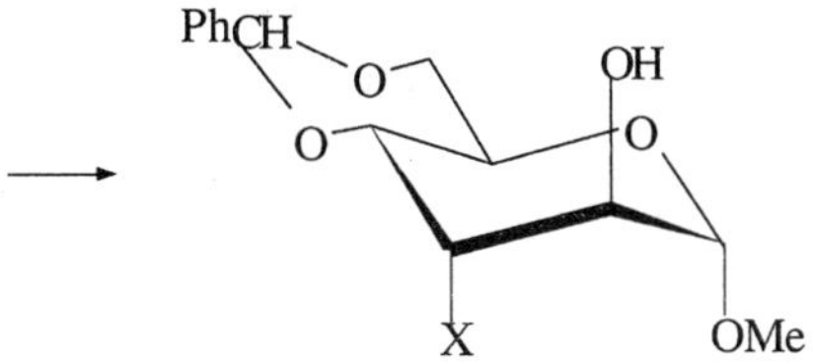

(7.6) **7.18** ⟶

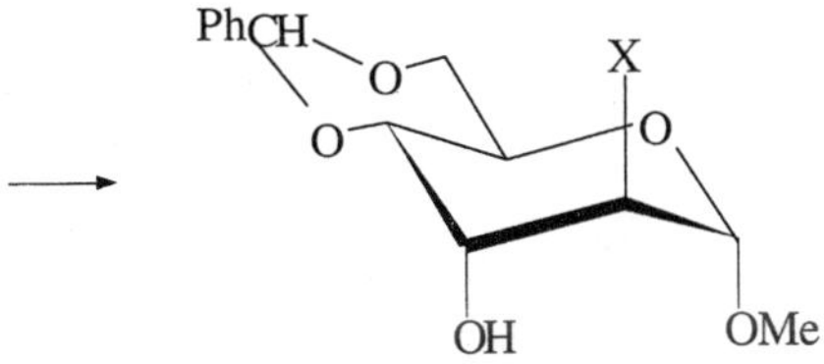

(7.7)

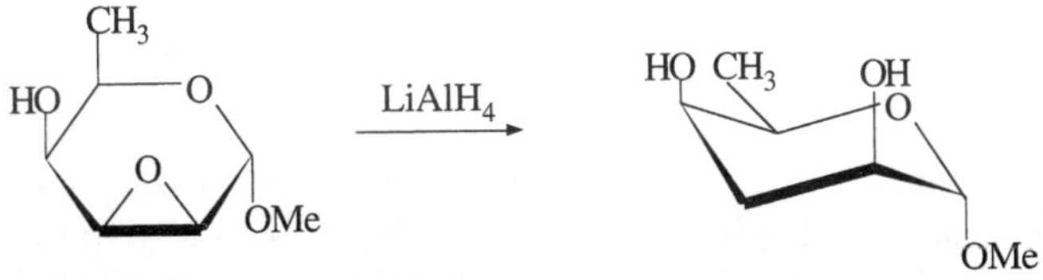

(7.8)

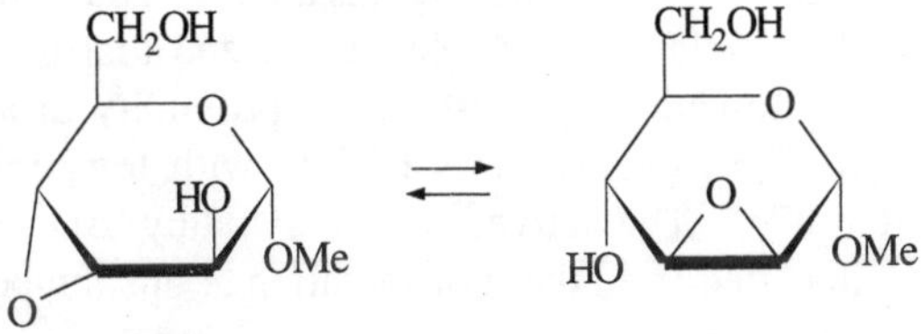

7.3 Cyclic acyloxonium ions (Paulsen 1971)

There are several ways to prepare cyclic acyloxonium ions. We will only discuss the reaction of antimony pentachloride with vicinal diacetates and vicinal chloro- or bromo-acetates because this technique has allowed a most interesting reaction in carbohydrate chemistry (reaction 7.9). The acyloxonium cation can react in two ways; the so-called *kinetic* reaction takes place at the most charged site as, for example with H_2O and OH^-, to give the unstable orthoester **7.20** which opens to the *cis*-hydroxyl ester **7.21**. The so-called *thermodynamic* reaction, with acetate in an acidic medium, choride, and bromide leads to the formation of the *trans* product **7.22** (X = AcO, Cl, Br).

(7.9) $CH_3COOCH_2CH_2X$ + $SbCl_5$ ⟶ 2-methyl-1,3-dioxolan-2-ylium $SbCl_5X^-$

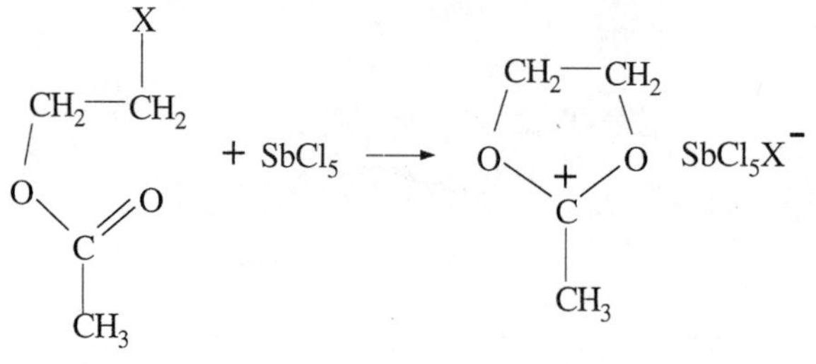

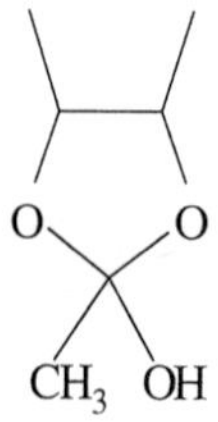

7.20 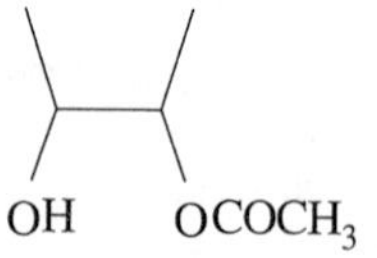 7.21 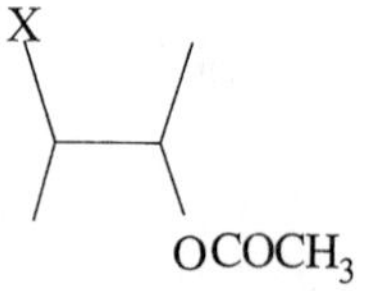 7.22

Glycerol tri-pivalate **7.23** (reaction 7.10) can react by cyclization with the participation of one or other of the terminal pivaloyl groups to give two identical acyloxoniums, **7.24** and **7.25**. These two cations are at equilibrium following the *trans* attack of the carbonyl ester on the acyloxonium ring. Interconversion is rather slow at room temperature. Two methyl signals are revealed in the proton NMR spectrum, that of the pivalate group, Me_3CCOO and that of the pivaloyloxonium group, shifted downfield because of the proximity of a positive charge. The rate of interconversion increases rapidly with temperature and coalescence is observed at 87°C. The activation free enthalpy, $\Delta G^{\ddagger} = 18.55 \pm 0.06$ kcal mol^{-1}, can be calculated from these observation at this temperature in nitromethane solution.

(7.10) $CH_2(OCOR)$-$CH(OCOR)$-$CH_2(OCOR)$ ⟶ 7.24 ⇌ 7.25

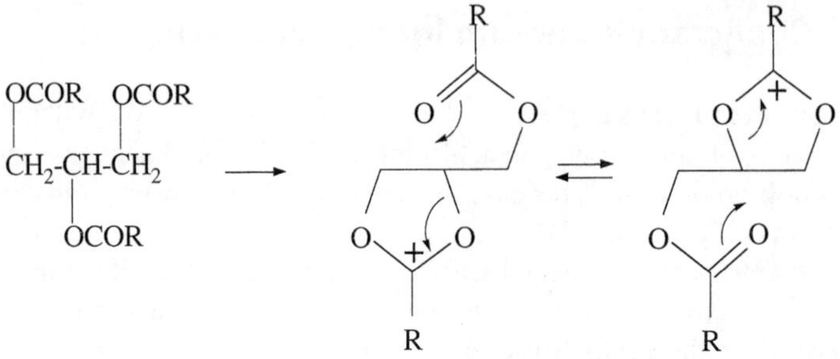

7.23 7.24 7.25

A clean reaction with pyranoses can only be observed if the first ionization is oriented at C-1. From glucose, β-pentaacetate **7.26** and tetra-*O*-acetyl-β-D-glucopyranosyl chloride **7.27** are favourably disposed for attack at C-2 by the acetate carbonyl group. The α-chloride anomer of **7.27** also reacts because of the ease with which it is ionized to the oxocarbenium ion. Any one of these three precursors in solution in dichloromethane, treated at 20°C, gives an equilibrium mixture of acetoxonium cations derived from four configurations, D-*gluco*, D-*manno*, D-*altro*, and D-*ido* (equilibrium 7.11). The mixture composition is determined by hydrolysis. We thus have the *cis* mode of decomposition which gives a mixture of partially acetylated sugars. The mixture is peracetylated and its composition determined by gas chromatography. At room temperature, the equilibrium between the acetoxonium cations derived from the four configurations correspond to the following proportions: D-*gluco*, 54%; D-*manno*, 13%; D-*altro*, 6–9%; D-*ido*, 21%. But since the hexachloroantimonate of the D-*ido* derivative is nearly insoluble in dichloromethane, it is separated by crystallization which displaces the equilibrium so that it can be isolated in 73% yield, thus giving easy access to this rare configuration.

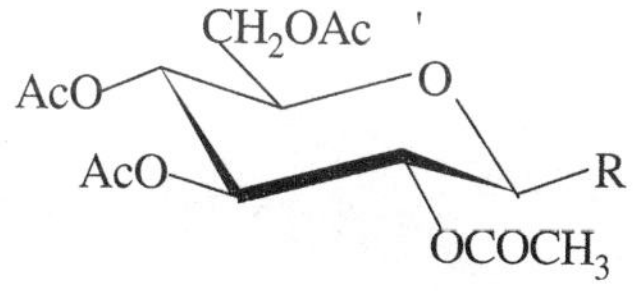

7.26 R=OAc

7.27 R=Cl

(7.11)

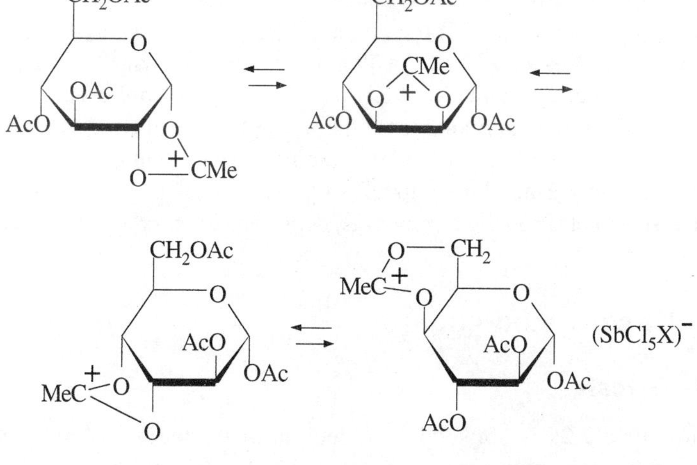

$(SbCl_5X)^-$

7.4 Nucleophilic displacements with participation

Acyloxonium ions (Section 7.3) are stabilized by heavy anions. The formation of acyloxonium ions is also assumed, this time as non-isolated intermediates, in the solvolysis of some tosylates. In reaction (7.12) with sodium acetate, the introduction of an acetyl group at position 6 and migration of benzoate to position 5 testify to the presence of a benzoxonium intermediate. Likewise, solvolysis of sulfonates presenting a *trans*-vicinal acetamido group involves an oxazolinium cation **7.28**, the analogue of an acyloxonium.

(7.12)

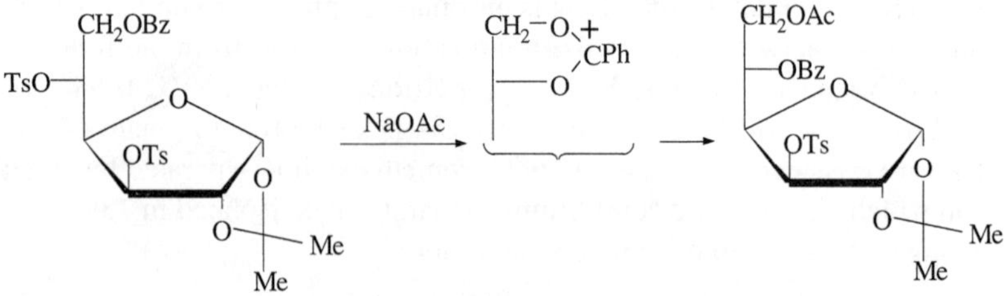

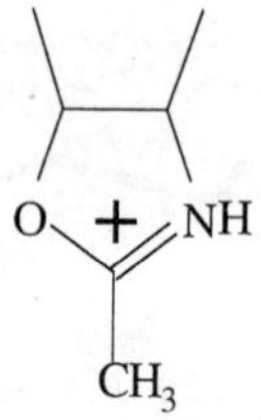

7.28

The opening referred to as '*cis*' on the side of the oxygen atom gives the same result as an S_N2 substitution of a sulfonate. But substitution is observed under conditions in which an S_N2 reaction without participation would be very slow. Thus reaction (7.13) gives a 66% yield in 40 h in refluxing methylcellosolve (124°C), a relatively energetic treatment which does not involve polar aprotic solvents (Jeanloz 1957). From the β-anomer of the same substrate in DMF, the nucleophile $PhCH_2SK$ gives a mixture of benzyl thioethers with retention and inversion of configuration (corresponding to the *trans* and *cis* openings of the oxazolidinium cation, respectively) and the oxazolidine, deprotonation product of this cation.

7.5 Unsaturated sugars

7.5.1 Glycals

The prototype **7.29** is obtained by reduction of the tetra-*O*-acetyl-β-D-glucopyranosyl bromide with zinc and acetic acid. We can suppose that the leaving

(7.13)

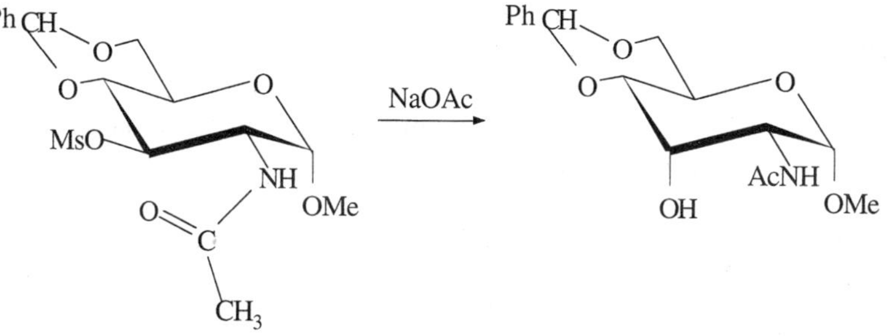

NaOAc

bromide leads to an oxocarbenium cation, reduced to the anion stage by zinc, and that the reaction ends in the C-1–C-2 elimination of acetate. The deprotected glycal **7.30** is obtained from triacetate **7.29** by alkaline methanolysis. This is general pathway that can also be used in the preparation of the D-galactal isomer. Glycals, easily accessible and capable of various reactions, have great synthetic potential (Ferrier 1980).

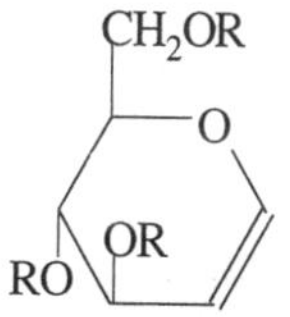

7.29 R=Ac

7.30 R=H

Glycals are enol ethers that give rise to electrophilic additions. With a non-symmetric reagent, the electrophilic part is added to C-2 in such a way as to create an oxocarbenium ion, which is relatively stable, at C-1. Thus water, alcohols, and carboxylic acids, with acid catalysis, give the 2-deoxy sugar (**7.31**, Z = OH), the 2-deoxy glycoside (**7.31**, Z = OR), and the ester (**7.31**, Z = OCOR), respectively. These conditions, which also catalyse anomerization, lead to the most stable anomer. 2-Deoxyglycosyl chlorides and bromides (**7.31**, Z = Cl, Br) are obtained from HCl and HBr. Chlorine and bromine each give the two possible halogenated derivatives, from, for example, tri-*O*-acetyl-D-glucal, α-D-gluco (**7.32**, 60%), and α-D-manno (**7.33**, 30%). Here again, the initial step, due to the bromine molecule or the electrophilic Br^+ cation, gives rise to an oxocarbenium cation centred at C-1 which, in the presence of silver salts, traps a molecule from an alcohol solvent to give an α-glycoside, a 2-bromo-2-deoxy derivative such as **7.34**. On the other hand, bromides **7.32** and **7.33** react with alcohols in the presence of silver carbonate to give the expected β-glycoside. The halogen atom is easily eliminated in all these compounds by free radical reduction, thus leading to 2-deoxy sugar derivatives. Methoxymercuration of **7.29** and **7.30** with mercury acetate in methanol leads to a carbon–mercury bond. Triacetylglucal **7.29** gives, after replacement of the ionic acetate by chlorine, β-glycoside **7.35** mixed with its α-D-*manno* isomer, the two products being 1,2-*trans*. The deprotected D-glucal gives methyl 2-acetoxymercuri-2-deoxy-α-D-mannopyranoside in high yield. Reduction with borohydride replaces the mercury by hydrogen. Finally, the peracids, which behave approximately as OH^+ cation sources, reconstitute pyranoses from glycals. This can be a practical procedure for configurational inversion of a sugar at C-2. Thus, we go from galactose to D-galactal **7.36**, then, by treatment with perbenzoic acid in aqueous solution, to a mixture of galactose and talose (Tipson and Isbell 1961). This rare sugar, **7.37**, is easily separated from galactose by crystallization. It is important to remember that direct inversion at C-2 is impossible in the galactose series. To conclude, let us recall the epoxidation of glycals with dimethyldioxirane given in Section 3.4. We will see another type of addition, azidonitration, discussed in Section 10.3.7.

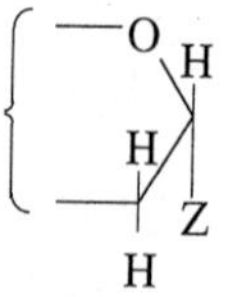

7.31

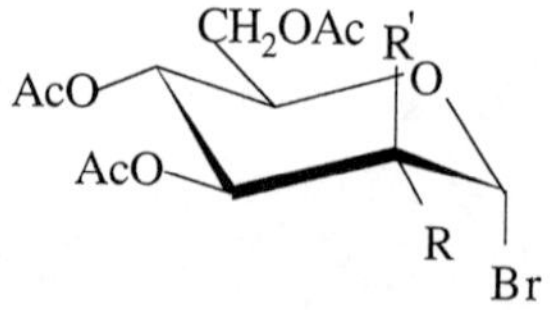

7.32 R=Br, R'=H
7.33 R=H, R'=Br

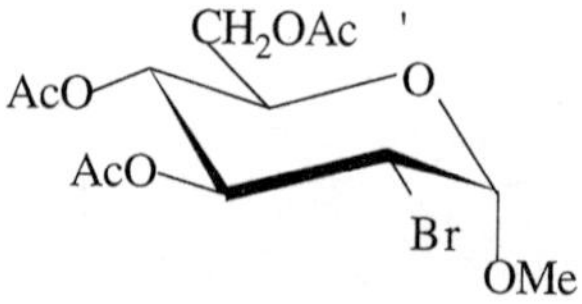

7.34

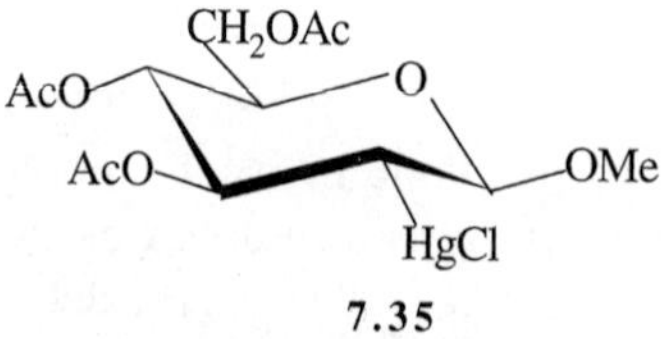

7.35

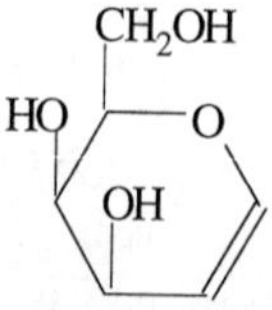

7.36

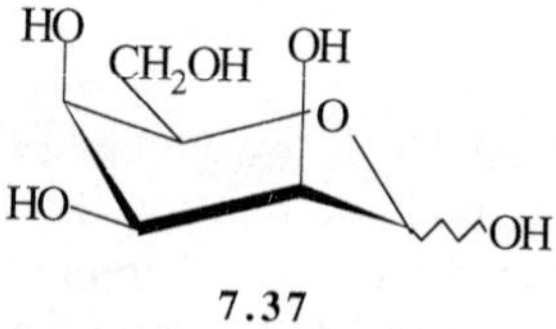

7.37

Compounds with hydroxyl groups—water, alcohols, phenols—are not added to the double bond of glycals and their esters in the absence of acidic catalysts. On the other hand, at a relatively high temperature, displacement of the acetyl group at position 3 of triacetylglycals is produced, with migration of the double bond to positions 2 and 3. Hence, methanol leads to methyl 4,6-di-*O*-acetyl-2, 3-dideoxy-α-D-*erythro*-hex-2-enopyranoside **7.38**.

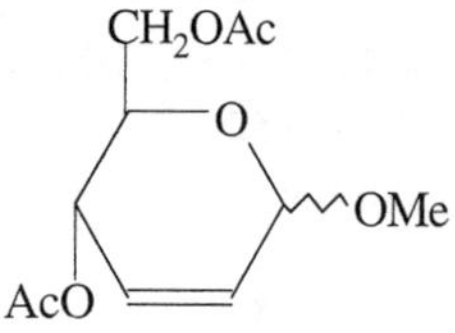

7.38

7.5.2 Ferrier reaction

We have seen above pyranoses with a double bond at C-1–C-2 or C-2–C-3. Examples of sugars with double bonds at all other positions are known. Here we will limit our description to one important reaction of pyranoses unsaturated at positions 5 and 6. They are prepared by elimination starting from 6-deoxy-6-iodo pyranoses. In one article Adams (1988) proposed treatment in DMF for 4 h at 80°C in the presence of a non-nucleophilic base, DBU **7.39**, but prior to this, other reagents were used. Refluxing tetrabenzoate **7.40** for a few hours in aqueous acetone in the presence of mercury chloride gives the substituted cyclohexanone **7.41** (Blattner *et al.* 1980). The starting product **7.40** is, like glucal, a vinyl ether. Working in very mild conditions, evidence is found for the formation of the opened intermediate **7.42** which then cyclizes as a δ-keto aldehyde (Blattner *et al.* 1985). The conversion to a cyclohexanone by replacing $HgCl_2$ by a catalytic amount (relatively high, 10–20 mol%) of palladium acetate or chloride was also carried out (Adams 1988).

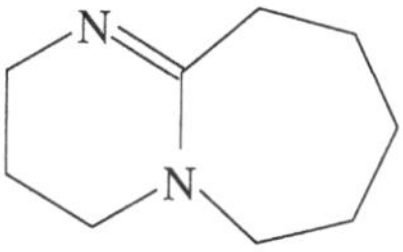

N N

7.39

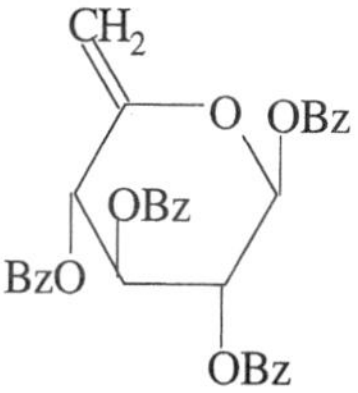

CH_2 O OBz OBz BzO OBz

7.40

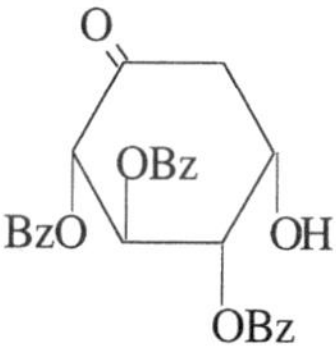

O OBz BzO OH OBz

7.41

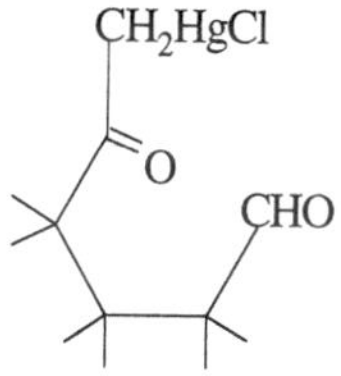

CH_2HgCl O CHO

7.42

7.6 Branched-chain sugars

7.6.1 General features

Two branched-chain sugars, apiose **7.43** and hamamelose **7.44**, are widespread in the plant kingdom. At least 40 others have been isolated as constituents of antibiotics, stimulating a great deal of effort in synthetic work. Controlling the branching conditions seems essential in order to use sugars as a chiral starting material. It is practical to classify branched-chain sugars into two groups. In the most abundant group, as this is the easiest to prepare, there is a heteroatom, most often the oxygen of a hydroxyl group, at the branch point, according to the structure > C(OH)–R. In the other group, there is a hydrogen following the structure > CH–R (Yoshimura 1984).

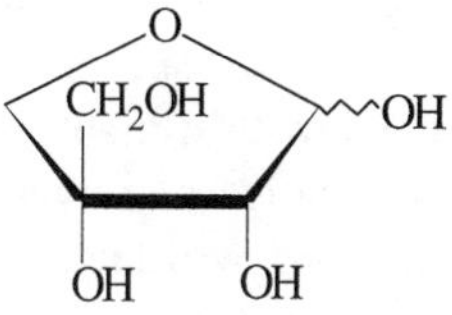

O CH_2OH OH OH OH

7.43

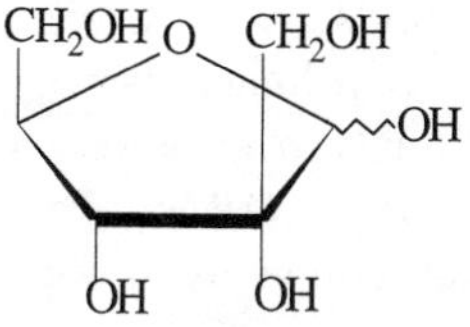

CH_2OH O CH_2OH OH OH OH

7.44

7.6.2 The >C(OH)–R family

Organometallics add in the usual fashion to carbonyl groups of furanose or pyranose rings. To avoid wasting the reagent or causing unwanted precipitations during the reaction, it is desirable to employ ether or acetal protection. The reaction is often stereospecific. Diacetal **7.45** gives with CH_3MgBr or CH_3Li only the α-D- *allo* product **7.46** resulting from *exo* attack. Reaction (7.14) shows a few steps of the synthesis of hamamelose, which consists in the *exo* addition of lithiated dithiane on the adapted ketone (78%). Liberation of the aldehyde group by mercury oxide and boron trifluoride (75%) is followed by reduction by lithium aluminium hydride (70%). Similar additions are observed with other carbon nucleophiles such as cyanide, nitromethane anion, etc.

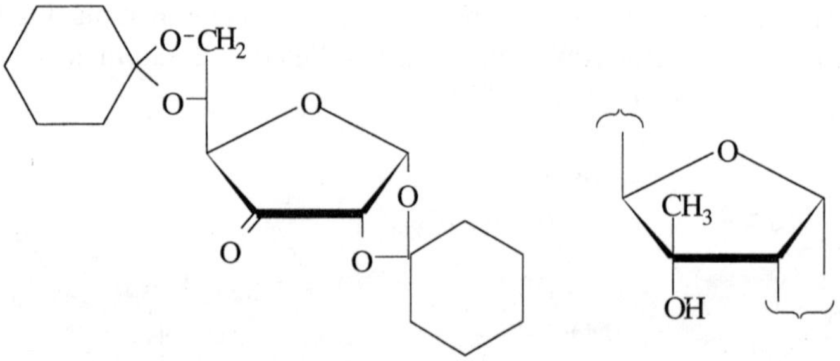

7.45 **7.46**

(7.14)

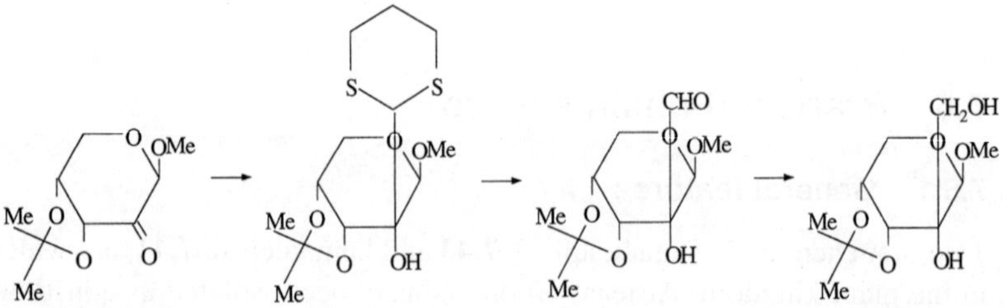

In another procedure for branching the Wittig reaction is used. Condensation of keto sugars with methylenetriphenylphosphorane gives sugars with an exocyclic double bond, such as furanose **7.47**. Conversion to vicinal diol **7.48** by OsO_4, followed by selective tosylation of the primary alcohol function, and reduction of sulfonate **7.49** by $LiAlH_4$ gives the final product **7.50**. This pathway is very long but ends with the *endo* isomer. Epoxidation of the exocyclic double bond by peracids gives a 'spiroepoxide' which can be hydrolysed to a glycol or reduced at its primary carbon end (reaction 7.15). A spiroepoxide is also obtained directly from a keto sugar by treatment with dimethylsulfoxonium methylide, $CH_2{=}S(O)Me_2$. Thus we have a vast choice of routes to this first family which allows us to hope that one or the other will lead to the desired isomer.

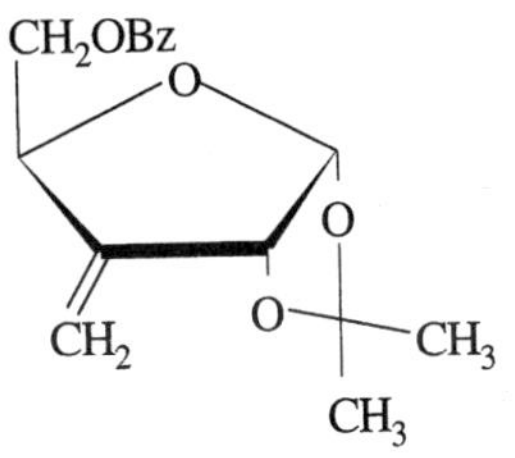

7.47

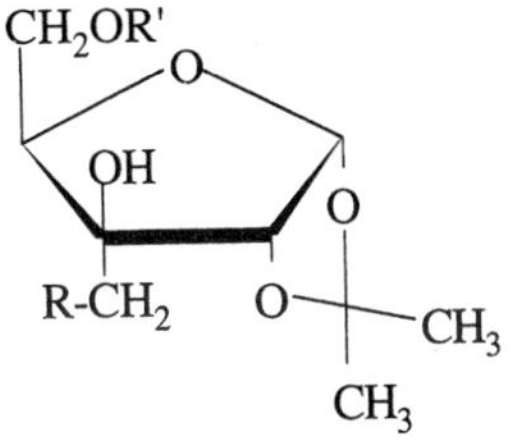

7.48 R=OH, R'=Bz
7.49 R=OTs, R'=Bz
7.50 R=H, R'=H

(7.15)

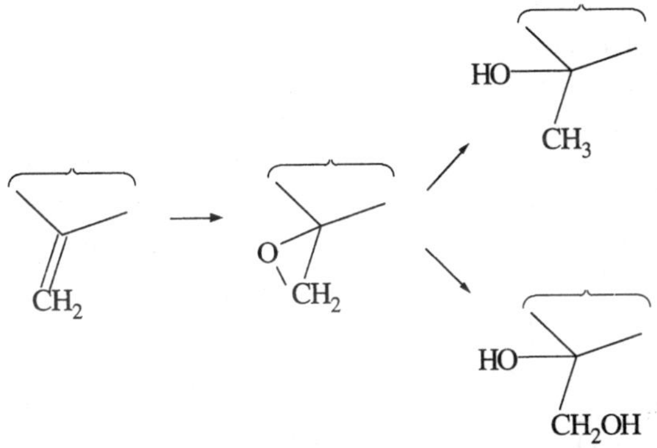

7.6.3 The > CH–R family

The synthesis of these products is becoming more and more important with the increasing use of sugar sequences in total synthesis. It is rare that the sequences to be prepared are not branched. A first solution consists in catalytic hydrogenation of exocyclic double bonds. Thus, ketone **7.51** (R = Me_3CPh_2Si), easily accessible from 2-deoxyglucose, condensed with phosphonate $(MeO)_2P(O)CH_2CO_2Me$ (Wittig–Horner reaction) gives the exocyclic double bond product **7.52**, which is then hydrogenated over palladium on charcoal. Branched-chain sugar **7.53** is obtained in at least 95% yield. Stereospecificity comes from the constraints imposed by the polyfunctional structure of the pyranose and the size of the R group. Product **7.53** is an intermediate in the synthesis of thromboxane B_2 (Hanessian and Lavallée 1981).

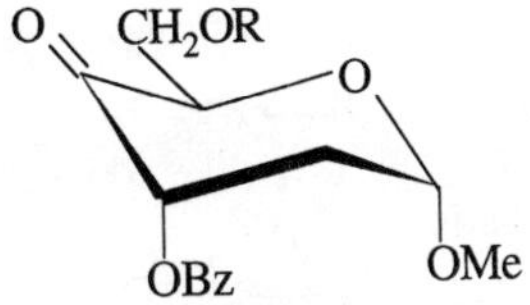

7.51

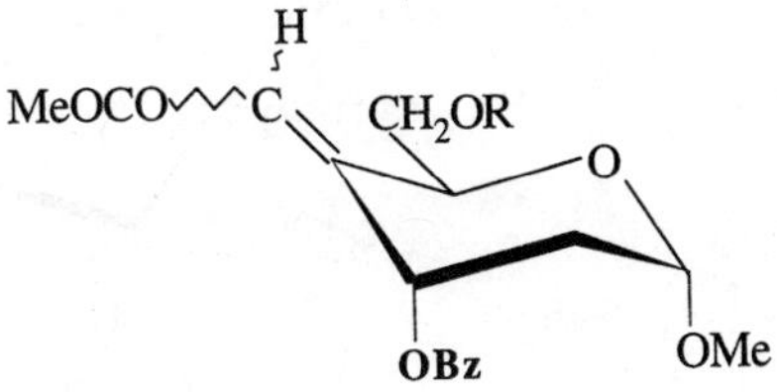

7.52

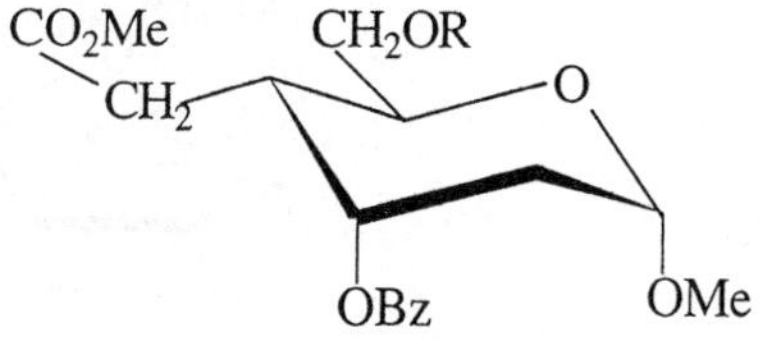

7.53

Another route is the deoxygenation of the tertiary hydroxyl group at the branch point. Sometimes the configuration facilitates a selective orientation of endocyclic dehydration to give a double bond, which can then be reduced (reaction 7.16). Radical deoxygenation of a benzoate or, better, a 4-cyanobenzoate, vicinal to a carbonyl group with tributylstannane (reaction 7.17) has also been described. In this case, there is inversion of configuration but this is not always so (Redlich *et al.* 1977; 1982). The result is probably due to the configuration of the transient radical at C-2 of the sugar. Other reactions introduce the > CH–R chain directly, such as the opening of oxiranes by carbon nucleophiles. The cuprate Me_2CuLi gives, with oxirane **7.18**, the diaxial opening product **7.54**. Similar reactions are observed with dialkylmagnesium (the presence of a halide anion should sometimes be avoided), Et_2AlCN, cyanide, and sodium ethyl malonate.

(7.16)

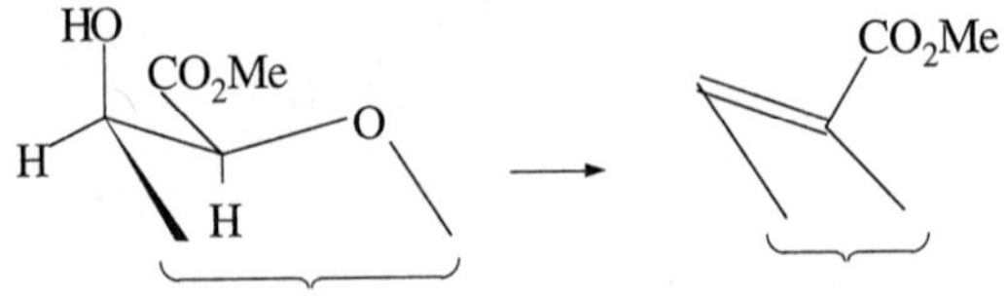

(7.17)

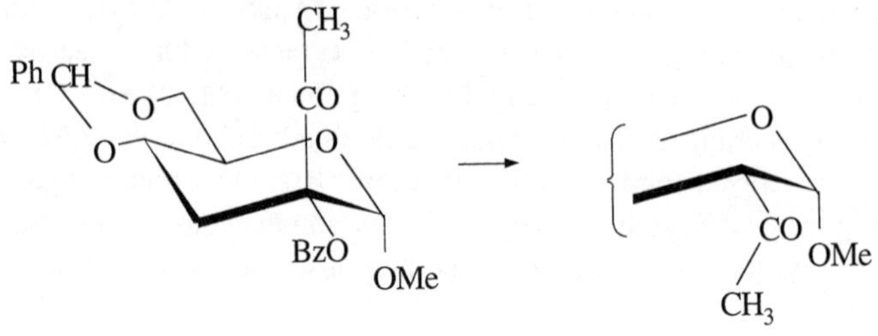

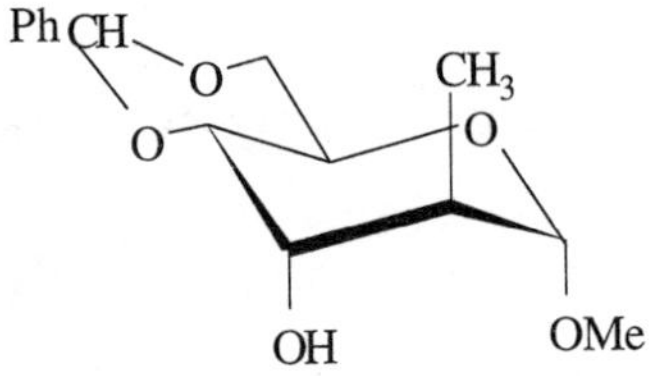

7.54

Until the present, we have only seen ionic reactions in the preparation of the two types of branched-chain sugars. A > CH–R type chain can also be introduced by a radical route. For example, the 2-bromo-2-deoxy derivative **7.55** is treated with allyltriphenylstannane, CH_2=CH–CH_2–$SnPh_3$, in benzene at 80°C and the chain radical reaction is maintained by slow addition (7 h) of azoisobutyronitrile, a classical initiator for chain reactions. Thus the branched-chain sugar **7.56** is obtained, mixed with 10% of its epimer at position 2, but isolable by crystallization in 39% yield (Korth *et al.* 1990). The mechanism of the chain reaction is given in Fig. 7.1 where the bromo sugar **7.55** is represented by RBr. The initiator transforms it into the $R^•$ radical which is added to allylstannane. The new radical is fragmented to give the branched-chain sugar R–CH_2–CH=CH_2 and the radical $Ph_3Sn^•$, which detaches Br from RBr and the reaction continues (Giese 1986). Other chains have been carried out by this method at positions other than C-2 and with other acceptors, such as methyl acrylate.

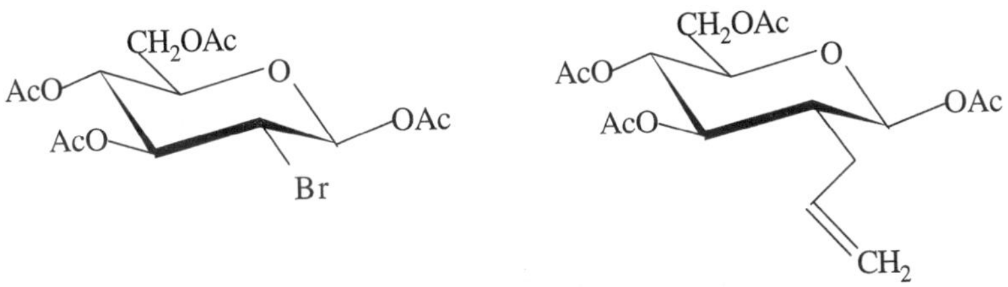

CH2OAc O AcO AcO OAc Br

7.55

CH2OAc O AcO AcO OAc CH2

7.56

SnPh3 R SnPh3 Ph3SnBr R• Ph3Sn• R RBr

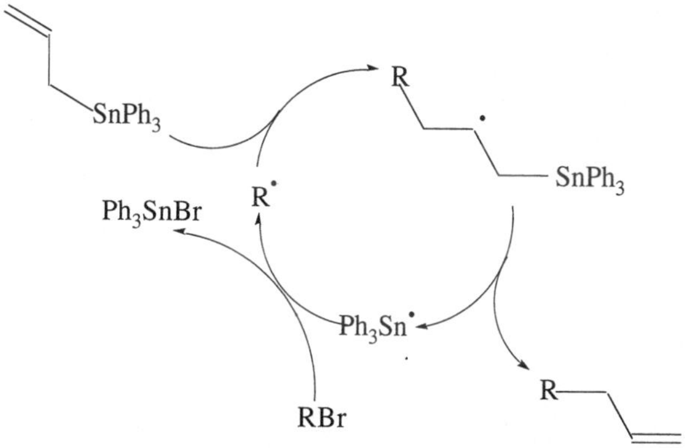

Fig. 7.1 Mechanism of the radical substitution of tetra-*O*-acetyl-2-bromo-2-deoxy-*β*-D-glucopyranose.

Finally, branched-chain sugars have been prepared by the sigmatropic reaction known as the Claisen rearrangement. Heating of vinyl ether **7.57** at 180°C gives dihydropyran **7.58**. Branching is on the same side of the ring as the vinyl ether group of the starting product and rearrangement is strictly stereospecific.

This type of reaction is useful in chiral synthesis for the final 'sugar' is deoxygenated at three positions, but the presence of a double bond in **7.58** allows the possibility of introducing *cis* or *trans* hydroxyl groups at C-3 and C-4 of the carbon atoms.

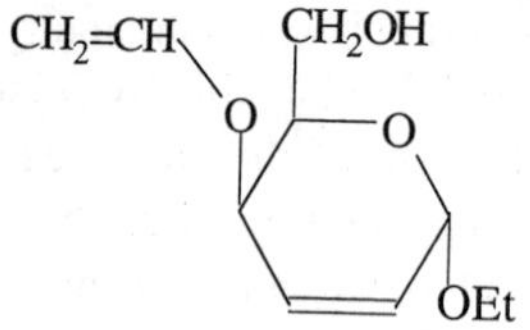

CH_2=CH CH_2OH O O OEt

7.57

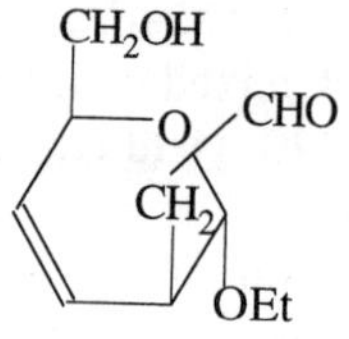

CH_2OH O CHO CH_2 OEt

7.58

7.6.4 Aldol condensation

Although aldol condensation is another type of ionic reaction, it deserves special consideration. It can lead to two classes of sugars. Condensation of an aldehyde in an α-position of a free carbonyl group under mildly akaline conditions still seems possible. This very simple reaction can also be quite efficient. Thus diacetal **7.59**, quantitatively prepared by treatment of mannose with acetone in an acidic medium, is at equilibrium with the carbonyl tautomer. Condensation with formaldehyde, used in aqueous solution in the presence of potassium carbonate, gives the hydroxymethylated sugar **7.60** in 86% yield (Ho 1979).

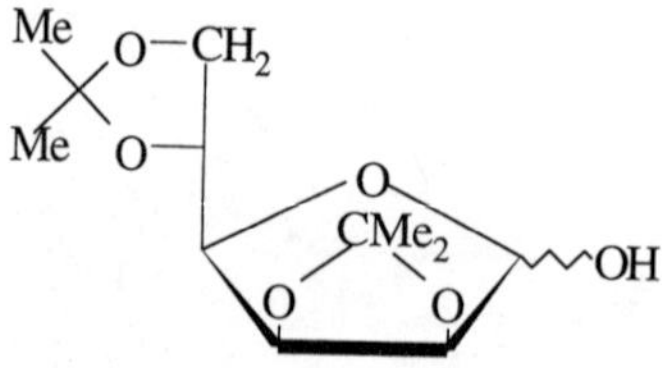

Me O–CH_2 Me O O CMe_2 O O OH

7.59

Me O–CH_2 Me O O CMe_2 O O OH CH_2OH

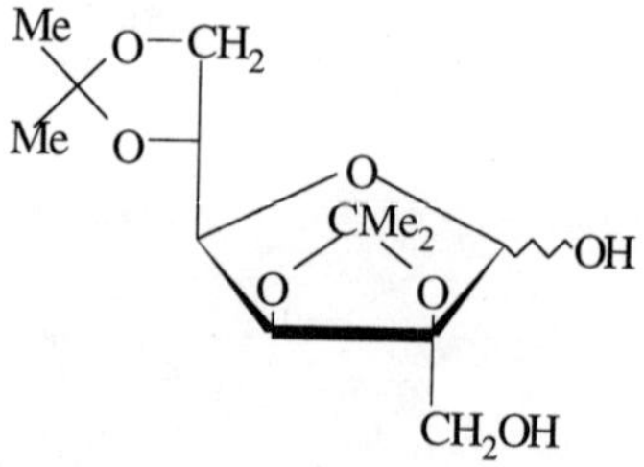

7.60

References

Adam, S. (1988), *Tetrahedron Lett.*, **29**, 6589–6592.

Alais, J. and David, S. (1990), *Carbohydr. Res.*, **201**, 69–77.

Augé, C., David, S., Gautheron, C., Malleron, A., and Cavayé, B. (1988), *New J. Chem.*, **12**, 733–744.

Baggett, N. (1988), The synthesis of monsaccharides. In *Carbohydrate Chemistry* (ed. J. F. Kennedy), Oxford University Press, Oxford.

Blattner, R., Ferrier, R. J., and Prasit, P. (1980), *J. Chem. Soc., Chem. Commun.*, 944–945.

Blattner, R., Ferrier, R. J., and Haines, S. R. (1985), *J. Chem. Soc., Perkin Trans I*, 2413–2416.

Ferrier, R. J. (1980), Unsaturated sugars. In *The carbohydrates, chemistry and biochemistry*, Vol. 1B. (2nd edn.) (ed. W. Pigman and D. Horton), p. 843, Academic Press, New York.

Giese, B. (1986), *Radicals in organic synthesis: formation of carbon–carbon bonds*, Pergamon Press, Oxford.

Gross, P. H., du Bois, F., and Jeanloz, R. W. (1967), *Carbohydr. Res.*, **4**, 244–248.

Hanessian, S. and Vatèle, J. M. (1981), *Tetrahedron Lett.*, **22**, 3579–3582.

Hanessian, S. and Lavallée, P. (1981), *Can. J. Chem.*, **59**, 870–877.

Ho, P. T. (1979), *Can. J. Chem.*, **57**, 381–383.

Jeanloz, R. W. (1957), *J. Am. Chem. Soc.*, **79**, 2591–2592.

Korth, H. G., Sustmann, R., Giese, B., Rückert, B., and Gröninger, K. S. (1990), *Chem. Ber.*, **123**, 1891–1898.

Lubineau, A. and Bienaymé, H. (1991), *Carbohydr. Res.*, **212**, 267–271.

Paulsen, H. (1971), *Adv. Carbohydr. Chem. Biochem.*, **26**, 127–195.

Redlich, H., Neumann, H. J., and Paulsen, H. (1977), *Chem. Ber.*, **110**, 2911–2921.

Redlich, H., Neumann, H. J., and Paulsen, H. (1982), *J. Chem. Res.*, (**M**), 352–372.

Tipson, R. S. and Isbell, H. S. (1961), *Methods Carbohydr. Chem.*, **I**, 157–159.

Williams, N. R. (1970), *Adv. Carbohydr. Chem. Biochem.*, **25**, 109–179.

Yoshimura, J. (1984), *Adv. Carbohydr. Chem. Biochem.*, **42**, 69–134.

8 Sugars in chiral synthesis

8.1 Asymmetric induction (Kunz and Rück 1993)

8.1.1 Enantioselective allylation and aldolization

The experiments we are about to describe bring into play chiral reagents constructed from a titanocene (titanium–carbohydrate complex) (Riediker and Duthaler 1989*a*; Duthaler *et al.* 1989; Bold *et al.* 1989; Riediker *et al.* 1989*b*). The starting compound is cyclopentadienyltitanium(IV) trichloride, $C_5H_5TiCl_3$, which is prepared from dichloride $(C_5H_5)_2TiCl_2$ by heating with $TiCl_4$. This trichloride reacts with 1,2:5,6-di-*O*-isopropylidene-α-D-glucofuranose ('diacetone glucose') in the presence of triethylamine to provide the bis-titanate **8.1**. The latter is a crystalline compound, hence its structure could be determined in the solid state. Its form is that of a three-legged piano stool with the cyclopentadiene ring being its seat. The two ligands have completely different orientations with respect to the cyclopentadienyl ring. Because of this, they create a chiral cavity around the titanium atom. A study of the NMR spectra of **8.1** in solution indicates neither inversion at the metal centre nor rapid ligand exchange. This suggests that the conformation in solution is close to the conformation in the crystal. Treatment of this complex with allylmagnesium chloride affords the organometallic allyltitanium **8.2**. It appears that the chiral cavity of the precursor **8.1**: is still present in **8.2**. Two other derivatives were prepared from monochloride **8.1**: the titanium enolate **8.3** by reacting with the lithium reagent $LiCH_2COOCMe_3$ derived from *t*-butyl acetate, and the titanium enolate **8.4** by reacting with the lithiated ethyl glycinate protected by the amino function **8.5**.

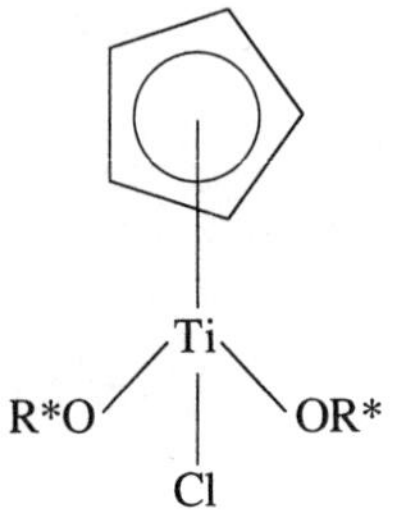

Ti
R*O
OR*
Cl

8.1

R*OH=
CH_3 O–CH_2
CH_3 O
O
OH
O
O
CH_3
CH_3

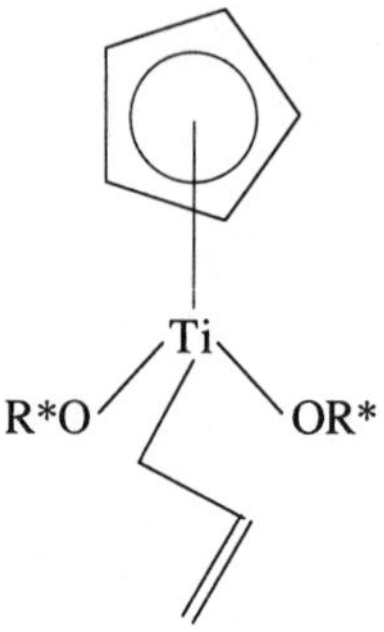

Ti
R*O
OR*

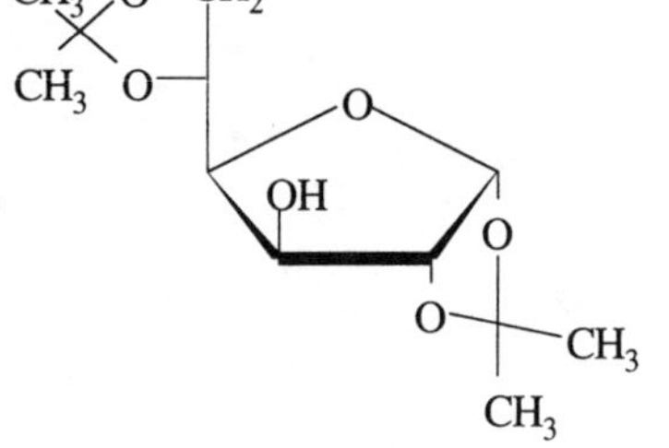

8.2

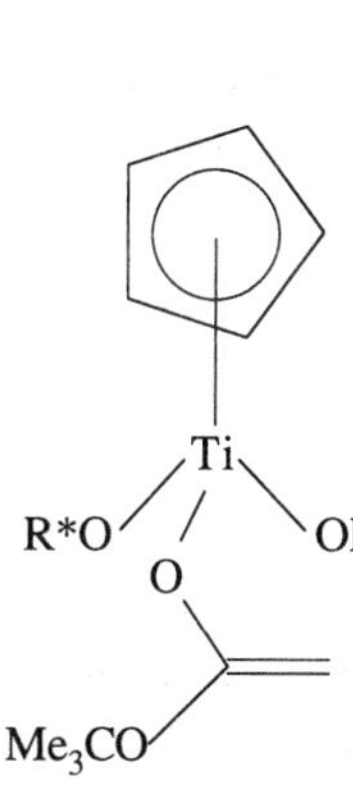

8.3 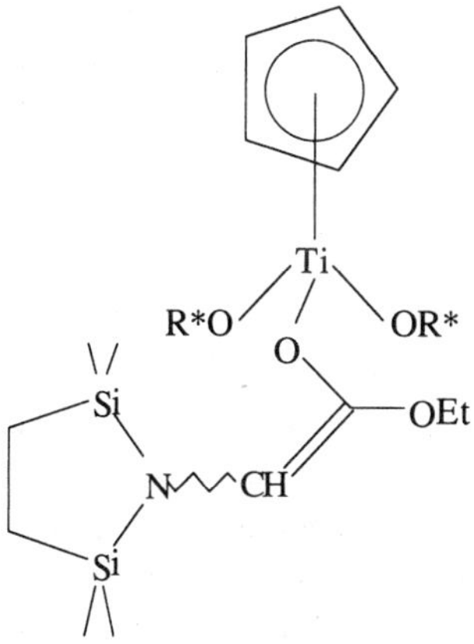 8.4 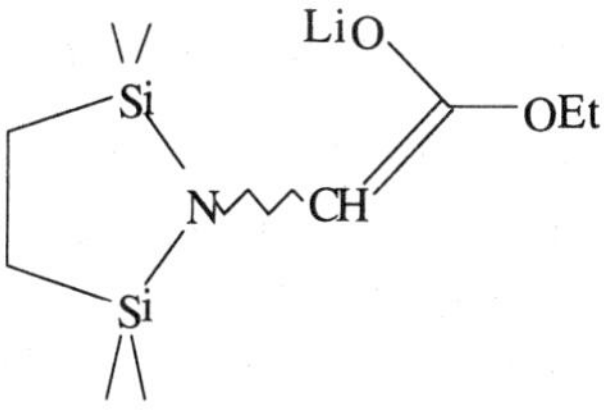 8.5

Allyltitanium **8.2** gives homoallylic alcohols with aldehydes (reaction 8.1). At –76°C, addition takes place preferentially on the *re* face of the carbonyl group. This reaction was carried out with 16 aliphatic and aromatic aldehydes. The average value of the enantiomeric excess (ee) of the addition products was 90.5%. Titanium enolate **8.3** gives an aldol condensation product with aldehydes, a β-hydroxyl ester (reaction 8.2). The average enantiomeric excess, measured from 13 examples, was 94%. Finally, the protected ethyl glycinate enolate **8.4** gives, as condensation product with aldehydes, β-hydroxyl amino acids, **8.6** (reaction 8.3). This reaction leads to D-*threo* products with two chiral centres. The average diastereoisomeric and enantiomeric excesses were 97 and 96%, respectively. Just as with allylation, the additions of enolates (reaction 8.2) and (reaction 8.3) take place preferentially on the *re* face of aldehydes. It is reasonable to attribute these enantioselectivities to the coordination of the aldehyde in the chiral cavity of the complex.

(8.1) **8.2** + R-CHO ⟶ + ····

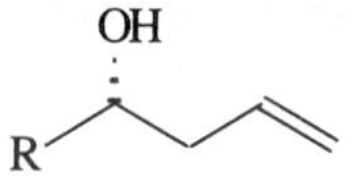

(8.2) **8.3** + R-CHO ⟶ + ····

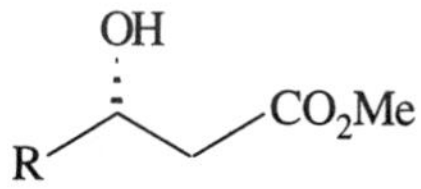

(8.3) **8.4** + R-CHO ⟶ + ····

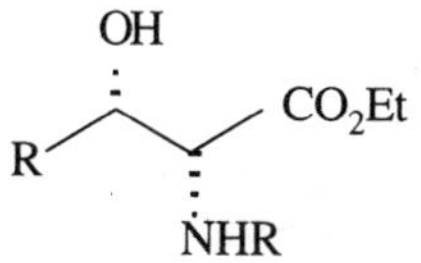

8.6

In every experiment, a polymer of cyclopentadienyl titanate $[(C_5H_5)Ti(OH)O]n$ is recovered as a by-product, which can be reconverted to the trichloride and then recycled. The other by-product, diacetone glucose, can also be recovered but is generally of little value.

8.1.2 Cycloaddition

2,3-Dihydro-6H-pyrans (Zamojski *et al.* 1982)

The cycloaddition of butyl glyoxylate $OHCCO_2C_4H_9$ to dienylic ethers gives 2, 3- dihydro-6*H*-pyrans, substituted at position 6 by an alkoxy group and at position 2 by an ester function, that can be considered as very reduced pyranoid analogues (Zamojski *et al.* 1982). Dienylic ethers can be prepared from sugars. The simplest of these preparations consists of the addition of the oxygen atom of the hydroxyl group from a partially protected sugar, such as **8.7**, to one of the triple bonds of butadiyne sold in a stablized form, $Me_2C(OH){-}C{\equiv}C{-}C{\equiv}C{-}C(OH)Me_2$. This addition gives ether **8.8** (*cis–trans* mixture), which can then be half-hydrogenated to **8.9** (*cis–trans* mixture), with the two constituents being separable. Cycloaddition of butyl glyoxylate creates two new chiral centres and, taking into account the chirality of the sugar, four diastereoisomeric dihydropyrans. Using the nomenclature convention for sugars, it is convenient to designate them as α-D, β-D, α-L, and β-L. The proportions of the mixture of dihydropyrans **8.11**, obtained by cycloaddition with the *trans*-dienylic ether **8.10**, are α-D (44%), β-D (4%), α-L (0%), and β-L (52%). They correspond to 56% of the *endo* addition. It is interesting to note the high degree of facial selectivity, as the addition proportion on the *si* face of the dienylic ether which corresponds to the total (α-D) + (β-L) reaches 96%. There is thus a high degree of protection of the *re* face against the approach of the dienophile by the perbenzylated glucose molecule. The α-D and β-L configurations would have been identical if an adapted dienophile had been used. Indeed, cycloaddition of ethyl mesoxalate, $CO(CO_2Et)_2$, to the *cis*-diene **8.12** prepared from diacetone glucose gives, in an overall yield of 85%, a mixture containing 92% of the *S*-adduct **8.13**. Dihydropyrans **8.11** and **8.13** and a number of analogues are transformable to authentic pyranoid sugars by appropriate functionalizations (David *et al.* 1978; 1979).

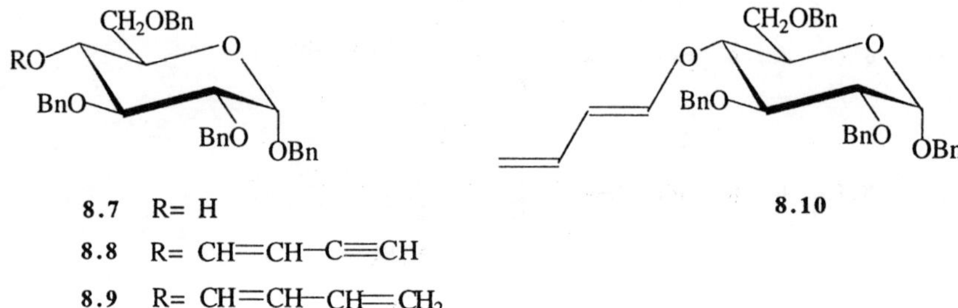

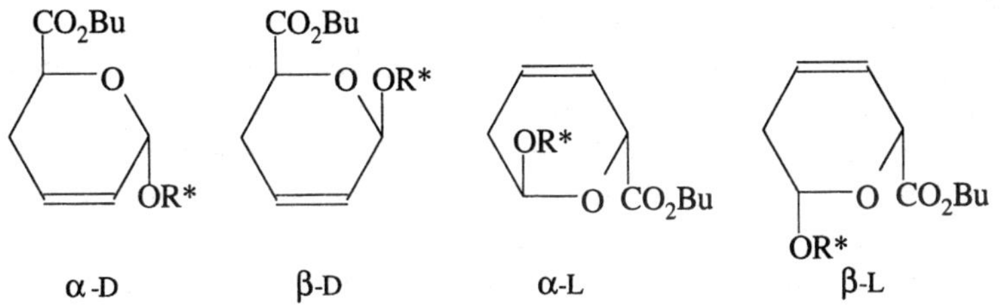

α-D β-D α-L β-L

8.11

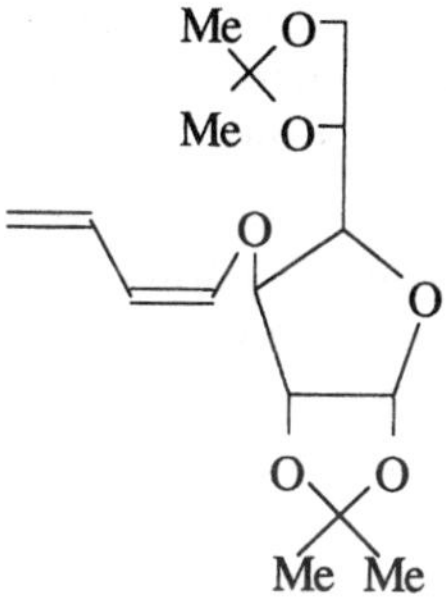

8.12 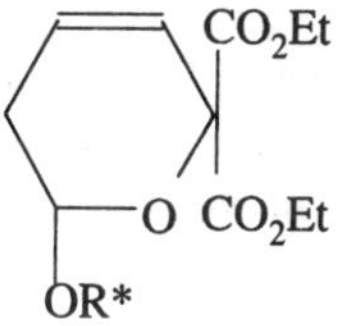 **8.13**

These first cycloadditions are only complete after 72 h at 60°C. When partners are more reactive and the cycloaddition can be catalysed by Lewis acids, it is possible to work at –78°C. Under these conditions, it may be possible to observe high asymmetric induction. Reaction (8.4) of cyclopentadiene with an acrylate derivative of 5-*O*-trimethylsilyl-α-D-xylofuranose carried out at –78°C in dichloromethane in the presence of $TiCl_4$ gives exclusively the *endo* adduct (1′*R*, 2′*R*). This is explained by the formation of an intermediate complex which contains both the dienophile and Lewis acid attached to the chiral sugar inductor (Kunz *et al.* 1987).

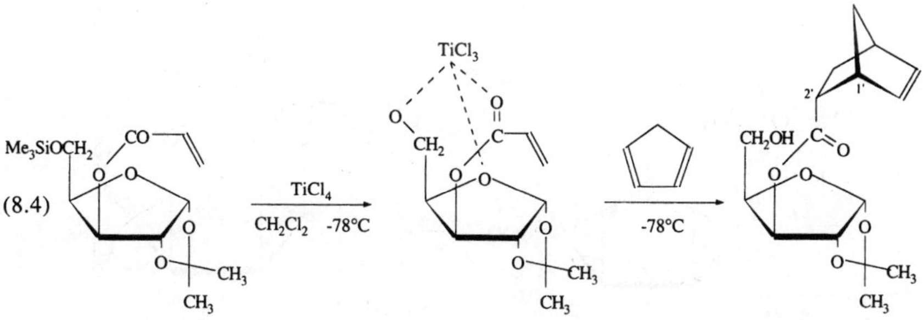

(8.4)

$TiCl_4$, CH_2Cl_2 -78°C; -78°C

3,6-Dihydro-2H-1,2-oxazines (Felber *et al.* 1986)

[4+2]-Cycloaddition of dienes to nitroso chloro derivatives in the α-position provides easy access to 3,6-dihydro-2*H*-1,2-oxazines non-substituted at nitrogen. This reaction was developed using enantiomerically pure nitroso chloro compounds derived from sugars. The preparation of one of them is outlined in reaction (8.5). The oxime of 'diacetone mannose' **8.14** is in tautomeric equilibrium with the cyclic hydroxylamine. Oxidation of the latter by sodium periodate at 80°C in slightly basic conditions (sodium acetate) gives hydroximino lactone **8.15**. Compound **8.15** can be considered to be the oxime of an ester carbonyl. The nitroso chloro derivative **8.16**, blue in colour, is finally obtained by chloration of **8.15** with *t*-butyl hypochlorite. The nitroso chloro derivative **8.16** is both more stable and much more reactive than its simple aliphatic analogues. With highly reactive dienes such as 1,3-cyclohexadiene, the reaction is complete in less than 15 min at –70°C. With *trans–trans*-2,4-hexadiene, the reaction lasts less than 4 h at –20°C. The end of the reaction is indicated by the disappearance of the blue colour of **8.16**. The primary cycloaddition product of cyclohexadiene undergoes an internal quaternization to give a product which is cleaved by HCl (reaction 8.6). In this experiment leading to **8.17**, as with other symmetrical dienes, dihydrooxazine is obtained with an enantiomeric excess equal to at least 96%.

(8.5)

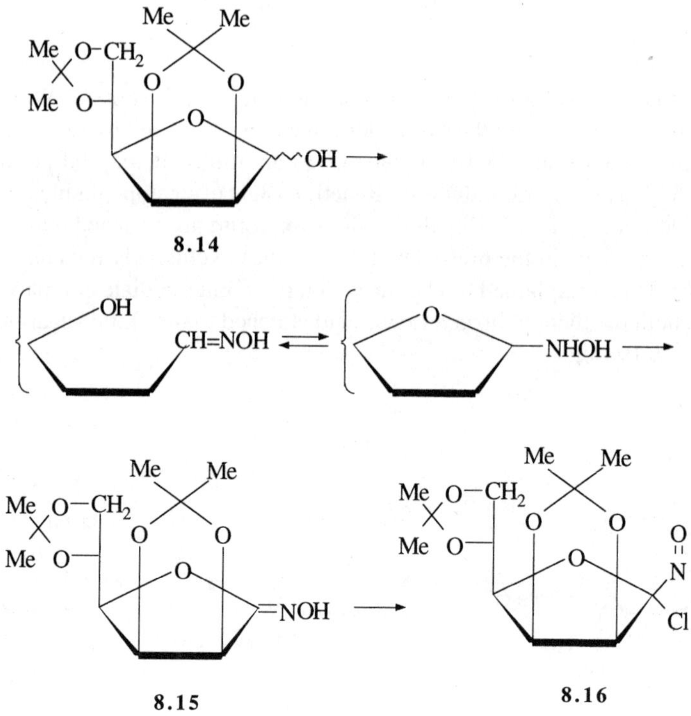

Me Me
Me O–CH_2
O O
Me O
O
OH →

8.14

OH
CH=NOH ⇌
O
NHOH →

Me Me
Me O–CH_2
O O
Me O
O
=NOH →

8.15

Me Me
Me O–CH_2
O O
Me O
O
O=N
Cl

8.16

(8.6)

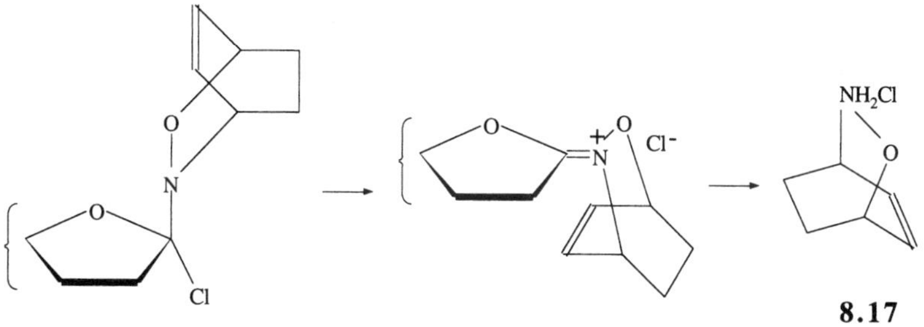

8.17

In crystalline furanose **8.16**, the chlorine atom adopts a pseudo-axial orientation and the nitrogen atom, a pseudo-equatorial one. The orientation of the N–O bond is given by the Newman projection along the C-1–N bond, **8.18**. A compact model indicates that one face of the N=O bond is very hindered. It is probable that the same conformation exists in solution and that this is what imposes the *endo* approach as shown in **8.19** in which the cyclohexadiene tries to avoid the sugar support of the nitroso chloro derivative.

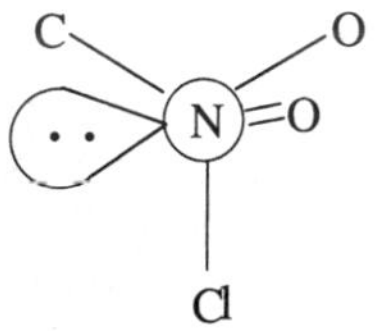

8.18 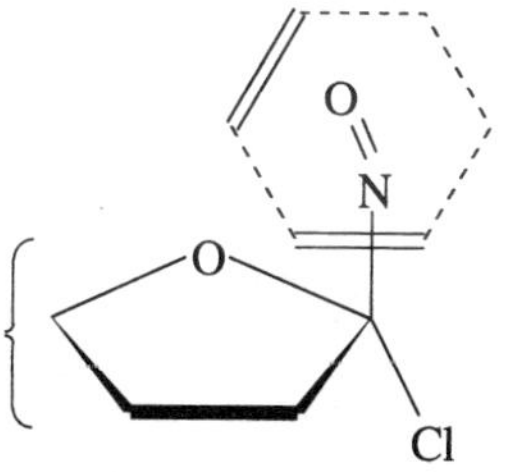 **8.19**

8.1.3 Ugi reaction

The Ugi reaction is the synthesis of an α-amino acid amide by reaction of an aldehyde with an amine and an isonitrile in the presence of zinc chloride and formic acid. β-D-Galactosylamine has been proposed as a chiral amine partner whose alcohol functions are protected by esterification with pivalate groups (reaction 8.7) (Kunz and Pfrengle 1988). The reaction is carried out in oxolane between –75 and –25°C. The diastereomeric mixture obtained corresponds to a D/L ratio close to 95:5. The pure D-isomer is obtained by recrystallization in heptane. It is known that the bond between a sugar molecule and nitrogen is easy to cleave in glycosylamines (see Section 3.6.2).

(8.7) R-CHO, Me_3C-NC; HCO_2H, $ZnCl_2$

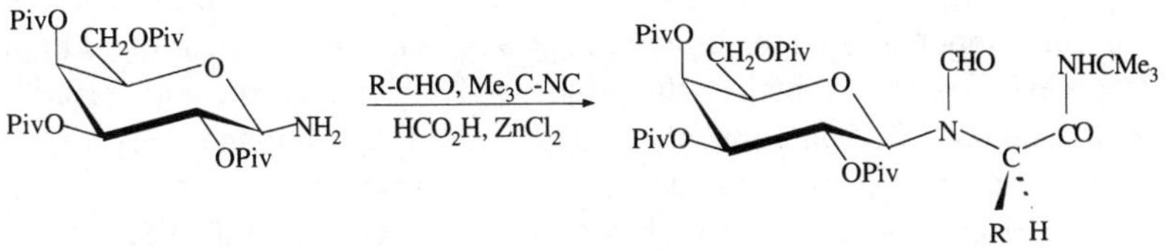

8.2 Sugars as starting materials in the synthesis of natural products

8.2.1 General

The synthesis of a natural substance can be achieved starting from racemic precursors by employing stereoselective or stereospecific reactions. The final product is racemic and the final synthesis must be completed by optical resolution. This approach now seems out of date, at least in academic laboratories. A second method is based on asymmetric synthesis. The starting products are achiral but the use of reagents and catalysts incorporating chiral structures allow the orientation of one or several favourable steps towards a chiral synthetic route. We have already seen examples of such reagents above in Section 8.1. The synthetic strategy that we are about to develop relies on the use of chiral starting compounds which are generally abundant. This is based on a body of principles whose use is facilitated by special computer-assisted programs, grouped under the heading '*the chiron approach*' (Hanessian 1983; Hanessian *et al.* 1990*a*; Hanessian 1993). The name *chiron* was coined from the expression 'chiral synthon'. It is well known that in total synthesis, the word *synthon* designates a small molecule made to fit exactly in place in multi-step total syntheses. Dissecting the target into synthons (on paper!) is practiced at predetermined positions where one hopes to be able to reattach the pieces along the synthetic pathway. In general the different functional groups present in the target (and also the experience of the chemist) will dictate the dissecting procedure. In the 'chiron approach', the guiding principle is to conserve the stereochemistry so that one tries to divide the synthetic target into fragments with minimal disturbance to the stereogenic centres. Naturally, a second condition comes into play, namely that the chirons to be stiched together must be either abundant natural products or easily accessible from them. To a certain extent, it is this simple access which determines the strategy. A quick look from an experienced chemist is not always sufficient enough to recognize the appropriate precursors of the target molecule. A software program has been proposed to help in recognizing cryptic relationships (Hanessian 1983; Hanessian *et al.* 1990*a*; Hanessian 1993). First, the molecule is observed from unusual angles as turned inside out, reversed, etc. From one class of compounds to another, chemists, imitated by journals specializing in the subject, consider different representational conventions. This diversity of representations slows down the mental transfer between the configurations of different chemical family members. The computer-assisted program allows one to go automatically from one conventional representation to another. This is illustrated in Fig. 8.1. The data furnished by the computer is the structure of a highly active natural compound, FK506, drawn in the way that is most often found in journals (top left). From this drawing, the program gives the absolute configuration, *R* or *S*, around each carbon atom. It divides up the large ring and lines up the chain along a vertical axis which makes it possible to observe a representation following the Fischer projection (top right), familiar to carbohydrate specialists, or the

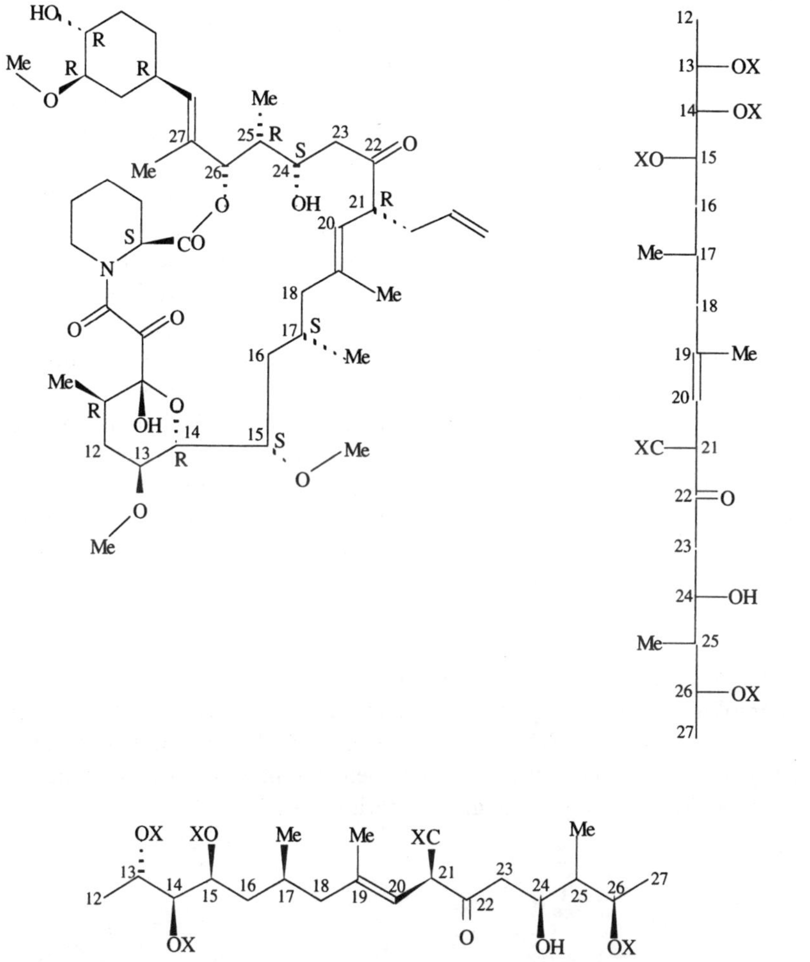

Fig. 8.1 Assigning *R* and *S* symbols and drawing two projections of compound FK506 by the 'chiron approach' program (Hanessian 1983; Hanessian *et al.* 1990*a*; Hanessian 1993) (reproduced with kind permission from the International Union of Pure and Applied Chemistry).

zig-zag representation (below). The Fischer projection is particularly practical for establishing a correlation with sugars. The program goes even further to suggest synthetic routes starting from accessible 'chirons'. Sometimes the correspondence between the chiron and target sequence is not very obvious. Sugars have too many alcohol functions making it necessary to deoxygenate them. More subtly, it may be better to conserve a function not present in the target until a latter synthetic step because this favours the stereospecificity of the reactions.

Since 1975, an annual report of the English Chemical Society, *Carbohydrate Chemistry*, has contained a chapter titled, *Synthesis* (from sugars) *of Enantiomerically Pure Non-Carbohydrate Compounds*.

8.2.2 Syntheses starting from sugars

(+)-Meroquinene (Hanessian *et al.* 1990*b*)

This piperidine, **8.20**, has an obvious relationship to quinine **8.21** for which it is a synthetic precursor. The starting compound is compound **8.22**, the peracetate of a 2-hydroxyglucal. Formula **8.22**, surrounded by arrows, indicates the modifications necessary to arrive at (+)-meroquinene. Tetraacetate **8.22** is very easily accessible because it is the elimination product of the tetra-*O*-acetyl-*α*-D-glucopyranosyl bromide **8.23** by the non-nucleophilic base, 1,5-diazabicyclo [5.4.0] undec-5-ene (DBU, **8.24**). It behaves strictly as a glycal in the presence of an alcohol and boron trifluoride etherate; the rearrangement gives the unsaturated glycoside **8.25** which, like **8.22**, is an enol acetate. Alkaline hydrolysis liberates the enolate whose keto tautomer undergoes readily acetate elimination at positions 3 and 4. On reacetylating, **8.26** is obtained. At this stage, the reader should note that **8.26** does not contain either of the two target stereogenic centres. There are only two asymmetric carbon centres remaining and, as we shall see, both are destined to disappear during the synthesis. But it is because of their presence at this point that the two permanent chiral centres will be established. Michael addition of the modified vinylcuprate, $(CH_2{=}CH)CuCN(MgBr)_2$, gives an enolate with which the methyl bromoacetate will instantaneously react. This gives nearly exclusively the *trans*-isomer **8.27**, isomerized to the *cis*-isomer **8.28** by triethylamine. The following operations are composed of deoxygenation at C-2, giving **8.29**, hydrolysis of protections, giving **8.30**, and cleavage of the 5,6-diol by periodate. Condensation of dialdehyde **8.31** with benzylamine, followed by reduction by cyanoborohydride at pH 4.3, gives piperidine **8.32**.

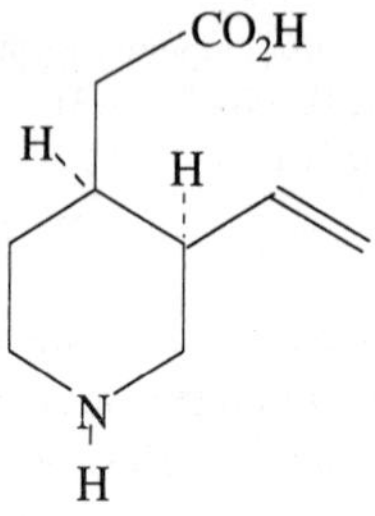

CO_2H H H N H

8.20

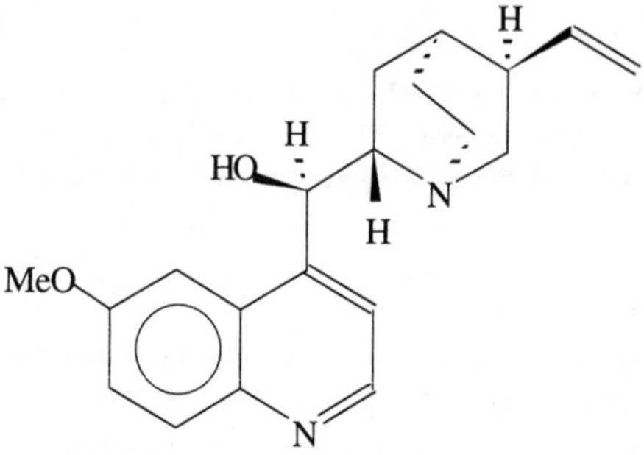

H H HO N H MeO N

8.21

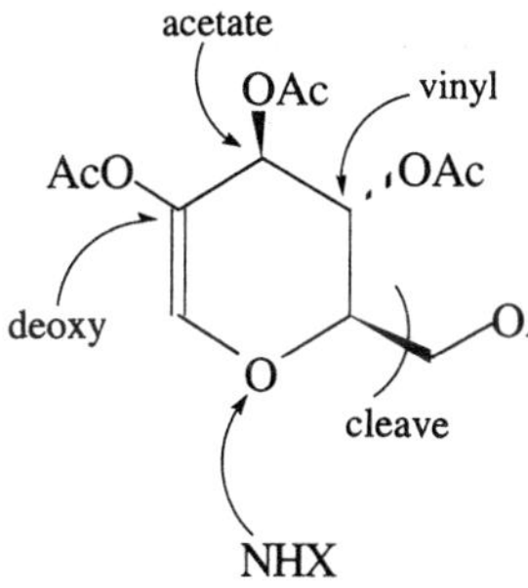

8.22

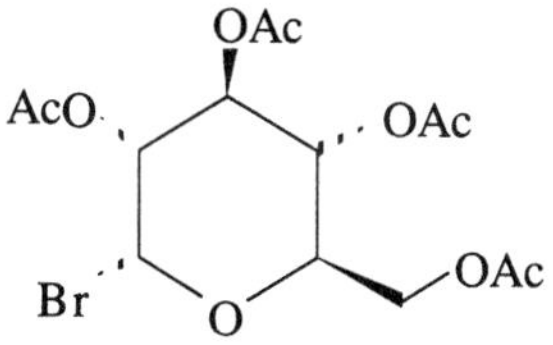

8.23

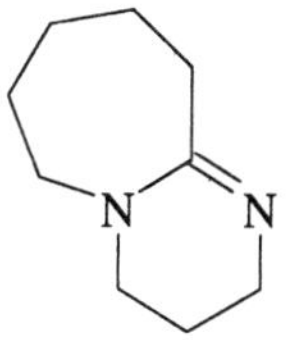

8.24

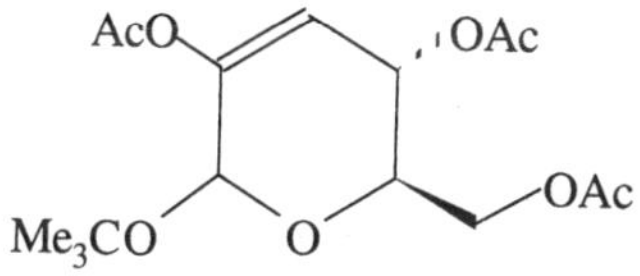

8.25

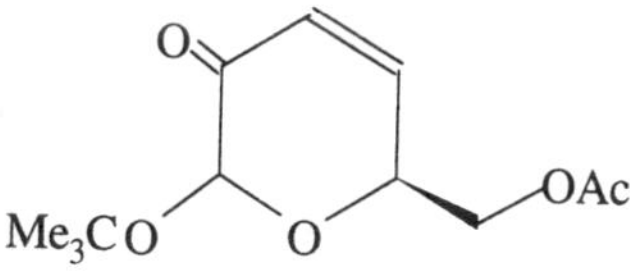

8.26

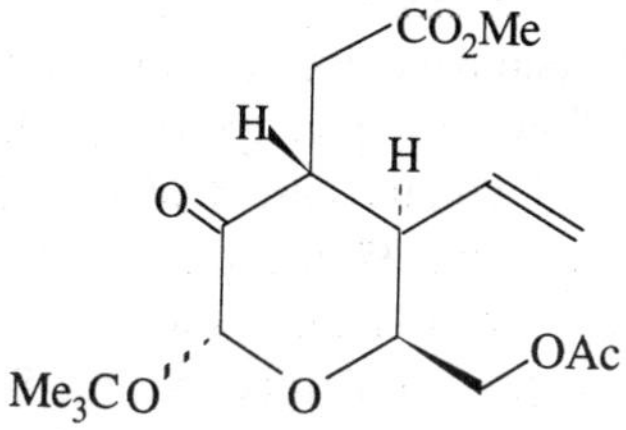

8.27

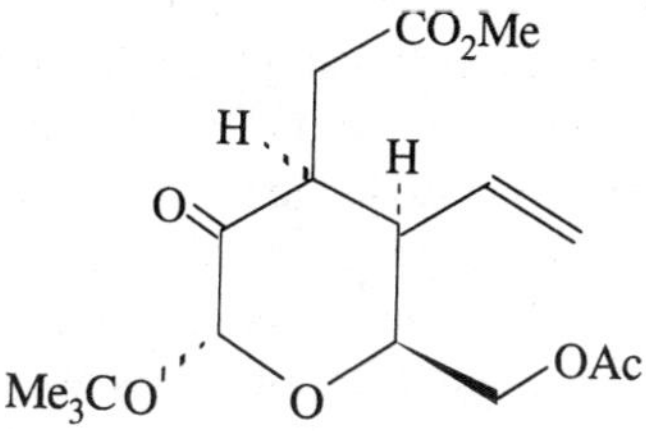

8.28

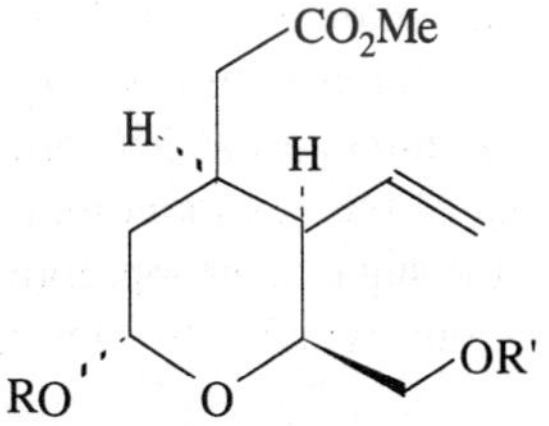

8.29 R= CMe_3, R'= Ac
8.30 R= R'= H

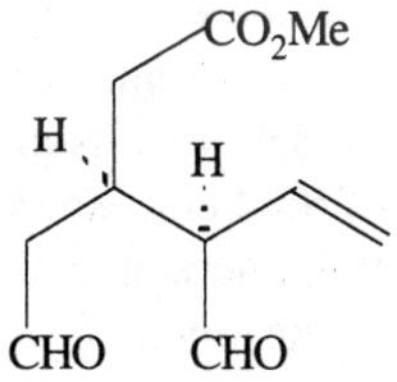

8.31

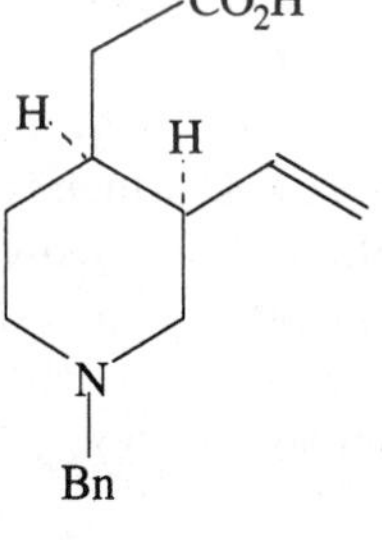

8.32

*(–)-Ajmalicine (**8.33**) and (+)-19-epiajmalicine (**8.34**)* (Hanessian and Faucher 1991)

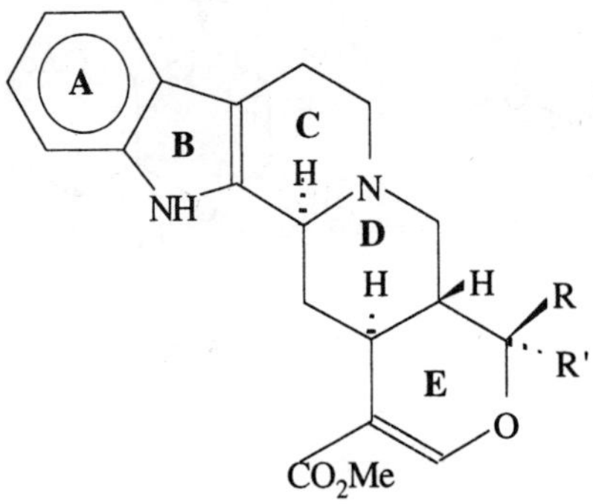

8.33 R= H, R'= Me
8.34 R= Me, R'= H

These two indole alkaloids possess important pharmacological properties. The starting product is the intermediate **8.29**, already used in the synthesis of (+)-meroquinene. The acetate function is hydrolysed by alkaline methanolysis to give the primary alcohol **8.35**. At this point we will temporarily go back to the traditional way of representing pyranoses in order to indicate the axial orientation of the vinyl group. Carbon 6 is deoxygenated by first replacing the hydroxyl group by chloride using the PPh_3–CCl_4 system, and the chloride is then reduced by a radical route with tributylstannane. The following step is the ozonization of the vinyl group to the axial aldehyde **8.36**. Treated with the base **8.24**, the –CHO substituent adopts the stable equatorial position as in **8.37**. Coupling with tryptamine **8.38** in the presence of the reducing agent ($NaBH_4$) gives **8.39** in which the ABDE rings of the two target alkaloids are seen. In this reaction, the Schiff base between the amine and the aldehyde is reduced to a secondary amine; the latter is cyclized to a cyclic amide by the ester function. Without giving more details, it is worth noting that the amide carbonyl group is well placed to allow cyclization of the indole nitrogen on the α-carbon atom, hence closing the C ring. The introduction of the CO_2Me substituent to the E ring requires the conversion of the glycoside to lactone **8.40** which can then be deprotonated to the α-position with the base Et_2NLi. Condensation of the anion with $CNCO_2Me$ gives **8.41** which is reduced to **8.42** by diisobutylaluminium hydride. The orientation of the methyl group of **8.42** corresponds to that of the *epi* alkaloid. (–)-Ajmalicine synthesis requires epimerization at C-5 of the sugar at an appropriate step. This method relies on the opening of the lactone which liberates the alcohol function at C-5. The latter is then inverted.

Natural substances of polypropionic origin

Molecules of a certain number of natural substances, often antibiotic, have sequences constituted by a linear carbon chain to which methyl or hydroxyl

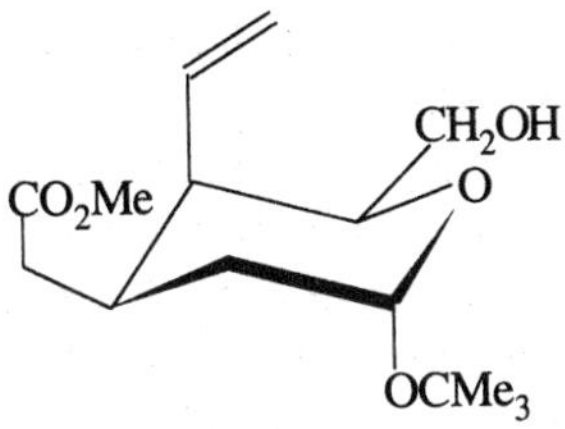

8.35

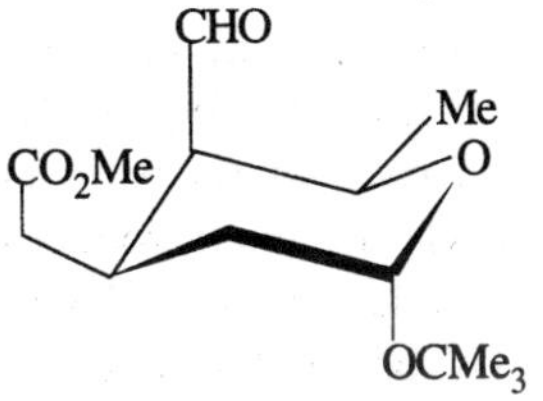

8.36

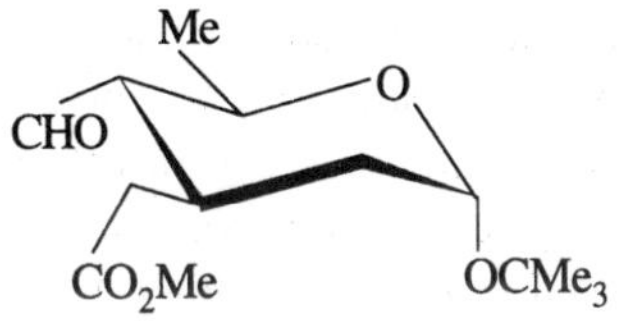

8.37

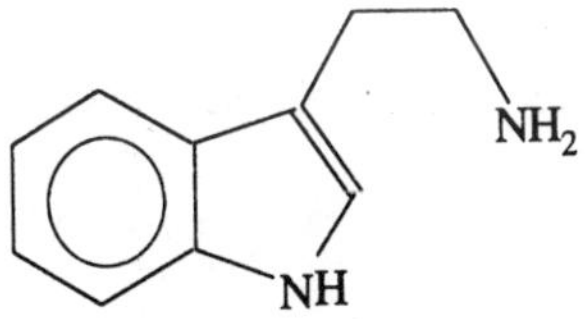

8.38

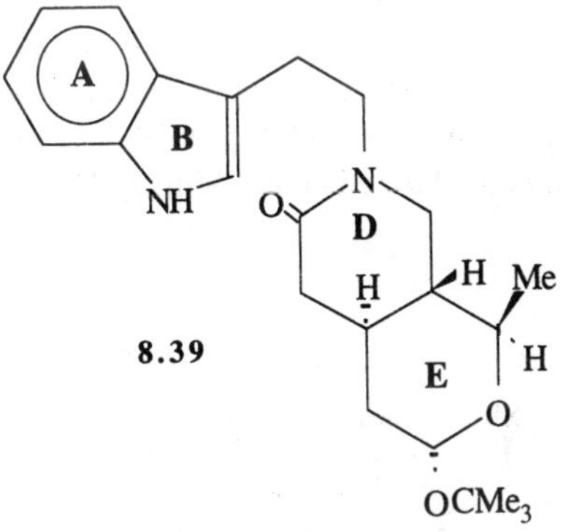

8.39

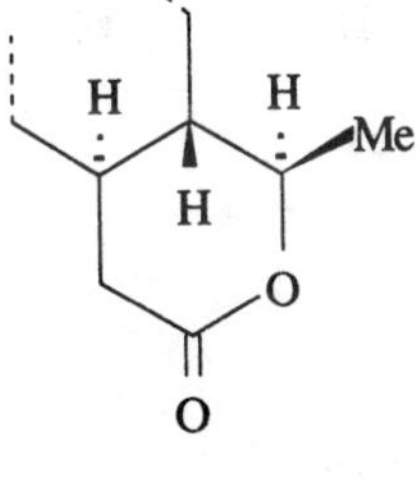

8.40

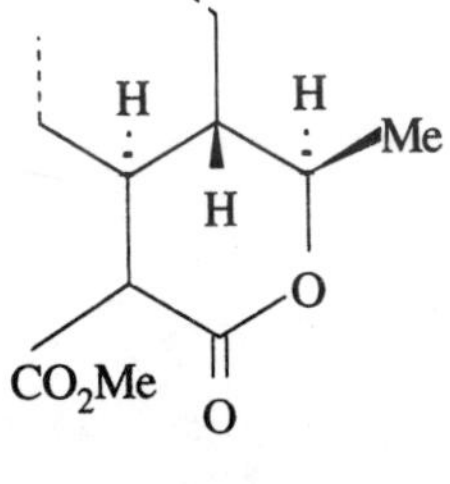

8.41

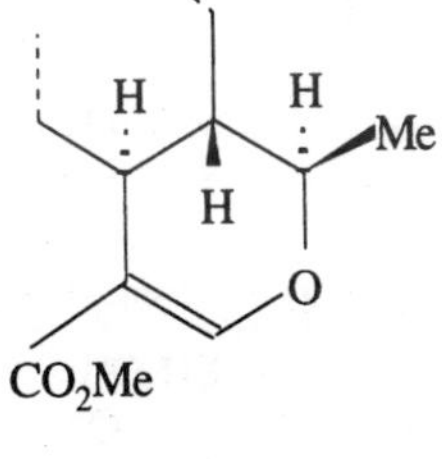

8.42

branched chains are alternated, as with fragment **8.43**. Most of them are probably derived from the condensation of biologically activated forms of propionic acid according to structure **8.44**. Deoxygenated sequences are also encountered.

To construct these sequences from sugar chains, one of the problems confronted is that of the stereospecific introduction of methyl branches. One possible chiron is D-ribonolactone **8.45**. Selective protection of the primary alcohol function and the derivatization of the diol system with the bis-imidazol thiocarbonyl **8.46** gives thionocarbonate **8.47**. Treatment of **8.47** with Raney nickel causes elimination and leads to the unsaturated lactone **8.48**. Michael addition of the bulky reagent $(MeS)_3CLi$ takes place *trans* to C-5. The enolate is oxidized directly with molybdene peroxide and, again for steric reasons, the hydroxyl group is introduced stereospecifically in the *trans* position to give, in one step, **8.49**. The sulfur substituent is reduced to a methyl group by Raney nickel. The carbon branch was thus introduced stereospecifically. From this we continue to **8.50** using classic transformations. We have used the numbering of the ribonolactone so that the reader may follow the sugar transformations. The epoxide is opened by the bifunctional organolithium reagent $LiCH(SPh)CO_2Li$. This introduces a carboxylic acid function which is lactonized on the secondary alcohol function due to the epoxide opening. Lactone **8.51** is obtained. The phenylthio function is located such as to allow the introduction of the double bond by elimination of the sulfoxide. Product **8.52** is an analogue of **8.48** except that the side chain has the opposite orientation. Also, Michael addition still takes place at the *trans* position of this side chain. Methyllithium gives the branched-chain product **8.53**. As before, oxidation can take place *in situ*. After the reductive opening of the lactone ring by $LiAlH_4$, **8.54**, the precursor of an ionomycin sequence, is obtained (Hanessian and Murray 1987).

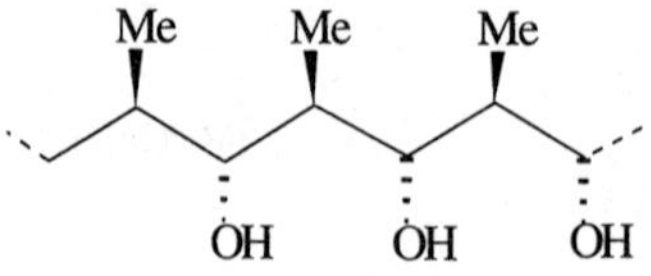

Me Me Me

OH OH OH

8.43

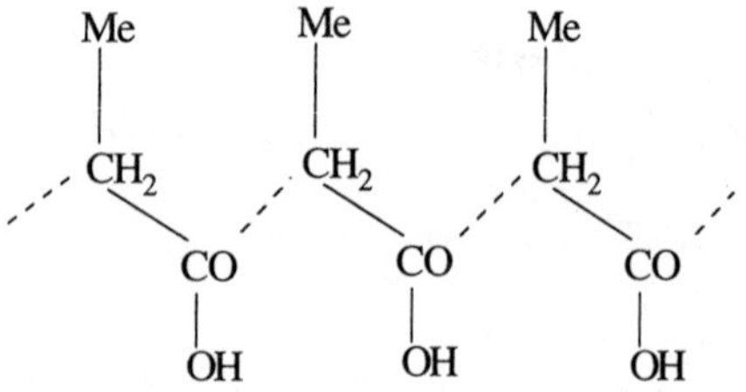

Me Me Me

CH_2 CH_2 CH_2

CO CO CO

OH OH OH

8.44

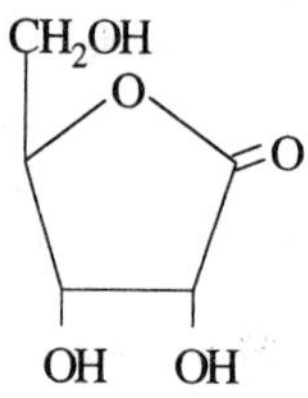

CH_2OH

O

O

OH OH

8.45

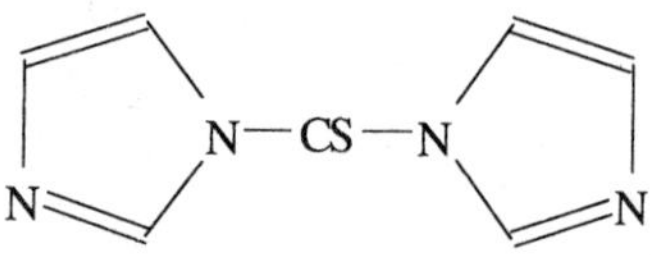

N—CS—N

N N

8.46

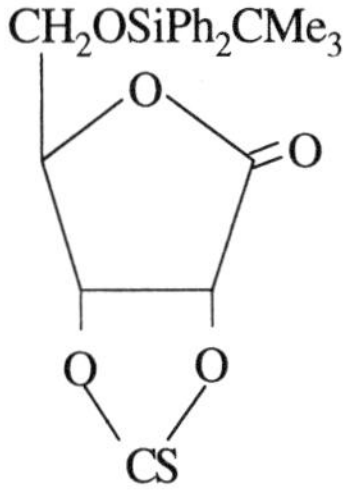

8.47 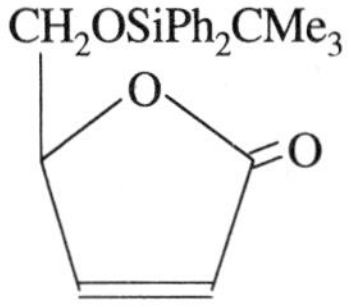 8.48 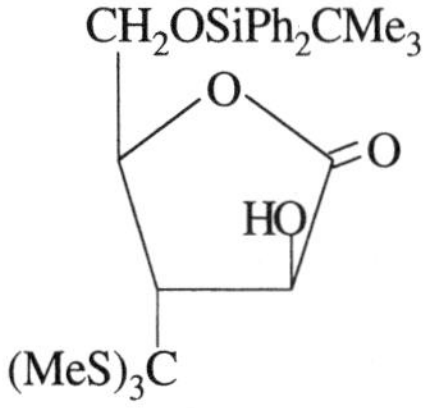 8.49

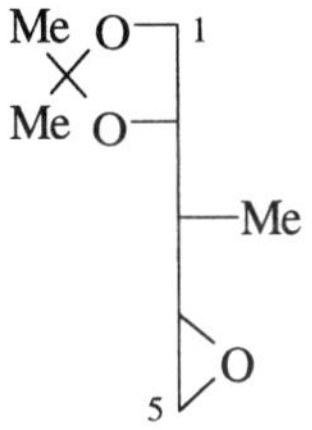

8.50 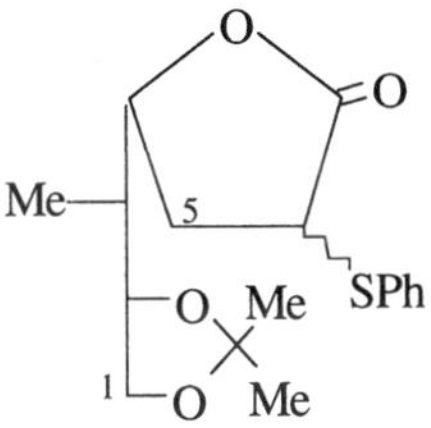 8.51 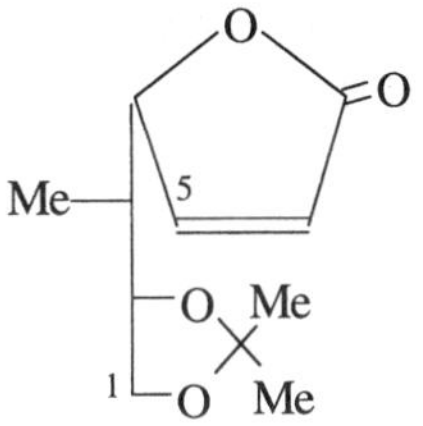 8.52

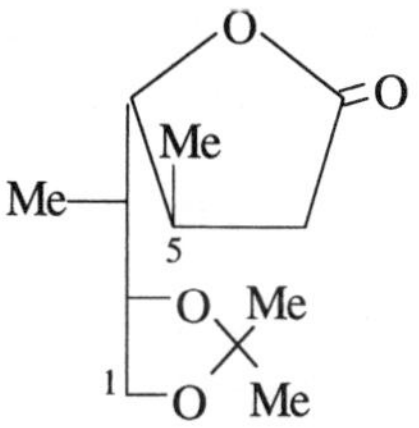

8.53 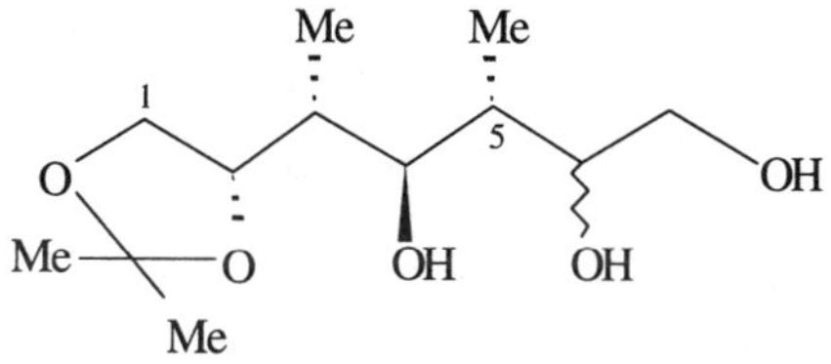 8.54

References

Bold, G., Duthaler, R. O., and Riediker, M. (1989), *Angew. Chem., Int. Ed. Engl.*, **28**, 497–498.

David, S., Lubineau, A., and Thiéffry, A. (1978), *Tetrahedron*, **34**, 299–304.

David, S., Eustache, J., and Lubineau, A. (1979), *J. Chem. Soc., Perkin Trans. 1*, 1795–1798.

Duthaler, R. O., Herold, P., Lottenbach, W., Oertle, K., and Riediker, M. (1989), *Angew. Chem. Int., Ed. Engl.*, **28**, 495–497.

Felber, H., Kresze, G., Prewo, R., and Vasella, A. (1986), *Helv. Chem. Acta*, 1137–1146.

Hanessian, S. (1983) *Total synthesis of natural products: The chiron approach*, Pergamon, Oxford.

Hanessian, S. and Murray, P. J. (1987), *Tetrahedron*, **43**, 5055–5072.
Hanessian, S. and Franco, J., and Larouche, B. (1990*a*), *Pure Appl. Chem.*, **62**, 1887–1910.
Hanessian, S, Faucher, A.-M., and Léger, S. (1990*b*), *Tetrahedron*, **46**, 231–243.
Hanessian, S. and Faucher, A.-M. (1991), *J. Org. Chem.*, **56**, 2947–2949.
Hanessian, S. (1993), *Pure Appl. Chem.*, **65**, 1189–1204.
Kunz, H. and Pfrengle, W. (1988), *Tetrahedron*, **44**, 5487–5494.
Kunz, H., Müller, B., and Schanzenbach, D. (1987), *Angew. Chem., Int. Ed. Engl.*, **26**, 267–269.
Kunz, H. and Rück, K. (1993), *Angew. Chem., Int. Ed. Engl.*, **32**, 336–358.
Riediker, M. and Duthaler, R. O. (1989*a*), *Angew. Chem., Int. Ed. Engl.*, **28**, 494–495.
Riediker, M., Hafner, A., Piantini, U., Rihs, G., and Togni, A., (1989*b*), *Angew. Chem., Int. Ed. Engl.*, **28**, 499–500.
Zamojski, A., Banaszek, A., and Grynkiewicz, C. (1982), *Adv. Carbohydr. Chem. Biochem.*, **40**, 1–112.

9 Oligosaccharides: configuration and analysis

9.1 Introduction and nomenclature

The molecules we have seen so far have been derived, in general, from a single simple sugar such as a pentose or a hexose. Although we have referred to them as sugars, which is a correct term, we will henceforth need to differentiate them from other sugars built from two, three, four, or a great number of simple sugars. Now is the moment to introduce terms whose meanings are obvious such as mono-, di-, tri-, tetra-, ... poly-saccharide.

Monosaccharide units are linked by oxygen bridges which arise as the result of the substitution of the hemiacetal hydroxyl group of one unit by the oxygen of a hydroxyl group bound to the other. If the latter is also a hemiacetal, these two functions protect each other, and no potential aldehyde is formed. The disappearance of the reducing power was easy to recognize with the old chemistry procedures; it was used very early on as a basis of classification. As a non-reducing disaccharide, we have already seen the example of ordinary table sugar, **3.21**. The trehaloses are three non-reducing disaccharides built solely from D-glucopyranosyl units, with fixed α,α-, α,β-, and β,β-anomeric configurations. The most common one is the α,α-isomer **9.1**, the storage substance of invertebrates and lower plants.

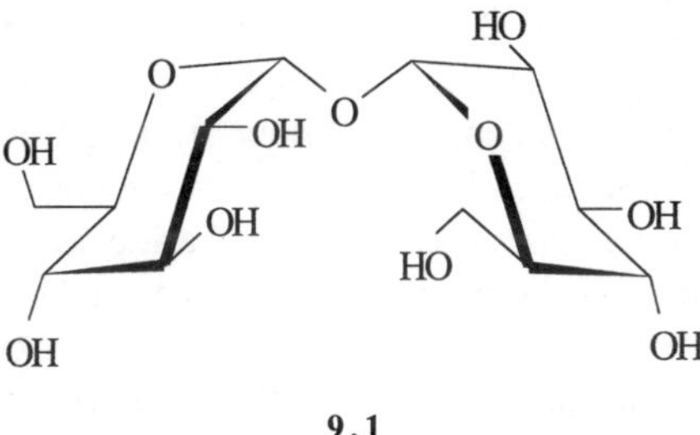

9.1

In the other type of bond, the bridging oxygen comes from an alcoholic hydroxyl group. It is encountered in a multitude of natural products. It allows the formation of both large and small sequences that the polymer chemist would call oligomers and polymers of polycondensation. A potential aldehyde function always remains at one end of these chains which is incorrectly referred to as the *reducing end* even if this is not so, because it is involved in a glycosidic bond with a molecule not belonging to the sugar family. Really long chains are periodic. For amylose **11.13**, and cellulose **9.2**, the repeating unit is the D-glucopyranosyl monosaccharide linked at positions 1 and 4. The anomeric configurations are all α for amylose and β for cellulose. Portions of more or less long repeating units are observed when the unit is a disaccharide, as in certain glycolipids or heparin and parent polysaccharides, or even a tri- or tetra-saccharide as in bacterial antigens. For the last 20 years, particular importance has been placed on a class of intermediate molecules imprecisely referred to as *oligosaccharides*. This name covers all degrees of condensation from two to a poorly defined maximum, perhaps about 20 monosaccharide units in all. This classification is not dictated by chemical properties but rather by the frequently observed activity of certain representatives in recognition phenomena. There is no obvious regularity about these structures, to which the rest of this work will be almost exclusively devoted.

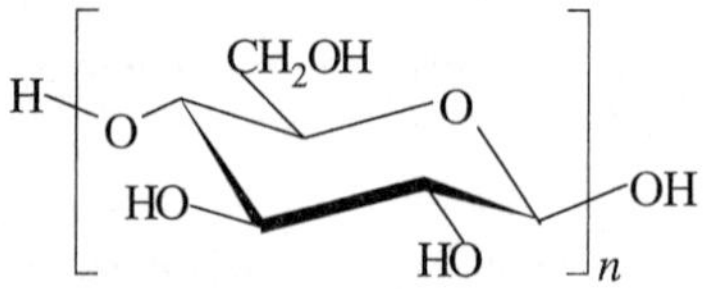

H–[O–(CH$_2$OH, HO, HO, O)]$_n$–OH

9.2

In the recommended nomenclature, an oligosaccharide is defined as a compound whose complete hydrolysis gives a restricted number of monosaccharides. A non-reducing disaccharide is described as a glycosyl glycoside. Some examples are ordinary table sugar or sucrose **3.21** (β-D-fructofuranosyl-α-D-glucopyranoside) and α,α-tetrahalose **9.1** (α-D-glucopyranosyl-α-D-glucopyranoside).

A reducing disaccharide is described as a glycosyl glycose, the non-reducing unit being considered as a substituent of the other, for example *N*-acetyllactosamine (see formula **9.9**, R = H, R′ = OH, n = 1) (2-acetamido-2-deoxy-4-*O*-β-D-galactopyranosyl-D-glucose).

Generally the non-branched chains are drawn starting from the non-reducing end on the left, to the reducing end on the right. One nomenclature system, more graphic than the preceding, *writes* the monosaccharide units in the same direction, with the glycosidic bond being described by the symbol (n→m), with n = 1 or 2. Thus *N*-acetyllactosamine is written as *O*-β-D-galactopyranosyl-(1→4)-2-acetamido-2-deoxy-D-glucose. To give another example, the tetrasaccharide glycoside studied in Section 11.5.3 is written as 4-nitrophenyl-*O*-α-D-glucopyranosyl-(1→4)-*O*-α-D-glucopyranosyl-(1→4)-*O*-α-D-glucopyranosyl-(1→4)-*O*-α-D-glucopyranoside. The branch monosaccharide units on the major chain are written in brackets.

These two nomenclature systems, the only ones which allow the position of the substituents on the constituent monsaccharides to be clearly indicated, are found particularly in articles on synthetic chemistry. They are awkward but can be simplified for free oligosaccharides. Each monosaccharide is represented by the following symbols: Glc for glucose, Gal for galactose, Man for mannose,

Fuc for fucose, GalNAc for *N*-acetylglucosamine, and NeuAc (the most common) for sialic acid. The ring dimension is indicated by the additive *p* or *f*. Thus *N*-acetyllactosamine is written as

β-D-Gal*p*-(1→4)-D-GlcNAc

and the free tetrasaccharide is written as

α-D-Glc*p*-(1→4)-α-D-Glc*p*-(1→4)-α-D-Glc*p*-(1→4)-D-Glc

Biochemists who encounter a limited number of sugars, all having the pyranoid form, use an even more simplified nomenclature in their articles. They remove the D, *L*, *p*, and *f* and place the anomeric descriptor after the sugar symbol which gives, for *N*-acetyllactosamine,

Gal-β-(1→4)-GlcNAc

If a formula presented in this style is puzzling, the reader may consult Table 9.1 to have an unambiguous translation of this nomenclature.

9.2 *Exo*-anomeric effect

Oligosaccharides are constructed from monosaccharide residues linked by oxygen bridges. Is this articulation rigid or semi-rigid? Is there free rotation around the two simple oxygen–carbon bonds? To be truthful, these questions have already appeared with simple glucosides (see Chapter 3), but we will examine them here because they have aroused the most interest within the framework of oligosaccharide chemistry. We will see that in order to understand the mechanism of cell recognition, it is very important to know if these molecules form a flexible or semi-rigid group. In a general case, the conformation around a simple bond is *gauche* (or *synclinal, sc*) or antiperiplanar (*ap*). In the anomeric region the C-1–O-5 bond is taken as a reference and conformations are exactly described by the torsion angles θ = O-1–C-1–O-5–C-5 and φ = R–O-1–C-1–O–5, with the crystallographic notation conventions. In the ideal conformation of a pyranoside, the equatorial anomer **9.3e** corresponds to $\theta = 180°$ (*ap*),

Table 9.1 Names of monosaccharide residues used in biochemistry and precise nomenclature.

Biochemistry	Precise nomenclature
Glc α (β)	α (β)-D-Glc *p*
Gal α (β)	α (β)-D-Gal *p*
Fuc α (β)	α (β)-L-Fuc *p*
GlcNAc α (β)	α (β)-D-GlcNAc*p*
GalNAc α (β)	α (β)-D-GalNAc*p*
NeuAc α (β)	α (β)-D-Neu 5Ac*p*

and the axial anomer **9.3a** to $\theta = 60°C$ (+*sc*). Figure 9.1 shows the three conformations of each anomer to be considered a priori (Tvarovša and Bleha 1989).

9.3e **9.3a**

On crystallized samples, θ and φ can obviously be measured without difficulty by X-ray diffraction analysis, but it is rare to obtain them with oligosaccharides. In any case, this does not give the conformation in solution with certainty, that is in conditions where their real activity is demonstrated.

Equatorial anomer

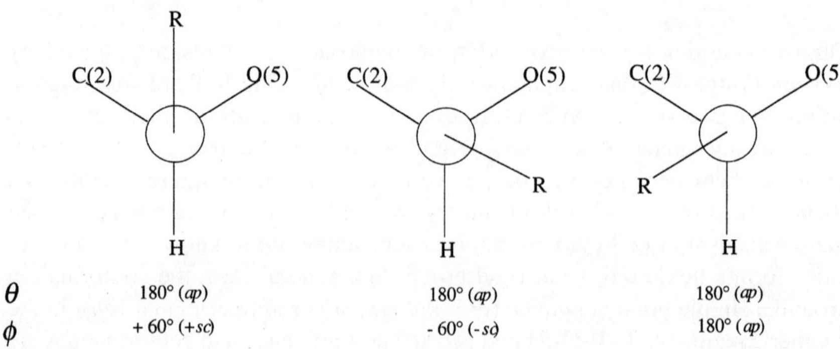

Axial anomer

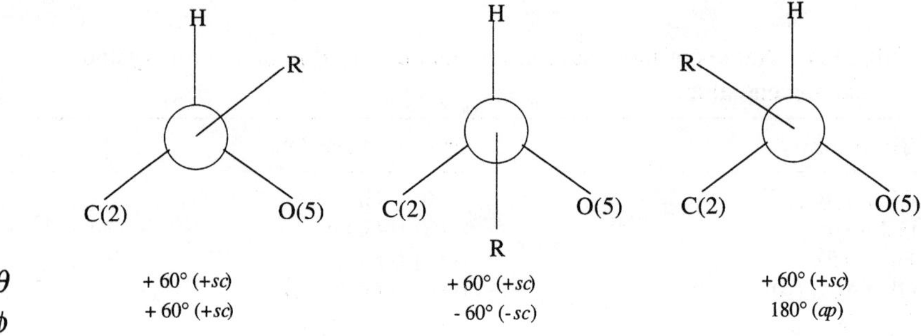

Fig. 9.1 Conformations to be considered for two anomeric glycosides.

Measurements in solution are achieved by ^{13}C NMR analysis which makes it possible to determine the ^3J coupling constant between the anomeric proton and the carbon atom of the glycosidic molecule junction (Fig. 9.2). Experimentally, the relationship between $^3J_{CH}$ and Φ has been established from presumably very rigid derivatives of sugars such as **9.4**. It is thus most likely that the torsion angles under consideration were hardly modified by the change to being in solution. The measurements are represented by equation (9.1) which gives graphically a curve of the Lemieux–Karplus type (Fig. 9.3) (Tvarovša *et al.* 1989).

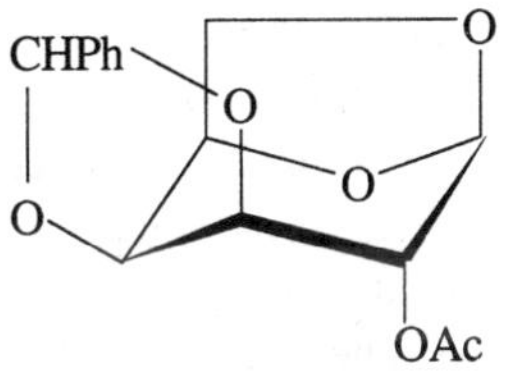

9.4

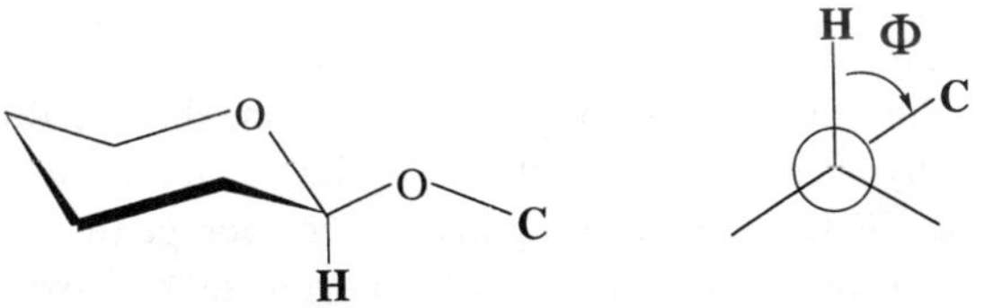

Fig. 9.2 The C and H atoms whose vicinal coupling constant is measured to the right, Newman projection along C-1–O -1.

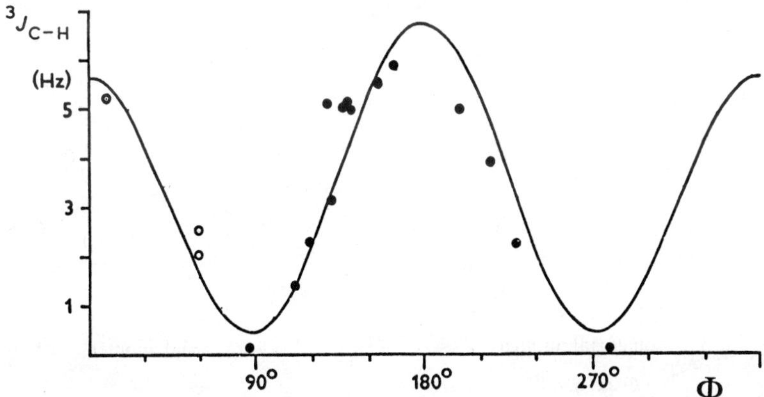

Fig. 9.3 Dependence of the ^3J-CH coupling constant of the Φ torsion angle defined in Fig. 9.2 (reproduced with permission of Elsevier Science).

$$^{3}J_{CH}/\text{Hz} = 5.7 \cos^{2}\Phi - 0.6 \cos \Phi + 0.5 \qquad (9.1)$$

A very similar equation had been previously derived from measurement of φ on 1,6-anhydro-β-D-glucopyranose, glycosides, and oligosaccharides in the solid state (Mulloy *et al.* 1988).

The proximity of carbon and hydrogen atoms can also be evaluated by measuring the spin-lattice relaxation rate of this proton (Dais and Perlin 1987).

Axial glycosides are found to exist uniformly in the (+*sc*, +*sc*) conformation. For example in the α-trehalose **9.1**, the four anomeric torsion angles correspond to +*sc* and the disaccharide possesses a C_2 axis, in solution as well as in the solid state. On the other hand, the known equatorial glycosides are divided into two groups, a major one being *ap*, –*sc* (three against one), and minor one, *ap*, *ap*. We may well wonder if these conformational preferences are due to something other than non-bonding interactions. The forbidden *ap*, *sc* and +*sc*, –*sc* conformations lead to the substituent molecule at O-1 in a position very close to the lower and upper surfaces of the pyranose. The preference for *ap*, –*sc* and +*sc*, +*sc* would come from the fact that the ring oxygen is less bulky than the tetrahedral carbon C-2. Figure 9.4 shows the conformations observed when the average of the φ values are taken from a great number of very diversified examples.

This may be interpreted as the tendency of the 2p orbital of the oxygen atom to align itself with the C-1–O-5 bond in a way as to interact with the corresponding antibonding orbital. More simply, we can assume that the R group naturally approaches the hydrogen group, as this is by far the least bulky. As usual, we can criticize the NMR spectra as only giving the average of several conformations. What obviously counts is being able to evaluate the barriers. Several calculations have been made on simple models. Perhaps the lesson to be learned is that angles and bond lengths do not vary independently of each other, thus each model must be optimized.

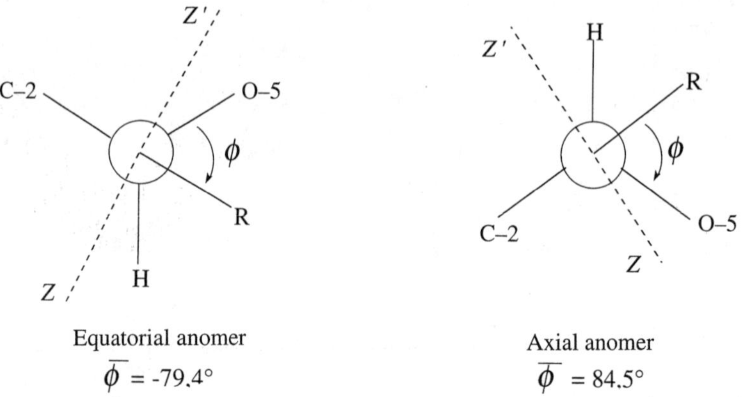

Fig. 9.4 Average of *exo*-anomeric conformations observed.

9.3 Determining sequences by chemical methods

9.3.1 Acidic hydrolysis

The first question concerns the nature and relative proportions of constituent monosaccharides. In principle, this is obtained by acidic hydrolysis (Biermann 1988) but, in practice, it must be carefully applied as there are a certain number of important specific cases. Hydrochloric, sulfuric, and trifluoroacetic acids are used whose 1 N solutions have a pH of 0.1, 0.3, and 0.7, respectively. When hydrolysis liberates monosaccharides fragile in an acidic medium, a delicate balance between the risk of incomplete hydrolysis and partial destruction of the hydrolysis product must be maintained. The fragile sugars are pentoses, deoxy sugars, and uronic and aldonic acids. When sialic acid is kept for 30 min at 90°C in 0.01 M HCl, 20% decomposition occurs. With neutral polysaccharides, decomposition can be limited to less than 9%. The acetyl groups of acetamides are hydrolysed and relatively stable protonated amino sugars are obtained.

Hydrolysis by methanolic hydrochloric acid leading to methyl glycosides would be less destructive but, in the worst case, a monosaccharide could be found in the product of methanolysis in four different chemical forms, the four methyl glycosides. Acetolysis, that is degradation by a mixture of acetic anhydride and sulfuric acid, transforms cellulose into the octoacetate of a disaccharide β-D-Glc*p*-(1→4)-Glc, but this is a preparative reaction rather than an analytical method. Acetolysis is sometimes recommended as an additional step in analytical work.

9.3.2 Enzymic hydrolysis

In Section 3.5.2 we have already discussed glycosidase enzymes. *Exoglycosidases* remove a monosaccharide unit located at a non-reducing end (there are often several since the chains are branched). They are named according to the unit they remove, such as neuraminidase (*N*-acetylneuraminic acid is another name for sialic acid), fucosidase, galactosidase, mannosidases, and aminohexosidase, and are normally specific to α- or β-configurations. In principle, they allow the residue by residue degradation from the non-reducing end and can be used in conjunction with the above-described methylation methods.

In Chapter 3 we did not describe the *endoglycosidases*. They catalyse the hydrolysis of a glycosidic bond in the middle of a chain and are specific to the configuration of the two sugars they separate (Rauvala *et al.* 1980). The endo-2-acetamido-2-deoxy-β-D-glucosidase cleaves the sequence called 'chitobiose')-β-D-GlcNAc*p*-(1→4)-β-D-GlcNAc*p*-(, present at the reducing end of certains glycoproteins. Another enzyme, an endo-β-D-galactosidase cleaves specifically the bond between galactose and *N*-acetylglucosamine in the very common sequence →3)-β-D-Gal*p*-(1→4)-β-D-GlcNAc*p*(1→.

9.3.3 Methylation analysis

In this analysis the idea is to etherify all free hydroxyl groups. Since ether functions resist conditions of acidic hydrolysis, the only free hydroxyl functions found in the fragments are those which were originally involved in the glycosidic bond and the one from the reducing end of the oligosaccharide. In a simple example where lactose **9.5** is permethylated to **9.6**, we observe that acidic hydrolysis of **9.6** gives a tetramethylated galactose **9.7** and a trimethylated glucose **9.8**, which establishes the arrangement of the two residues in the disaccharide. This commonplace idea runs into a few difficulties in practice. A solvent having contradictory properties is needed since the system is very hydrophilic at the beginning and quite hydrophobic at the end. A rather large oligosaccharide requires creating a great quantity of negatively charged alkoxide functions close to each other. The currently preferred solvent is dimethyl sulfoxide, CH_3SOCH_3, and the corresponding anion is used as base, obtained by adding sodium hydride or potassium *t*-butylate.

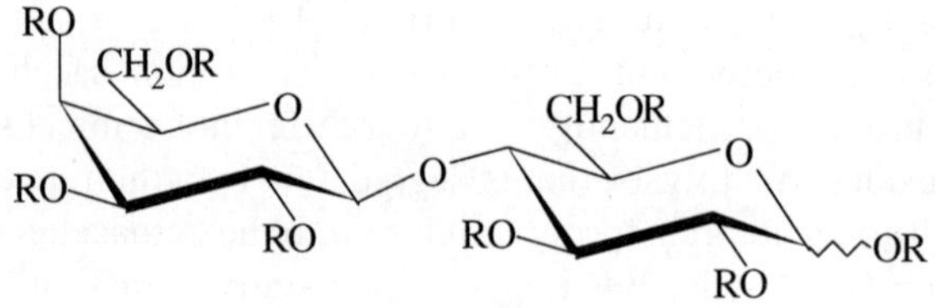

9.5 R= H
9.6 R= Me

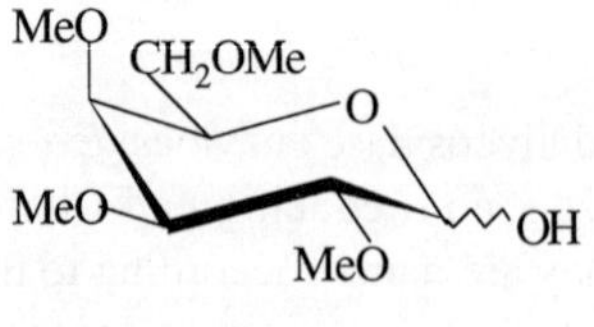

9.7 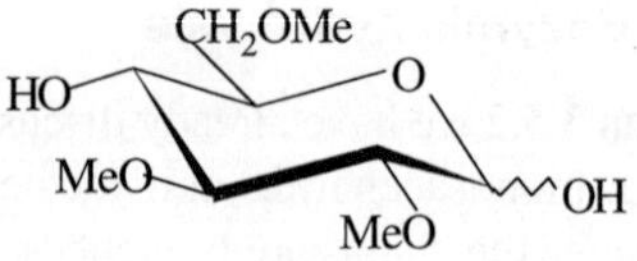 **9.8**

To establish a definite diagnosis, it is essential that the conversion of the hydroxyl groups to alkoxides be complete. An excess of $CH_3SOCH_2^-$ is checked by using the red colour that this anion gives with triphenylmethane. This technique methylates the amide nitrogens of the *N*-acetylated hexosamines as well.

The methods outlined in the last three paragraphs give finally a mixture of monosaccharides or derivatives of monosaccharides to be analysed. This analysis is carried out using the chromatographic techniques described in Chapters 1 and 3, with a possible adapted transformation to render the molecules volatile. The combination of gas chromatography and mass spectrometry techniques is particularly useful. The chemical operations have been miniaturized in order to treat very small quantities of the sample. This was necessary because it is rare to have really large oligosaccharides in notable quantities. They often come from human sources. We will now look at the spectroscopic methods which make it possible to go a little further in these investigations.

9.4 Determination of sequences by spectroscopic methods

9.4.1 FAB mass spectrometry

In one of the techniques of mass spectrometry (Dell 1987), abbreviated to FAB (*fast atom bombardment*), an accelerated beam of atoms or ions is fired at a target consisting of a solution of the sample to be analysed in a viscous solution (also called a matrix). By striking the target surface, the atoms transmit their kinetic energy to the molecules in the sample. Many of them are projected outside the target into the vacuum of the ion source and are ionized in notable proportions. In this way, gas-phase ions are produced without preliminary volatilization of the sample. Positive ions are formed, $[M + H]^+$ and $[M + \text{cation}]^+$, and negatives ones, especially $[M - H]^-$ as well as $[M + \text{anion}]^-$, depending on the nature of the viscous matrix containing the sample. Glycerol is used most often with polar molecules such as oligosaccharides and native glycopeptides, whereas for hydrophobic compounds as glycosphingolipids which tend to form aggregates in polar solvents, 1-thioglycerol CH_2SH $CHOHCH_2OH$ is preferred. It is useful to acidify the mixture with trace amounts of dilute HCl. The addition of ammonium thiocyanate to certain permethylated oligosaccharides brings about the formation of $[M + NH_3]^+$ and $[M + SCN]^-$ ions. Only the molecules present at the surface of the matrix are ionized by the atomic jet and impurities such as detergents must be avoided as they drive them away.

Strong peaks of pseudo-monomolecular ions and fragment peaks are observed. Non-derivatized sugars are used to determine M, but is is often useful to prepare, first, a derivative by permethylation (see Section 9.3.3) or peracetylation by treatment with a 2:1 (v/v) mixture of trifluoroacetic anhydride/acetic acid for 9 min at room temperature. The sample quantities needed for analysis are from 1 to 9 μg of free sugars, or 0.1 to 5 μg of derivatives. Three types of sugars give peaks above M = 4000: permethylated polysaccharides, permethylated glycosphingolipds, and natural acylated forms of mycobacterial polysaccharides.

Two modes of cleavage taken from Dell (1987) are shown in Fig. 9.5. In mode A, the major one, the charge is retained on the fragment on the side of the non-reducing end. In mode B, the charge is retained on the side of the reducing end. With derivatized oligosaccharides, the MNPQR sequence in which M is the monosaccharide residue on the non-reducing side, mode A is observed principally, according to the equation (9.2).

$$\begin{array}{l} \text{MNPQR} \rightarrow \text{M}^+, \text{MN}^+, \text{MNP}^+, \text{MNPQ}^+, \ldots \\ \qquad\qquad\qquad\qquad\quad \downarrow \qquad\quad \downarrow \\ \qquad\qquad\qquad\qquad\; \text{NP}^+ \quad\; \text{NPQ}^+ \end{array} \tag{9.2}$$

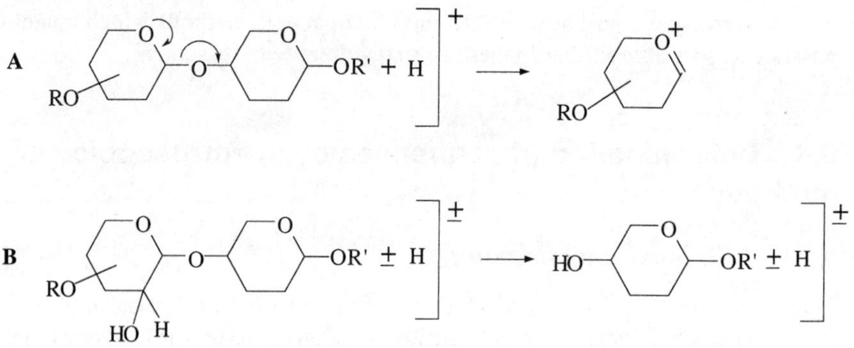

Fig. 9.5 Two modes of cleavage of oligosaccharide chains in FAB mass spectrometry.

The vertical arrows lead to fragments due to two cleavages. The fragment mass gives information on their composition since addition to the methylated fundamental structure Hex–HexNAc$^+$ of methylated fucose, *N*-acetyl or *N*-glycolyl neuraminic acid residues leads to different characteristic mass increments. No differentiation is made with isomers and compositions are given as hexoses, pentoses, deoxyhexoses, hexosamine, etc. Cleavage takes place preferentially, and sometimes even exclusively, on large permethylated oligosaccharides at each hexosamine residue, according to equation (9.3).

(9.3) $$\text{MN–HexNAc–QR} \rightarrow [\text{MN–HexNAc}]^+$$

FAB mass spectrometry is not restricted to glycolipids (see Section 13.1) although they are very well suited to this technique. A natural glycosphingolipid with 25 permethylated monosaccharide residues gives a [M + Na]$^+$ signal at 6184. Figure 9.6 gives the principal fragments of a permethylated ganglioside isolated from granulocytes (Fukuda *et al*. 1985). Besides the fragmentation types already described, we can observe the usual cleavage between the oligosaccharide and the ceramide lipid chain. The [M + H]$^+$ ion loses the acyl group to give an ion whose mass is characteristic of the ganglioside type, as in the case of Fig. 9.6, [M + H]$^+$ –238.

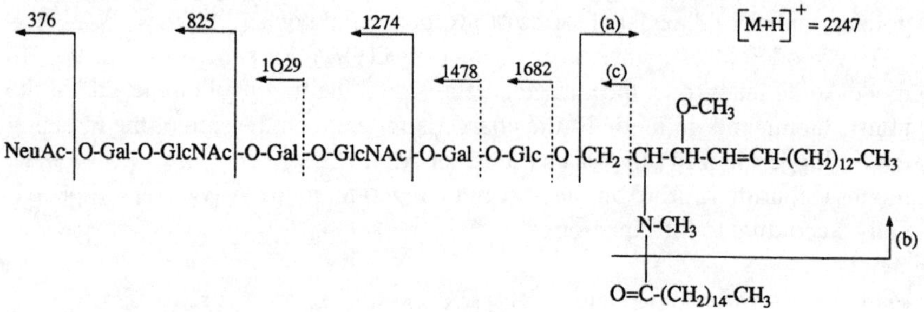

Fig. 9.6 Schematic representation of the fragmentation of a permethylated ganglioside from granulocytes (Fukuda *et al*. 1985) (reproduced with permission from Academic Press).

The tendency to cleave at the HexNAc residues is particularly interesting in the analysis of glycolipids. In these compounds a principal chain is formed from the residues β-D-Gal-(1→4)-β-D-GlcNAc which corresponds to the *N*-acetyllactosamine disaccharide. These dissacharide residues are linked by (1→3) bonds as shown in compound **9.9**. The non-branched hexasaccharide ($n = 3$) gives fragments at 3, 2, and 1 HexHexNAc units. On the other hand, the permethylated branched hexasaccharide **9.10** does not give a tetrasaccharide fragment (it is important to remember that the charge remains on the non-reducing side).

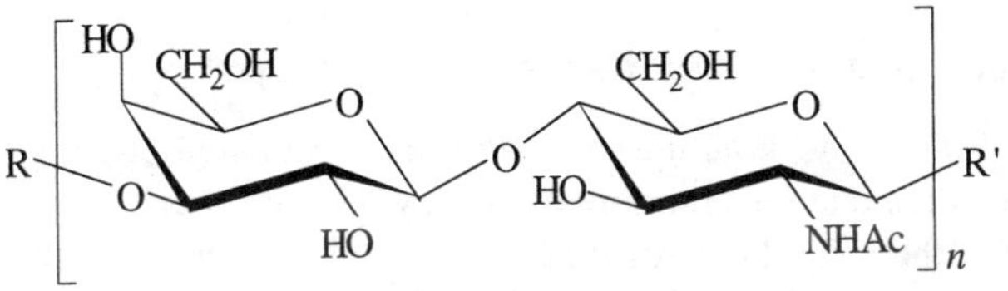

9.9

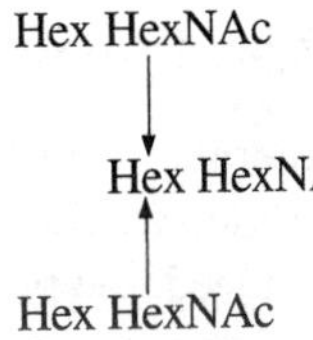

Hex HexNAc

Hex HexNAc

Hex HexNAc

9.10

In general, FAB mass spectrometry does not give the position of bonds. However, it is sometimes possible to recognize the presence of a fucose at position 3 of the β-D-GlcNAc residue. When the major ion fragment has a mass less than 900, further fragmentation can be observed according to equation (9.4). If OH-3 were not substituted in the oligosaccharide, the mass loss would correspond to CH_3OH, 32. If it were a fucose, the mass loss would be 206.

(9.4)

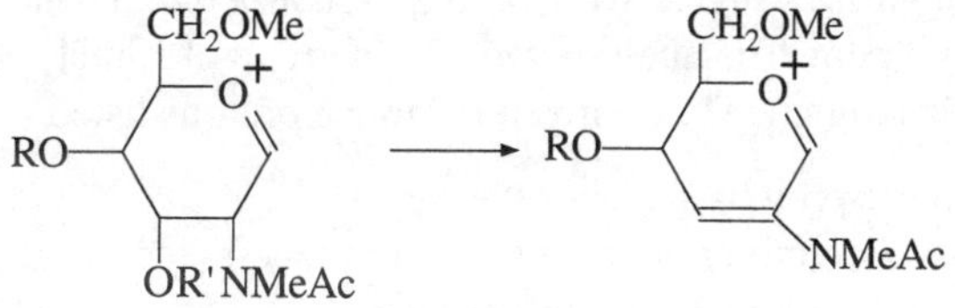

9.4.2 Electrospray

Certain recent models of mass spectrometer allow the use of a considerably easier injection technique. A solution of the sample to be analysed is injected directly into the apparatus by means of a syringe. Perfectly separated molecular peaks are observed. Here we will give the example of a sulfated pentasaccharide sodium salt, **9.11**. In Chapter 17 we will come back to this synthetic compound (Lubineau *et al.* 1994), which is the best known human-*E* selectin ligand today. Using the electrospray technique we observe the following peaks: $[M - Na^+]$ 932.2, $[M + Na^+]$ 978.3, and $[M + 2\,Na^+]$ 500.7.

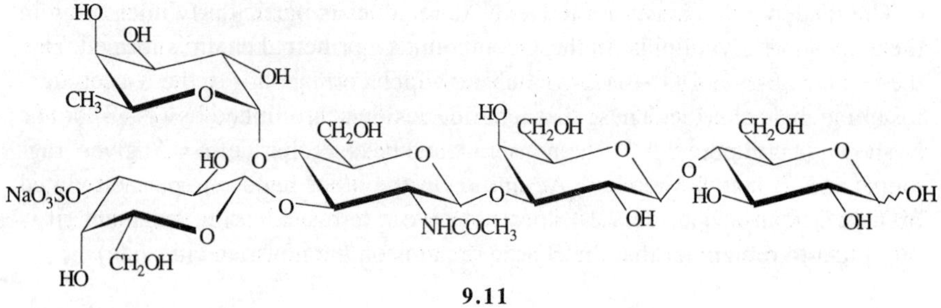

9.11

9.4.3 Proton nuclear magnetic resonance spectroscopy

Generally, in order to find out the structure of a naturally conjugated oligosaccharide, the chemist must operate on a mixture of closely related species. There is, indeed, the problem of the natural heterogeneity of these structures, even on a homogeneous protein support, and that of the artificial heterogeneity caused by the cleavage reagents as it is not conceivable to study the glycoconjugate in its entirety. Therefore, it is necessary to proceed, first, by fractionation, an especially difficult task with very similar structures. The analyses of proton NMR spectra such as those described in this section (Vliegenthart *et al.* 1983) give an indication of the sample's homogeneity since a mixture gives superimposed signals which can be attributed by considering the intensities.

Because the problem of structure is very complex, it is necessary to know the δ values to three decimal places. This implies a very high magnetic field (up to 14 telsa) for a frequency of 500 or 600 MHz. Working with computerized data, it is possible to increase the resolution or decrease the concentrations of a sample to 0.05 mM. Even at this frequency the non-anomeric skeletal proton signals of various sugars are superimposed as a broad band, unresolved between 3.4 and 4.0 ppm. The analysis relies on a group of signals, the *reporters*, located outside this region. They correspond to the protons listed below.

(1) anomeric protons;
Their ^3J coupling constants are significant (see Chapter 2). If H-2 is axial (Gal, GLcNAc, Fuc), we observe ^3J 2–4 Hz and 7–9 Hz for the α- and β-pyranoses, respectively. If H-2 is equatorial (Man), the difference is weaker, ^3J 1.6 and 0.8 Hz for the α- and β-anomers, respectively. However, the most interesting value is that of the δ.
(2) H-2 and H-3 protons of mannose;
(3) H-3 protons of sialic acids;
(4) H-5 proton and methyl of fucose;
(5) H-3 and H-4 protons of galactose;
(6) methyls of *N*-acetyl groups of amino sugars and sialic acid.

In the most favourable cases, the δ value depends on the type of residue, its anomeric configuration, the glycosidation site, the sequence of the sugar residues, and the position of the residue in the sequence. To this list was recently added (Hård *et al.* 1991) NH protons of the $-NHCOCH_3$ group. Naturally, they are not visible in D_2O nor in ordinary water at pH > 7. They are easily observed in H_2O at 27°C, pH 5.2. When an NH signal has been identified, the signals of certain protons close to the same monosaccharide residue can be detected by the nuclear Overhauser effect: the methyl group of CH_3CONH and the anomeric proton on a 2-acetamido-2-deoxy pyranose residue, the H-3 protons on a sialoside.

To make use of these spectroscopic data, we have recourse to a certain number of empirical rules, established by simple oligosaccharides whose structures have been well defined. Thus, the glycosidation of an oligosaccharide residue brings about small displacements of the recognition signals, sometimes also perceptible in neighbouring residues, in the order of 0.02–0.25 ppm. The use of these rules for determining unknown structures supposes additivity and that there are no important conformational changes which would greatly disturb the local diamagnetism. They are therefore the most certain in the series built from similar blocks. So that the reader may evaluate this, we will give an excerpt of this type of analysis involving the oligosaccharide chains of porcine thyroglobulin (de Waard *et al.* 1991). Thyroglobulin is the largest glycoprotein in the thyroid gland. The glucidic chain, linked to aspartic acid, is separated by hydrolysis catalysed by the N^4-peptide enzyme (*N*-acetyl-β-glucosaminyl)amidase asparagine F, whose bacterial origin is *Flavobacterium meningosepticum*, and the mixture of oligosaccharides is fractionated on a column. Among the products isolated, we will keep in mind the collection outlined in Fig. 9.7. The simplest unit is nonasaccharide **9.12** in which the framed residues and the sulfate are missing. The recognition signals of this nonasaccharide can be located, and the H-2 signal of the Man-4′ residue is drowned in the mass. There is also a decasaccharide in this collection with a supplementary GlcNAc residue. In this compound, the Man-4′ H-2 signal stands out from the mass and appears at δ 4.109 which indicates that the new residue GlcNAc is at position 2 of the Man-4′ mannose. The reporter signals confirm its end position, the one found in the 5′ frame. The addition of the D-Gal residue gives a undecasaccharide whose close analogue had already been found (Vliegenthart *et al.* 1983). The position of this residue is verified as being that found in the 6′ frame by displacement of +0.028 ppm of the GlcNAc-5′ H-1 signal. The following unit of this collection is a dodecasaccharide. It contains moreover the Gal residue whose α-anomer is proven by the signal of its anomeric proton, that of frame 7′, and its position by the effect of its presence on the Gal-6′ H-4 signal, which stands out from the majority to appear at $\delta = 4.185$ ppm. Characteristic signals of **9.13** are given in Table 9.2 so that the reader can assess, in concrete terms, the results of this type of analysis. There is also a collection of sulfated oligosaccharides found in this mixture. Sulfation displaces the geminal protons downfield. Here, the H-6 and H-6′ signals of GlcNAc-5 stand out from the majority (δ 4.306 and 4.440) which indicates the position in frame 8 for the sulfate group.

By these methods it could be shown that the enzymic galactosylation (Section 10.4.1) of tetrasaccharide **9.14** is selective for the GlcNAc terminal non-reducing unit bound at position 6 of galactose, for in the product pentasaccharide **9.15**, the anomeric proton of residue **E** is the only one displaced by 0.023 ppm (Augé *et al.* 1980).

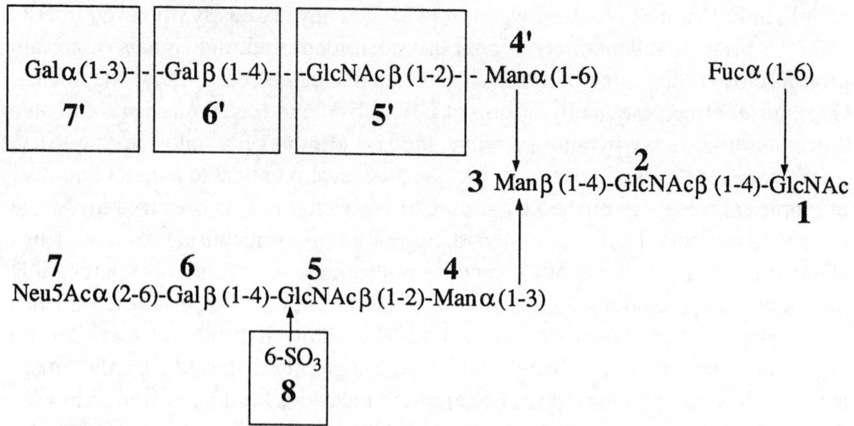

Fig. 9.7 Schematic representation of oligosaccharides obtained from pork thyroglobuline. The parts without framed units correspond to **9.12**. The other oligosaccharides separated are **9.12** + 5′, **9.12** + 5′ and 6′, **9.12** + 5′ + 6′ + 7′ (**9.13**), and the sulfated dodecasaccharide (**9.12** + 5′ + 6′ + 7′ + 8).

Table 9.2 Reporter signals of dodecasaccharide **9.13**.

Protons	Units	δ	Protons	Units	δ
H-1	1 α	5.180	H-2	4′	4.111
	1 β	4.692	H-3a	Neu 5Ac	1.722
	2 (α)	4.664	H-3e	Neu 5Ac	2.666
	2 (β)	4.669	H-4	6′	4.185
	3	4.772	H-5	Fuc (α)	4.098
	4	5.135		Fuc (β)	4.134
	4′	4.929	Me	Fuc (α)	1.209
	5	4.605		Fuc (β)	1.220
	5′	4.583	NAc	1	2.038
	6	4.445		2	2.097
	6′	4.544		5	2.069
	α-Gal	5.146		5′	2.048
	Fuc (α)	4.890		Neu 5Ac	2.029
	Fuc (β)	4.897			
H-2	3	4.257			
	4	4.195			

GlcNAcβ(1-6)
Galβ(1-4)-Glcβ0Me
GlcNAcβ(1-3)

9.14

F E
Galβ(1-4)-GlcNAcβ(1-6)
B A
Galβ(1-4)-Glcβ0Me
GlcNAcβ(1-3)
C

9.15

As another example of oligosaccharide structure elucidation, we now present an outline of the determination of the sequence of the repeating unit of the polysaccharide from the cell wall of the bacterium *Streptococcus gordonii* 38 (Reddy *et al.* 1994). The derivation of sequence **9.16** relied mainly on 2D NMR spectroscopy. A review of the theory and practice of 2D NMR would be outside the scope of this book. On the other hand, we feel it is very useful for the reader to have an idea of what can be done. This may be best achieved by looking first at the proton spectrum of polysaccharide **9.16**, reproduced in Fig. 9.8, which has been obtained by the so-called phase-sensitive double quantum filtered correlation spectroscopy (DQF-COSY). In this figure, the *x*, *y* coordinates in the plane are chemical shifts, while the *z* coordinate, positive or negative, would correspond to the intensities of the peaks, generally shown only by contour lines. The diagonal of the square gives the same pattern as the 1D experiment, while off-diagonal cross-peaks are the signals of coupled protons. The coordinates of these

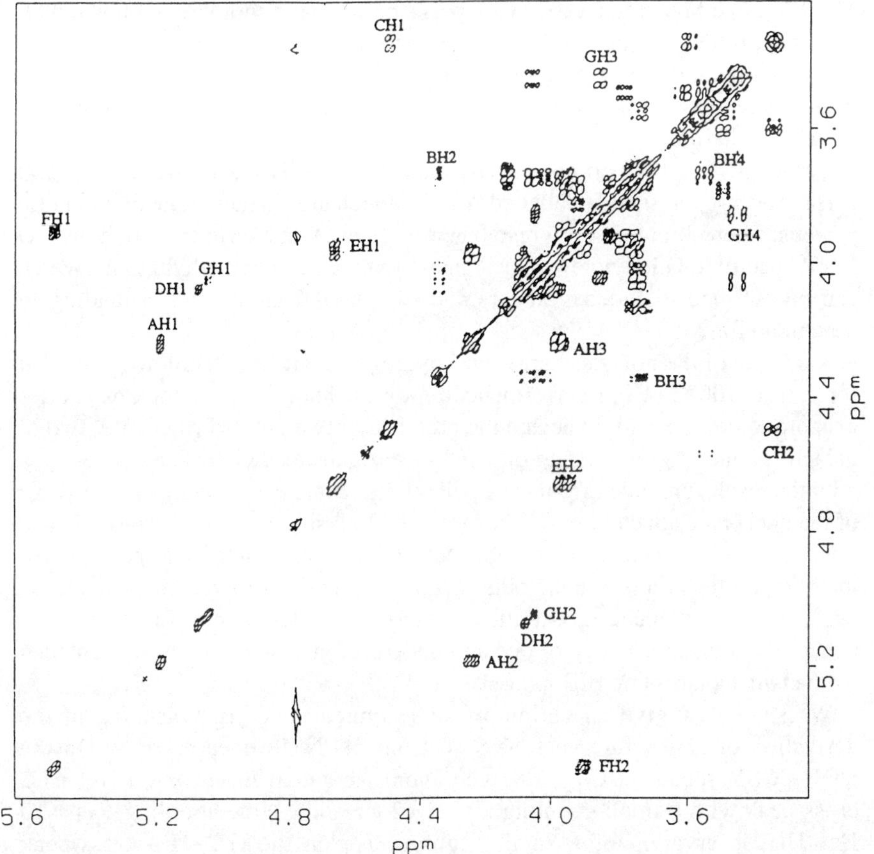

Fig. 9.8 Phase-sensitive DQF-COSY spectrum of the polysaccharide from *S. gordonii* at 500 MHz. (from Reddy *et al.* (1994) *Glycobiology*, **2**, 183–92; reproduced with kind permission of Oxford University Press)

$$\left[\rightarrow 6)\text{-}\alpha\text{-D-GalNAc}p\text{-}(1\rightarrow 3)\text{-}\beta\text{-L-Rha}p\text{-}(1\rightarrow 4)\text{-}\beta\text{-D-Glc}p\text{-}(1\rightarrow 6)\text{-}\beta\text{-D-Gal}f\text{-}(1\rightarrow 6)\text{-}\beta\text{-D-GalNAc}p\text{-}(1\rightarrow 3)\text{-}\alpha\text{-D-Gal}p\text{-}(1\rightarrow PO_4^-\text{-}\right]_n$$

with α-L-Rha$p$-(1→2)- attached at the β-L-Rha$p$ residue.

9.16

cross-peaks are the chemical shifts of the coupled protons, and the values of the coupling constants may be extracted from the signals. This is a homonuclear COSY experiment, but heteronuclear COSY is also possible in order to know which proton is connected to which carbon in an organic molecule. In this case, the *x*, *y* coordinates are the proton and carbon chemical shifts, respectively.

In sugars, couplings are most often very small between protons which are neither geminal nor vicinal. One way to make the corresponding cross-peaks appear is by magnetization transfer from one proton, obviously chosen because it is well resolved (homonuclear Hartman–Hahn spectroscopy, HOHAHA or TOCSY). In a favourable case, cross-peaks between the anomeric proton and all the other protons in a pyranose can be observed.

Finally, the nuclear Overhauser effect can be utilized to construct a 2D spectrum where cross-peaks correspond to protons close in space (phase-sensitive NOESY spectrum).

Let us now go back to polysaccharide **9.16**. The ^{1}H NMR spectrum at 500 MHz shows seven signals, labelled **A** to **F**, which are characteristic of anomeric protons. There are also seven resonances typical of anomeric carbon in the ^{13}C NMR spectrum. Other noteworthy signals are those of two methyl groups which can be ascribed to 6-deoxy sugars and two methyl groups corresponding to acetamido functions.

A sample of the polysaccharide was hydrolysed with 4 N trifluoroacetic acid for 2 h at 100°C. High-performance anion exchange chromatography, after evaporation of the acid, indicated the presence of two units of rhamnose, two of galactose, one of glucose, and one of *N*-acetylgalactosamine. The discrepancy with the results predicted from the NMR spectra came from incomplete cleavage of the phosphate attached to C-6 under the hydrolysis conditions employed. The missing residue was identified as galactosamine 6-phosphate by a special chromatography technique. On the other hand, mild acid hydrolysis induced cleavage of the polysaccharide at the anomeric phosphate ester function, with concomitant disappearance of one anomeric proton signal, which was replaced by the two signals of an α/β-mixture.

We shall now give an outline of some typical mode of reasoning in the derivation of a structure such as **9.16** from 2D NMR spectroscopy. On the DQF-COSY spectrum (Fig. 9.8), the anomeric signal **A** showed a H-1–H-2 cross-peak with a small coupling, $J_{1,2}$ = 3.2 Hz and a large coupling, $J_{2,3}$ = 10 Hz. The observation of a small coupling $J_{3,4}$ on the H-2–H-3 cross-peak completed the identification of the residue as α-*galacto*. Knowing the resonance of H-2, it was possible to find that of the geminal C-2 carbon of residue **A** from the ^{13}C-decoupled, ^{1}H-detected, multiple quantum correlation spectrum

[^{1}H(^{13}C)-HMQC]. The chemical shift of C-2, 50.91 ppm, identified residue **A** as the acetamido sugar, α-GalNA*cp*.

Identification of residue **B** as *rhamno p* is typical of the utilization of HOHAHA. Magnetization transfer from the anomeric proton **B** led to the appearance of only one cross-peak, the H-1–H-2 signal, but transfer from one of the well-resolved methyl resonances at 1.351 ppm disclosed the entire connectivity up to the same H-2 resonance. This showed that **B** was one of the rhamnose units, and gave the chemical shifts of all its protons. The NOESY showed the proximity of H-1 with H-2, H-3, and H-5, compatible only with the β-*rhamno* configuration.

The sequence of the repeating unit of polysaccharide **9.16** could also be derived from NOE spectroscopy. This indicated the proximity of the anomeric protons of units **A**, **B**, **C**, **D**, **E**, **F**, and **G**, with protons H-3 of **B**, H-4 of **C**, H-6 of **D**, H-6 of **E**, H-3 of **F**, and H-2 of **B**, respectively.

Finally, the anomeric resonance at 5.068 ppm showed a direct correlation to the ^{13}C resonance at 108.66 ppm in the HMQC spectrum, indicating that galactose residue **D** is the β-furanoside form (Beier *et al.* 1980).

Table 9.3 gives the NMR chemical shifts for all the protons (to three decimals) and carbon atoms of the repeating unit of polysaccharide **9.16**, in D_2O at 25°C.

Table 9.3 NMR chemical shifts of the polysaccharide from *S. gordonii* 38 in D_2O at 25°C.

Assignment		Residue						
		α-GalNAc **A**	α-Rha **G**	β-Rha **B**	β-Glc **C**	β-Gal$_f$ **D**	β-GalNAc **E**	α-Gal **F**
^{1}H[a]	H1	5.180	5.040	4.925	4.489	5.068	4.657	5.495
	H2	4.233	4.041	4.332	3.341	4.070	3.956	3.895
	H3	3.973	3.850	3.727	3.596	4.080	3.745	3.964
	H4	4.078	3.449	3.546	3.604	3.989	3.944	4.242
	H5	4.330	4.046	3.476	3.490	4.010	3.815	4.128
	H6	4.01	1.265	1.351	3.783	3.741	3.764	3.73
	H6′	4.07			3.947	4.053	3.913	3.73
	NAc	2.087	–	–	–	–	2.045	
^{13}C[b]	C1	95.81	100.92	101.51	103.53	108.66	104.01	96.58
	C2	50.91	71.29	73.05	74.02	81.78	53.37	68.02
	C3	68.00	71.10	79.63	76.83	77.53	71.63	79.68
	C4	68.80	72.75	71.75	78.07	83.93	68.78	70.05
	C5	70.44	69.47	73.42	75.60	70.51	74.58	72.38
	C6	65.18	17.49	17.70	61.80	72.08	68.08	61.89
NAc	CH_3	23.12	–	–	–	–	23.16	

[a] ^{1}H NMR chemical shifts are with reference to internal $Me_3SiCH_2CH_2CH_2SO_3Na$ (DSS) with acetone as the internal standard (2.225 ppm downfield from DSS).

[b] Carbon chemical shifts are with reference to internal acetone (31.07 ppm).

(from Reddy *et al.* (1994) *Glycobiology*, **2**, 183–92; reproduced with kind permission of Oxford University Press)

The capability of the [$^1H(^{13}C)$-HMQC] experiment in providing accurate assignments of strongly coupled proton resonances of carbohydrates is illustrated by the H-3 (3.596 ppm) and H-4 (3.604 ppm) signals of unit **C** which gave well-separated cross-peaks because of the chemical shift difference of the geminal carbon atoms, C-3 and C-4.

9.5 Potential efficiency of oligosaccharides to store and transmit information

Two deoxyribonucleotides, let us say *d*A and *d*C, can only give two distinct combinations, *d*A*d*C and *d*C*d*A. This is the same with two amino acids where there are two possible dipeptides, CysAl and AlCys. In both cases, two identical molecules can only give a single condensation product. On the other hand, the reader may verify that the association of two glucose molecules can give 11 distinct disaccharides if restricted to pyranoid tautomers. The number increases to awesome proportions if oligosaccharides derived from different monosaccharides are considered. Four different nucleotides can only give 26 distinct tetranucleotides, while four different monosaccharides can give 35 560 distinct tetrasaccharides (Sharon and Lis 1993). Nature has at its disposal four letters which allow 35 560 different words to be written! As long as these molecules can be recognized by specialized proteins, a very small number of basic elements is sufficient enough to store and transmit the most varied information, in the most compact form possible, to living cells. This will be developed at length in the rest of this work.

References

Augé, C., Mathieu, C., and Mérienne, C. (1980), *Carbohydr. Res.*, **151**, 147–156.

Beier, C. R., Mundy, B. P., and Strobel, G. A. (1980), *Can. J. Chem.*, **58**, 2800–2804.

Biermann, C. J. (1988), *Adv. Carbohydr. Chem. Biochem.*, **46**, 251–271.

Dais, P. and Perlin, A. S. (1987), *Adv. Carbohydr. Chem. Biochem.*, **45**, 125–168.

de Waard, P., Koorevaar, A., Kamerling, J. P., and Vliegenthart, J. F. G. (1991), *J. Biol. Chem.*, **266**, 4237–4243.

Dell, A. (1987), *Adv. Carbohydr. Chem. Biochem.*, **45**, 19–72.

Fukuda, M. N., Della, A., Oates, J. E., Wu, P., Klock, J. C., and Fukuda, M. (1985), *J. Biol. Chem.*, **260**, 967–982.

Hård, K., Spronk, B. A., Hokke, C. H., Kamerling, J. P., and Vliegenthart, J. F. G. (1991), *FEBS Lett.*, **287**, 108–112.

Lubineau, A., Le Gallic, J., and Lemoine, R. (1994), *Bioorg. Med. Chem.*, **2**, 1143–1151.

Mulloy, B., Frenkel, T. A., and Davies, D. B. (1988), *Carbohydr. Res.*, **184**, 39–46.

Rauvala, H., Finne, J., Krusius, T., Kärkkäinen, J., and Järnefelt, J. (1981), *Adv. Carbohydr. Chem. Biochem.*, **38**, 389–416.

Reddy, G. P., Abeygunawardana, C., Bush, C. A., and Cisar, J. O. (1994), *Glycobiology*, **4**, 183–192.

Sharon, N. and Lis, H. (1993), *Sci. Am.*, 74–81.

Tvarovša, I. and Bleha, T. (1989), *Adv. Carbohydr. Chem. Biochem.*, **47**, 45–123.

Tvarovša, I., Hricovíni, M., and Petráková, E. (1989), *Carbohydr. Res.*, **189**, 359–362.

Vliegenthart, J. F. G., Dorland, L., and van Halbeek, H. (1983), *Adv. Carbohydr. Chem. Biochem.*, **41**, 209–374.

10 Chemical transformations and synthesis of oligosaccharides

10.1 Oligosaccharide reactions

Reducing oligosaccharides have two different hydroxyl functions, namely the alcohol and the hemiacetal. The alcohol functions are divided into two fairly well differentiated categories, primary and secondary. More subtle differences exist between the secondary alcohol functions, depending on their being axial or equatorial, or combined in a vicinal diol, etc. With respect to monosaccharides, there is nothing fundamentally new here and what is known about the chemistry of oligosaccharides relies on corresponding selective reactions.

For example, it is possible to describe the preparation of a very important disaccharide, *N*-acetyllactosamine **10.1** starting from lactose, a cheap starting material. The oxime of the potential aldehyde function, that is to say R–CHOH–CH=N–OH, is simultaneously acetylated and dehydrated to a peracetylated nitrile R′–CHOAc–CN. Alkaline methanolysis deacetylates and causes elimination of HCN to give R–CHO, the *O*-*β*-D-galactopyranosyl-(1→3)-D-arabinose disaccharide. The latter is readily isolated in the form of the *N*-benzylglycosamine derivative **10.2**. Addition of HCN gives nitrile **10.3** which is converted to *N*-acetyllactosamine by catalytic hydrogenation over palladium and *N*-acetylation with acetic anhydride in methanol (Kuhn and Kirschenlohr 1956). In a more recent adaptation (Alais and Veyrières 1981), the use of HCN is avoided by progressively adding acetic acid to cyanide dissolved in the reaction medium. Nitrile **10.3** is recovered by filtration, fortunately obtained in the crystalline state, thus opening the way to large-scale preparation.

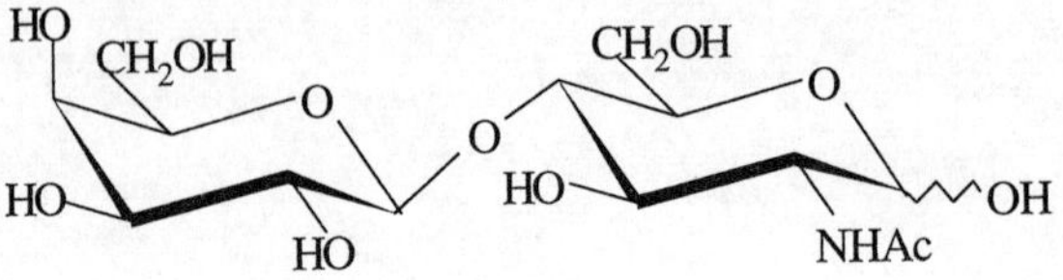

HO CH$_2$OH O HO HO O CH$_2$OH O HO NHAc OH

10.1

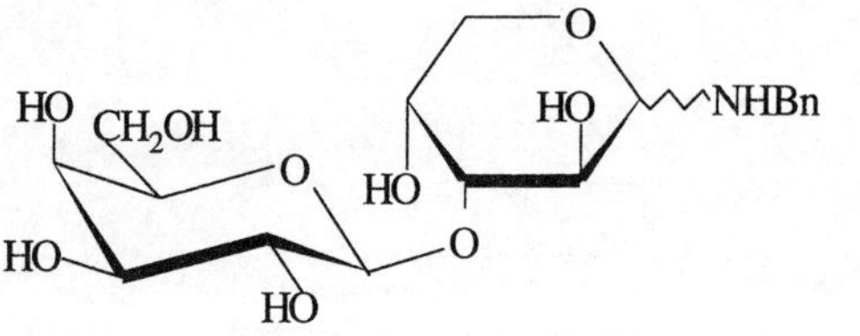

HO CH$_2$OH O HO HO O O HO HO NHBn

10.2

OH CN HO HO NHBn O

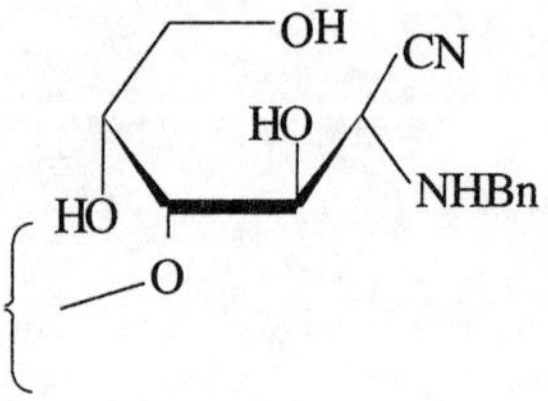

10.3

The aldehyde function can be selectively oxidized from lactose to a carboxyl group to give lactobionic acid. It can also be reduced to alcohol. Further on we will see that alkaline conditions are utilized to detach certain oligosaccharides from their protein support. These conditions cause degradation of the oligosaccharide, particularly if there is substitution at position 3, via the well-known elimination of β-alkoxy aldehydes from the carbonyl tautomer as outlined in formula **10.4**. The answers consist then in carrying out the cleavage with a mixture of hydroxide and sodium borohydride, whereby the hemiacetal is reduced to an alcohol and there is no longer elimination.

The primary alcohol functions are protected by traditional reagents: triphenylchloromethane, *t*-butyldimethylsilyl chloride, pivaloyl chloride. An axial–equatorial *cis*-diol system is selectively protected, if there is only one, by acetalation with acetone. In this way the hydroxyl groups are protected at positions 3 and 5 of the galactose unit of methyl lactoside **10.5** by preparing the isopropylidene derivative **10.6**. While on the subject, it is worth noting a reaction in which each lactose unit behaves as if it were alone; treatment of glucose with acetone and 2, 2-dimethoxypropane, by acid catalysis, gives, among others, **10.9** (Stevens 1972). Under the same conditions, the lactose gives compound **10.10** in which six out of eight hydroxyl groups are blocked (Hough *et al.* 1979). With $(Bu_2SnO)n$, lactoside **10.5** gives a stannylene, probably **10.7**, which, upon treatment with allyl bromide, leads to the monoallylic ether **10.8**. This is the only substitution product isolated in 70% yield, in spite of there being seven free hydroxyl groups in the starting lactoside (Alais *et al.* 1983). Vicinal diols in oligosaccharides are cleaved by periodate which allows the selective destruction of the monosaccharide units which possess them. Their disappearance from the total hydrolysate is then observed.

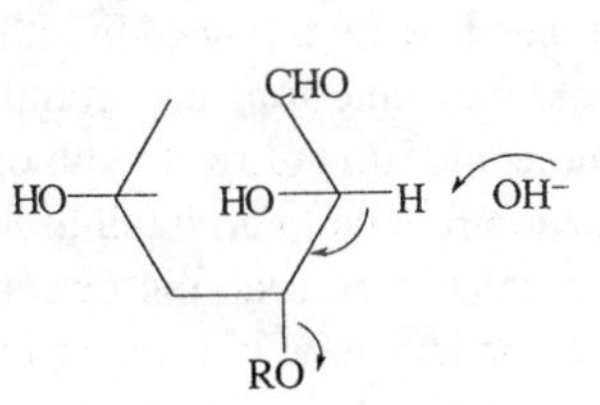

CHO, HO, HO, H, OH⁻, RO

10.4

R'O, CH_2OH, O, RO, HO, O, CH_2OH, O, HO, OMe, OH

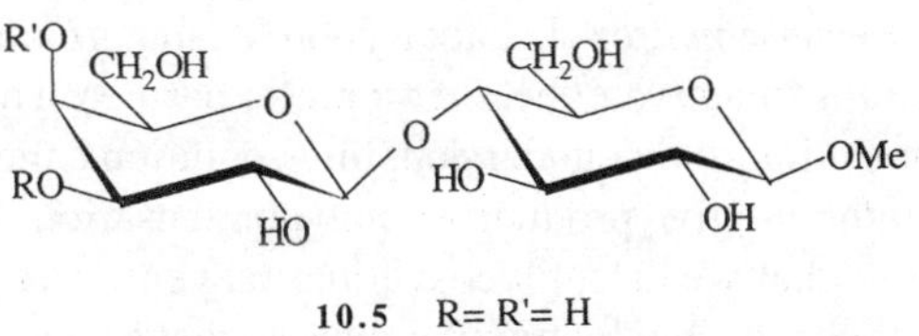

10.5	R= R'= H
10.6	R, R'=CMe_2
10.7	R, R'= $SnBu_2$
10.8	R= Allyl, R'= H

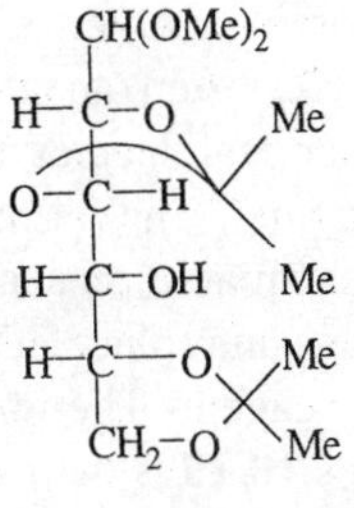

$CH(OMe)_2$, H–C–O, Me, O–C–H, H–C–OH, Me, H–C–O, Me, CH_2–O, Me

10.9

Me, O, CH_2OH, O, Me, O, HO, $CH(OMe)_2$, H–C–O, Me, O–C–H, H–C–O, Me, H–C–O, Me, CH_2–O, Me

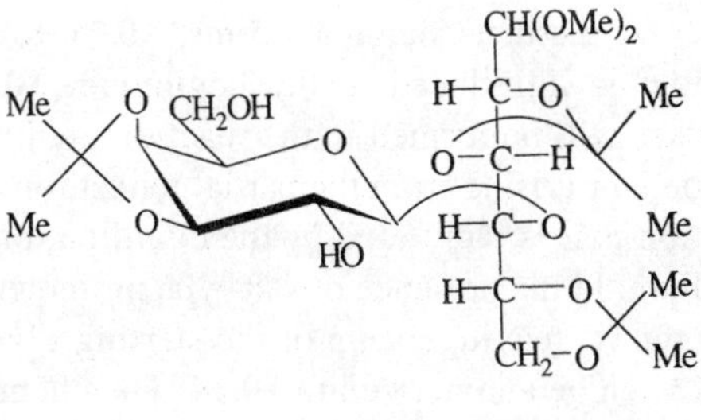

10.10

Below (Section 10.3.4) we will see the use of configurational inversion at C-2 of a monosaccharide unit in order to have access to the β-D-*manno* configuration. One of the most important transformations of oligosaccharides is the activation of the reducing end to achieve convergent syntheses of higher oligosaccharides, examples of which will be given in the following sections.

As previously mentioned, these reactions are predictable extensions of reactions with monosaccharides. But perhaps the chemistry of oligosaccharides could be a more specific field. How is the reactivity of each hydroxyl group modified by the remainder of the complex sequence, by its configuration, and its conformation? Are there functional groups which have lost all reactivity or else, on the contrary, unexpected preferred sites such as the active sites of proteins? This area has not yet been explored with really complicated oligosaccharides.

10.2 Non-enzymic coupling reactions: general principles

10.2.1 Reactions paths of the glycosyl donor

The coupling reaction is the most fundamental operation in the synthesis of oligosaccharides. It requires building an oxygen bridge with the hemiacetal carbon atom of one sugar and the alcohol function of another (Paulsen 1982; Schmidt 1986; Garegg and Lindberg 1988; Boons 1996). Considering the very high stability of the C–O bond in alcohols, compared to the lability of the C–O bond in a hemiacetal, in the great majority of cases the reaction involves nucleophilic substitution of the alcohol oxygen on the hemiacetal carbon.

Because hydroxyl is not a good leaving group, it needs to be replaced by other groups to activate the electrophilic partner. The alcohol functions also require protection so that under coupling conditions, the molecules do not react with one another to give products of polycondensation. These protected activated molecules that we call 'glycosylating reagents' are not capable of reacting directly with the alcohol hydroxyl groups so that Lewis acids or salts having more or less marked Lewis acid character must be added in proportions varying from catalytic to high molar excess, depending on the techniques used. These are known as the 'promoters'.

Under these conditions, we can predict that the glycosylating reagent will have six distinct chemical forms, **10.11** to **10.16**. Moreover, even if a pure reagent is introduced at the beginning, **10.11** or **10.12**, often an anomeric mixture will be formed in the reaction medium. The chemical species **10.13** is an ion pair arising from the partial ionization of the α-anomer. The formation of this ion pair is facilitated by the coordination of X^- with an acid present in the medium. In the presence of salt-type promoters, the anion can replace in this ion pair the X^- leaving group of the starting glycosylating reagent. The ionic pair **10.13** can be anomerized to **10.14**. The chemical species **10.15** is the free oxocarbenium ion. The formation of a carbenium ion at C-1 is favoured by the participation of the ring oxygen. However, if there is a participating acyloxy group at C-2, the intermediate has the **10.16** structure. The two problems with glycosidation, which incidentally are not completely independent, are that the coupling yield needs to be acceptable and that the glycosidic bond corresponds to the desired anomer. Reactions with participation involving intermediate **10.16** must be immediately separated. They are often fast and give exclusively the 1, 2-*trans* anomer as predicted in excellent yield. When there is no participation, we can expect that each intermediate, from **10.11** to **10.14**, will react by the S_N2 mechanism with inversion of configuration. We thus have the equatorial bond with **10.11** and **10.13**, and the axial bond with the other two. Unfortunately these intermediates are generally present simultaneously, whatever the departing glycosylating reagent. Little selectively is expected from a reaction with the oxocarbenium ion **10.15**, but there is evidence that is is associated with molecules of solvents in diethyl oxide or acetonitrile solution. Such complexation may favour one face. The general conclusion is that the nucleophilic partner in the medium has the choice between a certain number of reacting paths whose activation energies may not be very different.

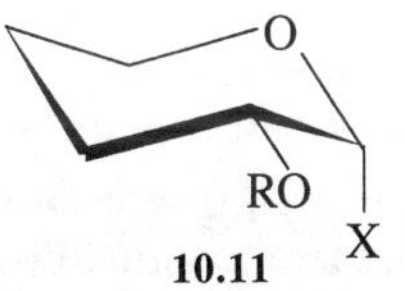

10.11 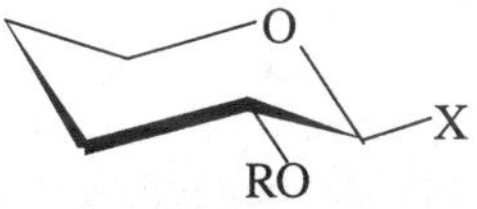 **10.12** 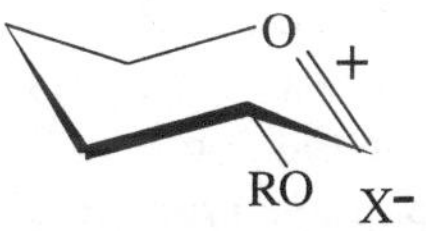 **10.13**

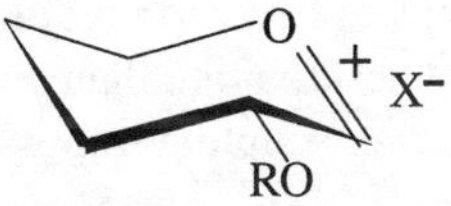

10.14 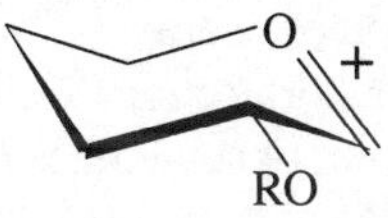 **10.15** 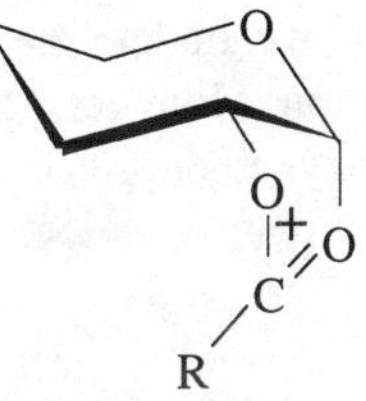 **10.16**

10.2.2 Effect of the acceptor configuration

The acceptor is confronted by a mixture of potentially glycosylating entities at equilibrium, six in the worst case, from **10.11** to **10.16**. Choosing one of them, which will alter the subsequent reaction pathway and the final outcome, depends on two factors: the rate at which it disappears by coupling and its reappearance rate resulting from the equilibrium. The reactivity of the hydroxyl group vis-à-vis one of the glycosylating entities depends on its position in the acceptor molecule and on the steric and electronic effects of the protecting groups. The steric effect (explained by the repulsion of filled orbitals in the theory of molecular orbitals) sometimes seems obvious; condensation of the peracetylated bromide **10.17** with the protected *N*-acetylglucosamine **10.18** gives the β-disaccharide

(1,2 *trans*) in more than 78% yield (Augé and Veyrières 1976). This reaction, carried out in the presence of $Hg(CN_2)$, is a typical example of glycosylation with participation. The same donor with phthalimidoglucosamine **10.19** gives an α/β mixture in low yield which can be interpreted as the effect of the bulky phthalimido group. However, the reaction, in principle S_N2, with trichloracetimidate gives 70% of the pure β-disaccharide.

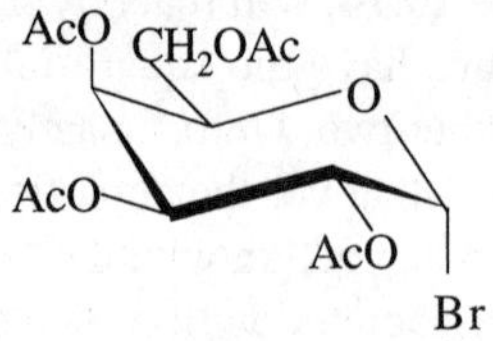

10.17

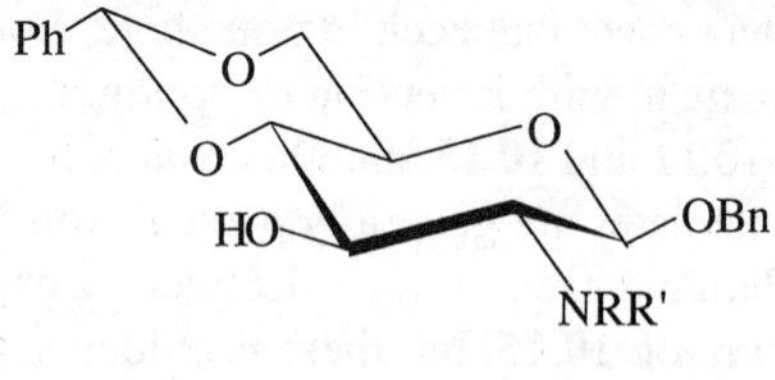

10.18 R= H, R'= Ac
10.19 R,R'= Phthalyl

It is necessary to bring in electronic factors (modifications of the shape and energy level of the HOMO of the free pair of the oxygen atom) to explain the sometimes considerable effect of modifications far from the coupling site, not only in the donor but in the acceptor as well. For example (Paulsen 1982), the coupling of bromide **10.20** with the benzyl rhamnoside **10.21** gives a ratio of 19:81 of an α/β mixture, typical of a moderately selective reaction without participation. But the ratio is inverted (81:19) with trichloroethyl rhamnoside **10.22**. It is clear that the HOMO orbital of rhamnoside **10.22**, essentially the 2p free pair of O-4, has undergone a modification of energy and perhaps even a certain delocalization. More generally, the acetate protections are more deactivating than are the benzyl ether protections, to the point of sometimes rendering coupling impossible.

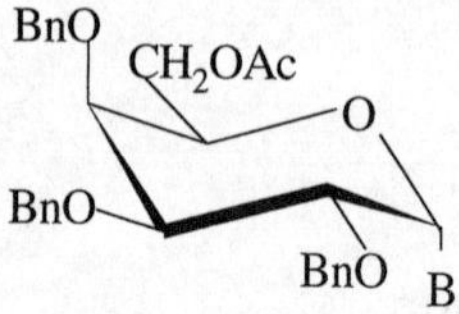

10.20

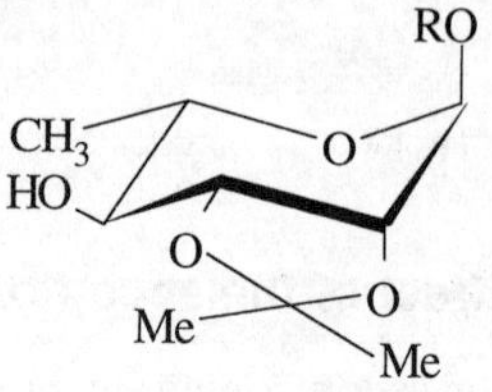

10.21 R= Bn
10.22 R= CH_2CCl_3

An attempt to separate the electronic and steric effects relies on the 'stereodifferentiation' notion (Spijker and van Boeckel 1991). To simplify this discussion, we will speak about the steric effect as an interaction between solids. Let us assume that two chiral entities RXYZ and R′X′Y′Z′ approach each other to create a transition state as in Fig. 10.1. We imagine an ideal case whereby mutual adaptation is perfect: each bump in RXYZ corresponds to a hole in

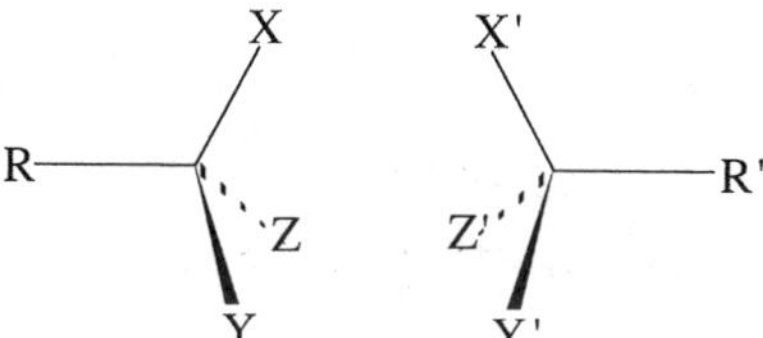

Fig. 10.1 Steric interaction of two chiral solids.

R′X′Y′Z′ and vice versa, so well that the two entities can approach each other enough to establish a bond. Let us switch X and Y: these branches are different so that the adapation cannot be as good. This switching around gives the enantiomer and does not change the energy of the frontier orbital. The lowering of reactivity is originally steric. An observation of this kind allows the steric and electronic contributions in the reactivity to be separated. This reasoning is not entirely rigorous. It is only valid if the frontier orbital is not chiral. Whatever may be, the practical results are suggestive. The reaction of the perbenzoylated D-fucopyranosyl bromide with the acceptor **10.24** (Fig. 10.2) in the presence of silver triflate and 2,6-di-*t*-butylpyridine gives a mixture of disaccharides (87%). We would expect a reaction with participation giving only the β-anomer; in fact, we have a 2:1 α/β mixture. Figure 10.2 shows that the transition state with the participating intermediate **10.23** is quite unfavourable. But the enantiomer bromide of L-fucopyranosyl gives an intermediate, **10.25**, in which the β-approach is more favourable, to give a 78% yield and a reversed α/β ratio of 1:8.4.

Thus, glycosidic coupling, in spite of all the considerable improvements of the last 20 years, is still a tricky reaction, the outcome of which cannot be safely predicted.

10.3 Carrying out coupling reactions

When carrying out coupling reactions, reagents and solvents should be carefully dried. Traces of water may compete with the acceptor for the donor molecule, and this may be disastrous because of the much lower molecular weight and much higher mobility of the water molecule.

Although most of the techniques described in Sections 10.3.1 to 10.3.6 may be utilized, with proper adjustment to the coupling of aminated sugars, we prefer to deal with all the problems of this family together in one separate Section, 10.3.7.

10.3.1 Reactions with participation

Reactions with participation are the least uncertain. They lead to the 1,2-*trans* anomer according to reaction (10.1). The *gluco* and *galacto* configurations give 1,2-diequatorial glycosides and the *manno* configuration gives the 1,2-diaxial glycosides. The simplest of the glycosylating reagents are the equatorial anomers

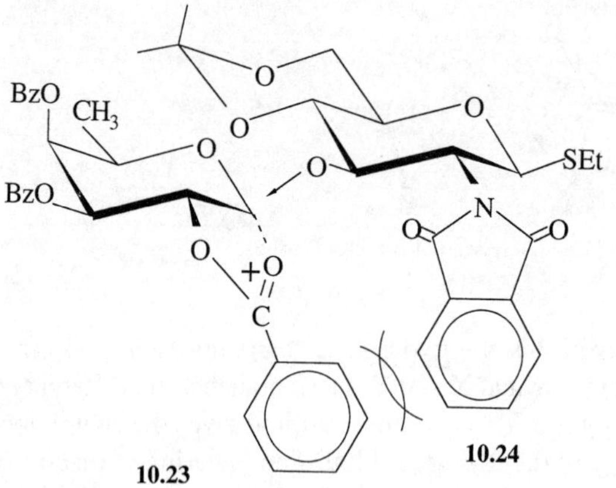

Fig. 10.2 The source of diastereoselectivity in a coupling reaction.

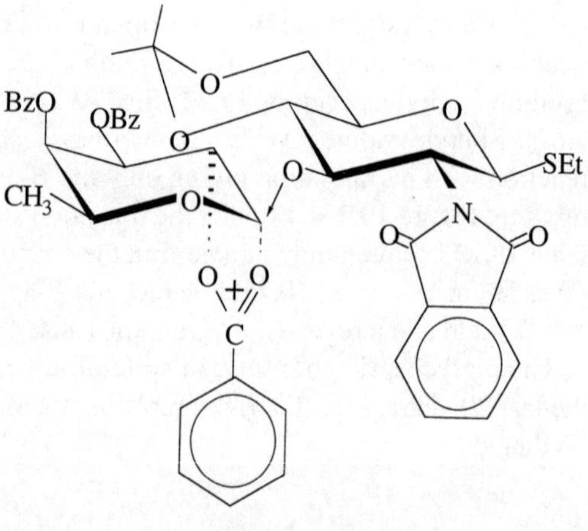

of pyranose peracetates such as *β*-D-*gluco* derivative **10.26**. The latter, in the presence of trimethylsilyl trifluoromethanesulfonate (triflate) $CF_3SO_3SiMe_3$ as promoter, leads to *β*-D-*gluco* pyranosides. This reaction has its limits, however, due to the need to prepare this unstable anomer and the high cost of the promoter. Treatment of the peracetylated pyranoses with HCl or HBr gives the stable glycosyl halides, **10.27** or **10.28**, with axial halogen. The 'universal promoter' for the coupling of these glycosyl halides seems to be silver triflate (Hanessian and Banoub 1977). It is very efficient and sometimes the coupling reaction is completed in a few minutes, even at –70°C. Dichloromethane is used with tetramethylurea as proton acceptor. Preparing large quantities at the beginning of a synthetic sequence requires using classic promoters ($HgBr_2$ or $HgCl_2$ in toluene–acetonitrile or nitromethane mixtures) or tin triflate in dichloromethane (Lubineau and Malleron 1985).

(10.1)

Glycosylating reagent + G-OH →

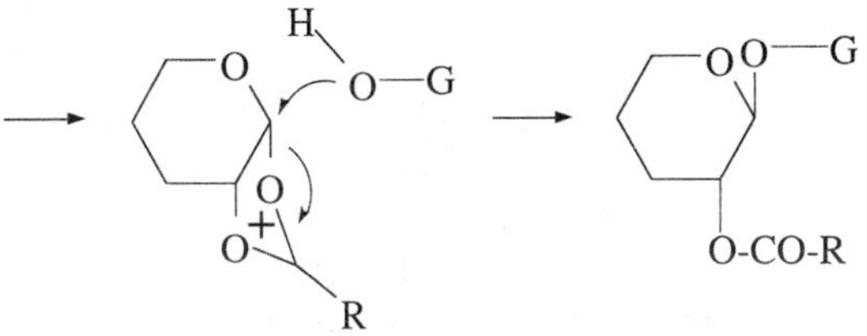

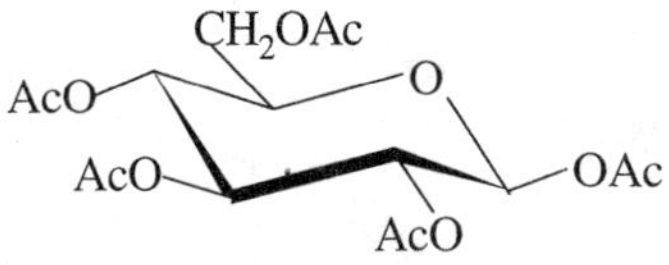

10.26

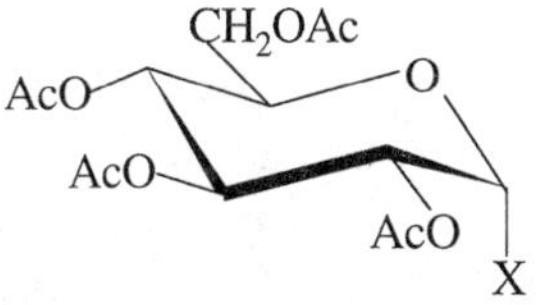

10.27 X= Cl
10.28 X= Br

10.3.2 S_N2 reactions

In the presence of silver salts, glycosyl halides attack the amides on the oxygen atom to give imidates such as **10.29**. Imidates without participating groups react with alcohols with inversion of configuration leading to 1,2-*cis* compounds (Pougny *et al.* 1978). Trichloroacetimidates (Schmidt and Kinsy 1994) are easily accessible by addition of alcohols to tricholoroacetonitrile according to equation (10.2) in the presence of strong bases. With hemiacetal pyranoses the equatorial imidate is formed first, under kinetic control, and anomerizes thereafter to the axial imidate. This behaviour is readily explained. Deprotonation displaces the bonding electrons of the proton in the direction of the ring, a fact which increases the bulk of the oxygen atom and favours equatorial conformation. However, anomerization of the 1-oxide ion substituent is possible. Starting from the *β*-imidate, retroreaction, anomerization, and renewed trichloroacetonitrile addition slowly leads to the thermodynamically stable axial imidate. Thus the addition of 2,3,4,6-tetra-*O*-acetyl-D-glucopyranose to CCl_3CN in the presence of K_2CO_3 in CH_2Cl_2 solution gives 78% of the *β*-imidate **10.30** in 2 h at room temperature, but after 48 h under the same conditions, the *α*-imidate is obtained in quantitative yield.

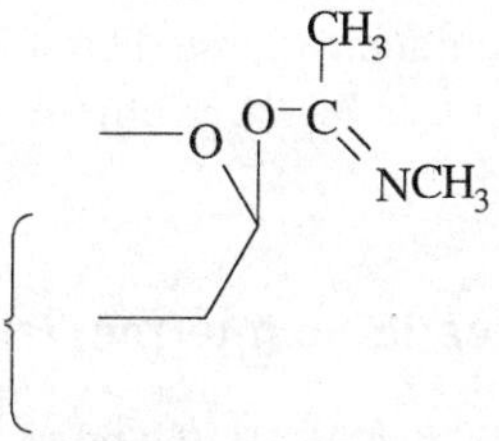

10.29

$$CCl_3{-}CN + R{-}OH \rightarrow R{-}O{-}C(CCl_3){=}NH \tag{10.2}$$

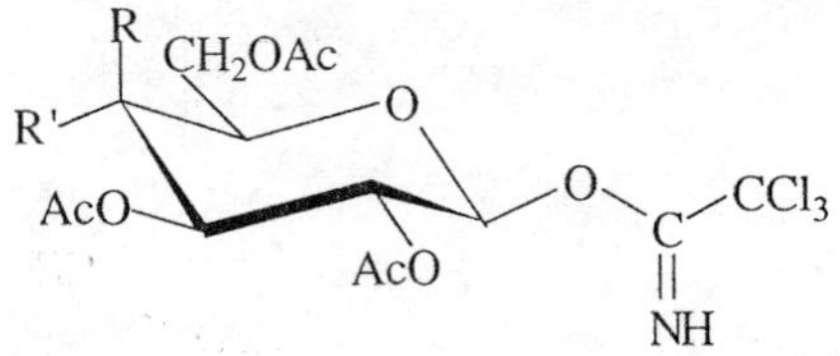

10.30 R = H, R' = OAc

10.31 R = OAc, R' = H

α-Trichloroacetimidates are stable derivatives which act as glycosyl donors in the presence of boron trifluoride etherate in CH_2Cl_2 (reaction 10.3). β-Glycosides are obtained in excellent yields, at temperatures as low as –70°C, with or without participation. Noteworthy in this methodology is the simplicity of the base and acid catalysts.

$$\text{Glycosyl–O–C(=NH)CCl}_3 + \text{G–OH} \xrightarrow{BF_3.Et_2O} \text{Glycosyl–O–G} + CCl_3CONH_2 \tag{10.3}$$

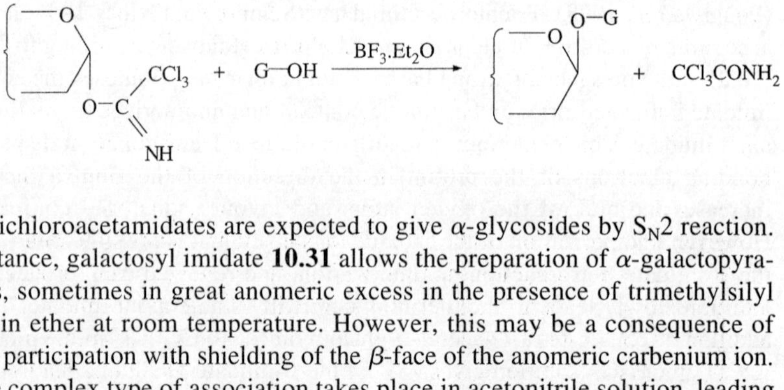

β-Trichloroacetamidates are expected to give α-glycosides by S_N2 reaction. For instance, galactosyl imidate **10.31** allows the preparation of α-galactopyranosides, sometimes in great anomeric excess in the presence of trimethylsilyl triflate in ether at room temperature. However, this may be a consequence of solvent participation with shielding of the β-face of the anomeric carbenium ion. A more complex type of association takes place in acetonitrile solution, leading to variable results. Problems occur with protected fucoyl imidates because of their high reactivity that leads to decomposition in the reaction medium. This difficulty was overcome by an inverse procedure, the glycosyl donor being added to a mixture of acceptor and catalyst. This allowed the preparation in high yields of versatile building blocks for the syntheses of Le^a, Le^x, Le^y, and H blood group epitopes (Section 16.3.1).

10.3.3 Reactions with cationic intermediates

The 'halide-assisted' reaction is applied to a halide without group participation at position 2. The intermediate **10.16** is therefore excluded. Work is carried out in a non-polar solvent to minimize the appearance of the ionic intermediate **10.15**. The remaining chemical forms **10.11** to **10.14** are at equilibrium, and this equilibrium is hastened by addition of a mineral salt to furnish a common ion X^-. The most reactive intermediate in glycosidation is the equatorial ionic pair **10.14** which leads, in this case, to the 1,2-*cis* glycoside axial anomer. If **10.14** appears at a rate clearly higher than that of the coupling, the entire glycosidation will be diverted to this route, whatever the starting anomer. Sometimes the hydroxyl partner is not reactive enough for condensation to proceed in an acceptable fashion (there is no promoter in the medium!). In the latter case, silver triflate is the most efficient promoter, with selectivity being as high as the hydroxyl group's reactivity is low.

These methods were essentially developed for the introduction of an α-D-galactopyranosyl derivative which corresponds to a 1,2-*cis* glycosidation. To effect this goal, an especially brillant method was recently published (Kahne *et al.* 1989), but this will be discussed later (see Section 10.3.5) to avoid anticipating its mechanism.

Besides the pyranosyl chlorides and bromides, the use of fluorides is becoming more popular in glycosidations without participation. The substituent at position 2 is protected by a benzyl group. The axial–equatorial 1,2-*cis* fluoride gives the axial–equatorial 1,2-*cis* glycoside in solution in diethyl ether in the presence of silver perchlorate and tin(II) triflate. Fluorides, just as bromides, can be prepared from thioglycosides. Using thioglycoside is a means of temporarily protecting the hemiacetal function, for the C-1–S bond can remain intact through a certain number of transformations in the rest of the molecule. This is also true for methyl glycosides; however, thioglycosides are transformed into glycosylating reagents in a manner which is much more practical than methyl glycosides, with fewer risks to the remainder of the molecule. Equation (10.4) shows how the same phenyl thioglycoside can be converted to a glycosylating or glycosylable partner by treatment with either $Et_3N–SF_3$ or Bu_4NF.

$CH_2OSiCMe_3Ph_2$ AcO BnO BnO O F ← $CH_2OSiCMe_3Ph_2$ AcO BnO BnO O SPh → CH_2OH AcO BnO BnO O SPh

(10.4)

More generally, thioglycosides are activated by the so-called 'thiophilic' reagents. Among these we find chlorine and bromine; displacement by the halide counterion yields the pyranosyl halide. Iodonium dicollidine perchlorate is a milder reagent. When the counter anion is a poor nucleophile, in the presence of an alcohol, there is direct formation of an *O*-glycoside, a reaction which is observed in the presence of *N*-bromosuccinimide. Another efficient activator is dimethyl(methylthio)sulfonium triflate, $Me_2S(MeS)^+OTf^-$. The reactivity of thioglycosides can be controlled by the nature of the protecting groups and the size of the anomeric leaving moiety (Boons 1996).

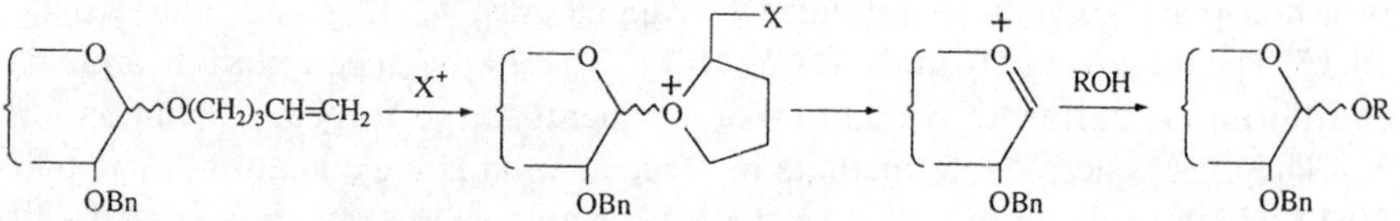

Fig. 10.3 Mechanism of the activation of pentenyl glycosides for glycosidic coupling.

Pentenyl glycosides with a benzyl protection at C-2 are activated by iodonium dicollidine perchlorate and converted to glycosylating agents by the mechanism depicted in Fig. 10.3 (Mootoo *et al.* 1988). There is no activation with an acetyl protection at C-2. Thus, a pentenyl 2-*O*-acetylglycopyanoside with a free hydroxyl group behaves as an acceptor. But it may be converted to a glycosyl donor in the next step after de-*O*-acetylation and *O*-benzylation.

10.3.4 Creating the equatorial–axial 1,2 bond

This is the configuration of β-mannosides. It is possible to imagine that they are obtained by S_N2 substitution of the α-*manno* halides, axial and easily accessible with non-participating protection at O-2. But these derivatives are not very reactive, perhaps because of the steric repulsion of the axial group at position 2. However, β-mannosides could be prepared directly (Garegg *et al.* 1983) with an insoluble promoter, the silver derivative of a natural zeolite cation exchanger, for example, according to equation (10.5). Another technique (David *et al.* 1989), which offers a high degree of security, consists in preparing the β-D-*gluco* glycoside, generally accessible in high yield by coupling with participation and inversion of configuration at C-2. In this method, after the glycosidation reaction with β-acetate **10.32** in the presence of trimethylsilyl triflate, generally effected in high yield, the di- (or tri-)saccharide **10.33** undergoes alkaline methanolysis followed by benzylidenation. Thus **10.34** and **10.35** are obtained successively. A particularly efficient leaving group is necessary to achieve substitution at C-2. The conversion of **10.35** to an imidazoylsulfonate (Hanessian and Vatèle 1981) which is displaced by a benzoate gives **10.36** in very high yield.

(10.5)

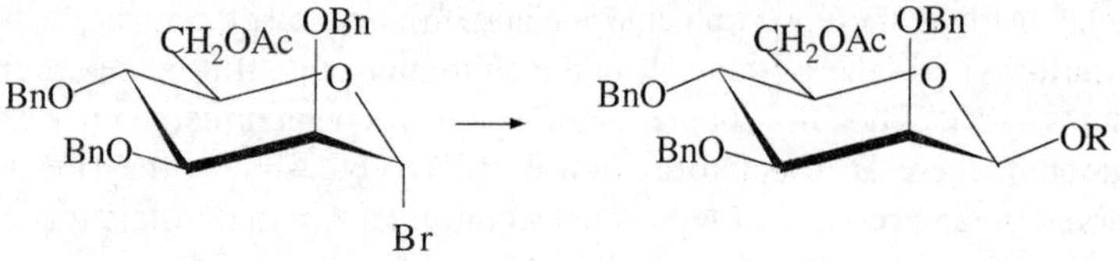

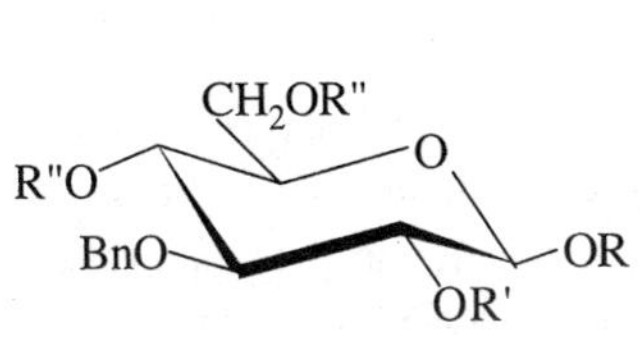

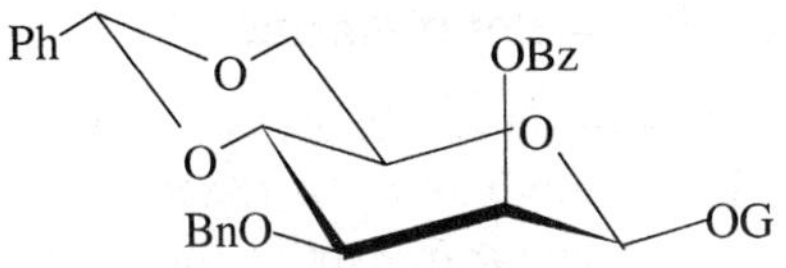

10.32 R = R'= R"= Ac
10.33 R = G R'= R"= Ac
10.34 R = G R'= R"= H
10.35 R = G R'= H R", R"= PhCH

10.36

10.3.5 1,2-*Cis* glycosidation without participation using sulfoxides

This recent method, first described by Kahne *et al.* (1989), was tested by Sarkar and Matta (1992). For example, sulfoxide **10.37**, readily obtained by oxidation of the perbenzylated phenylthio galactoside by means of peracid, can be used as a glycosylating reagent. Coupling takes place at –76°C in dichloromethane in the presence of trifluoromethanesulfonic anhydride and a hindered base. Only the α-glycoside is formed in a nearly quantitative yield. Thus coupling with the methyl β-D-galactopyranoside protected by benzylation, except at position 3, gives disaccharide **10.38**, characteristic of the blood group B (see Chapter 16) isolated in 90% yield. There are still very few examples of this method.

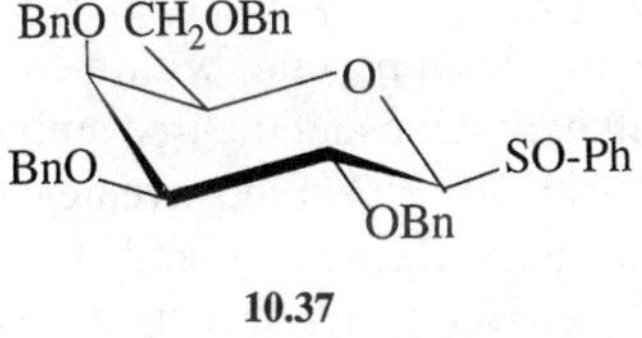

10.37

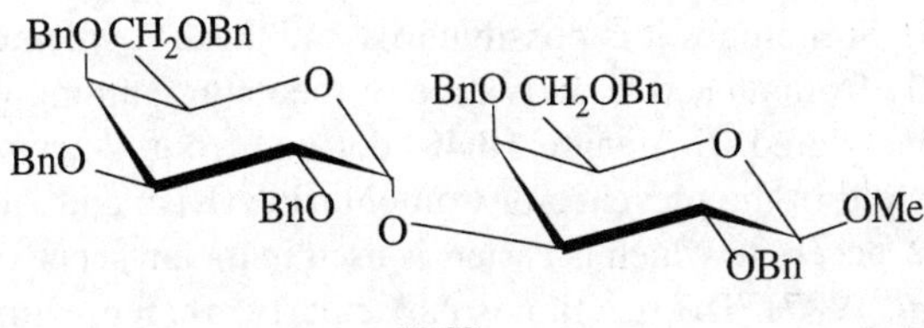

10.38

10.3.6 Other methods

1,2-Anydro-hexopyranoses are readily obtained from glycals by oxidation with dimethyldioxirane (Section 3.4). Mixing with primary alcohols gives straightaway the *trans*-glycosides. However, secondary alcohols do also react, in the presence of zinc chloride. The reaction may be conducted at –78°C in ether–oxolane solution. In reaction (10.6), the disaccharide glycal product, obtained in 81% yield, may in turn be activated and coupled in two steps (Danishevsky *et al.* 1995).

O CH_2OR O CO O O + HO CH_2OR O BnO $ZnCl_2$ -78° O CH_2OR CO O O O OH BnO CH_2OR O

(10.6)

Mannose is converted to methyl β-D-mannoside with methyl sulfate in an alkaline medium. This coupling is the opposite of all those we have considered thus far; the bridge oxygen comes from the glycosyl donor. This type of reaction has been used in the preparation of disaccharides, with the use of sugar triflate as electrophile (Schmidt 1986).

10.3.7 Coupling of amino sugars

To make glycosides of amino sugars, the chloride derivative of phthalimido glucosamine **10.39** is used preferentially. This β-anomer is much more reactive than the α-anomer in glycosidation reactions and is readily obtained (Nilsson *et al.* 1990) by treatment of a mixture of anomeric acetates (phthalimido analogues of **10.40** and **10.41**) with dichloromethyl methyl ether in the presence of BF_3 etherate. Choride **10.39** is employed with silver triflate as promoter or, in the case where a maximum yield is not so important, with mercury salts. The phthalimido group is hydrolysed with hydrazine and the free amine is acetylated to obtain the true natural structure. This is somewhat inconvenient. Peracetylated methyl thioglycosides of GlcNAc and GalNAc could be converted to *N*,*N*-diacetyl derivatives which were activated with dimethyl(methylthio)sulfonium triflate. After coupling, mono-*N*-deacetylation could be performed with methanolic sodium methanolate in quantitative yield (Castro-Palomino and Schmidt 1995). Sometimes it is possible to avoid these supplementary steps by starting directly from an active derivative of the natural amine which is, as we all know, *N*-acetylated. β-Acetate **10.40**, derived from *N*-acetylglycosamine, gives a good yield in the presence of iron chloride (Kiso and Anderson 1985). Chloride **10.42**, access to which is easier, is used in the presence of tin(II) triflate (Lubineau *et al.* 1987). The reaction is particularly efficient with primary alcohols, as in the preparation of the β-D-GlcNAc*p*-(1$\rightarrow$6)-D-Gal sequence, frequently found in natural oligosaccharides.

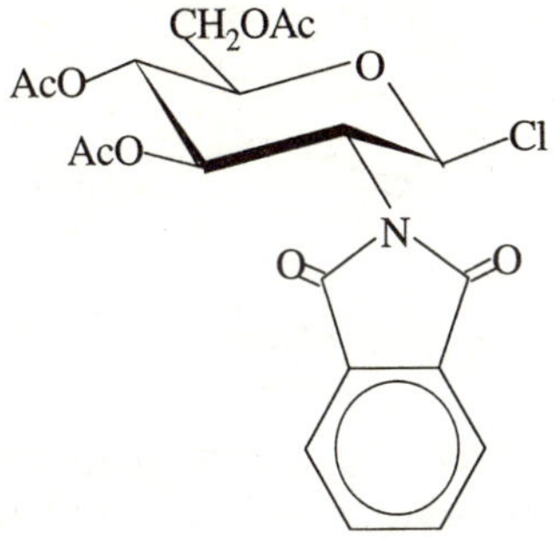

CH2OAc
AcO
O
AcO
Cl
N
O
O

10.39

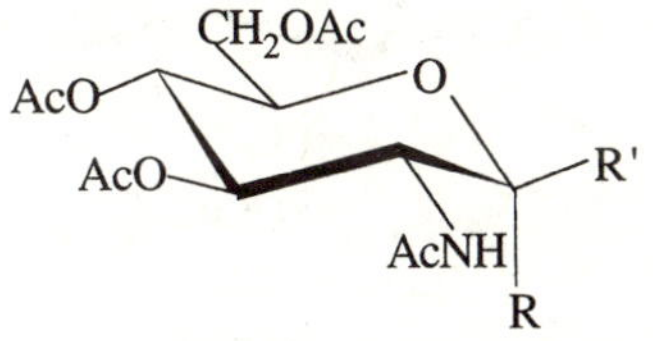

CH2OAc
AcO
O
AcO
R'
AcNH
R

10.40 R = H, R'= OAc
10.41 R = OAc, R'= H
10.42 R = Cl, R'= H

In non-amino glycosylating reagents, non-participating protection of O-2 is provided by a benzyl group. In the amino glycosylating reagents, the acetamido group is replaced by an azido group which is reduced to an amine and *N*-acetylated after glycosidation. The usual preparation of azides at position 2 begins with glycals. For example, reaction (10.7) describes the preparation of a glycosyl halide, **10.45**, used to introduce a 2-acetamido-2-deoxy-α-D-galactopyranosyl unit. Galactal **10.43** is treated with a mixture of cerium ammonium nitrate $(NH_4)_2Ce(NO_3)_6$ and sodium azide. Addition to the double bond gives **10.44** and the intermediate nitrate is treated successively with lithium iodide and ammonium chloride (Lemieux and Ratcliffe 1979).

(10.7)

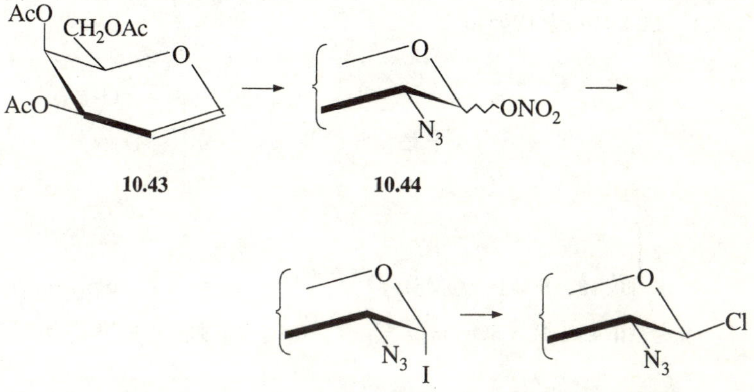

AcO CH2OAc
O
AcO
→
O
ONO2
N3
→

10.43 **10.44**

O
N3 I
→
O
Cl
N3

10.45

An alternative to azidonitration has recently been reported (Czernecki and Ayadi 1995). Glycal **10.43** reacted with (diacetoxydiodo)benzene $PhI(OAc)_2$, sodium azide, and diphenyldiselenide PhSeSePh, in dichloromethane solution at room temperature gives the cyrstalline azido phenylselenide **10.46** as the only product in 92% yield. Hydrolysis with *N*-iodosuccinimide then gives hemiketal **10.47** (87%). A few modifications allow the preparation of analogues with a benzyl protection. These hemiketals may be activated by conversion to imidates.

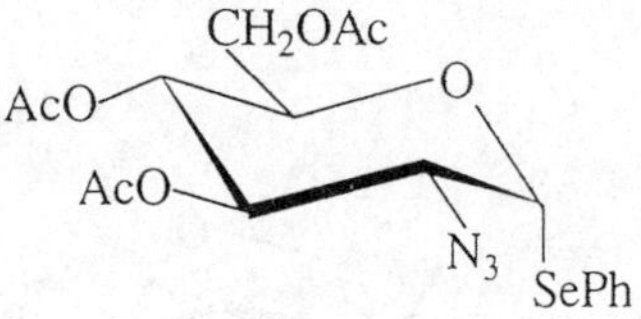

10.46 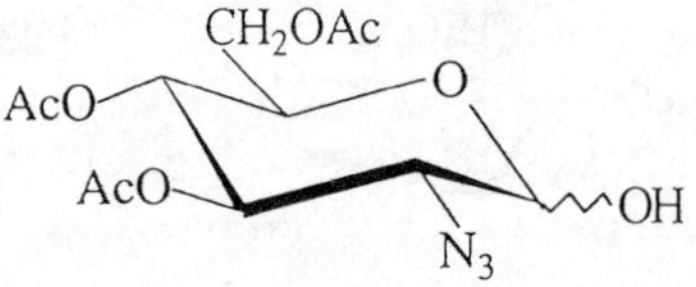 10.47

Reactions with the β-imidates of 2-azido-2-deoxy-D-galactropyanose derivatives gave high yields of 1,2-*cis*-α-D-glycosides which could afford ultimately 2-acetamido-2-deoxy-α-D-galactopyranosides (Schmidt and Kinzy 1994).

A radically new approach to 1,2-*trans*-2-acetamido-2-deoxy glycosides has been reported. The opening by an oxygen nucleophile of an aziridine bridge spanning C-1 and C-2 in a pyranose ring should give 1,2-*trans*-2-amino-2-deoxy glycosides. On treatment with sodium methanolate, potential aziridine precursors such as **10.48**, prepared from benzylated glycals, give the *trans*-glycosides **10.50** (R′ = Me) (Lafont and Descotes 1988). The amino glycoside is then obtained by hydrazinolysis. Treatment of protected glycals with benzenesulfonamide and iodonium dicollidine perchlorate gave compounds **10.49**. Again, the corresponding aziridines were not isolated but the expected product of opening **10.51** was obtained when silver tetrafluoroborate was added to a mixture of **10.49** and a tributylstannyl ether, R′–OSnBu$_3$, in oxolane solution at –78°C. De-*N*-sulfonylation was achieved with sodium and ammonia. In this way, the oligosaccharides of the Le^y and Le^b family were prepared (Section 16.1.2) (Danishefsky *et al.* 1995).

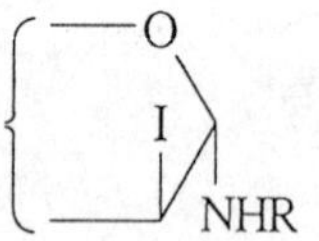

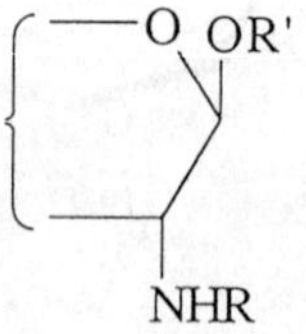

10.48 R = PO(OMe)$_2$ 10.50 R = PO(OMe)$_2$

10.49 R = SO$_2$Ph 10.51 R = SO$_2$Ph

10.3.8 Thiooligosaccharides

Thiooligosaccharides are analogues of oligosaccharides in which the interglycosidic oxygen is replaced by sulfur. The interest in these artificial products stems from their particular behaviour in enzymic chemistry such as the resistance to enzymic hydrolysis and inhibition or induction of glycosidases. They are prepared by nucleophilic substitution of an activated ester (triflate) of the acceptor by sulfur of an activated 'thioglycose'. Success is due to the highly nucleophilic character of the sulfur; the synthesis of real oligosaccharides in a similar fashion has been envisioned (Schmidt 1986) but has not yet been generalized. Naturally, there is inversion of configuration on the acceptor which must be consequently selected. Thus, the sodium salt of 2,3,4-tri-*O*-acetyl-1-thio-β-D-xylopyranose **10.52** leads to '4-thioxylobiose' **10.53**. This reaction (10.8) is complete in a few hours at room temperature in the presence of a sodium complexation agent in 92% yield after isolation (Defaye *et al.* 1992). From here we go on to the free thiodisaccharide by alkaline methanolysis.

(10.8)

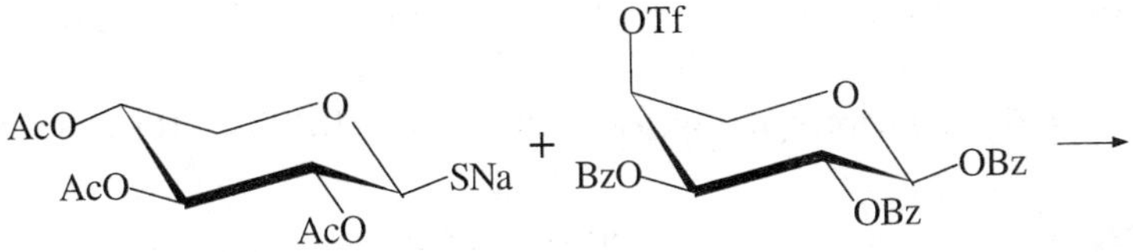

AcO, AcO, AcO, O, SNa + OTf, O, BzO, OBz, OBz →

10.52

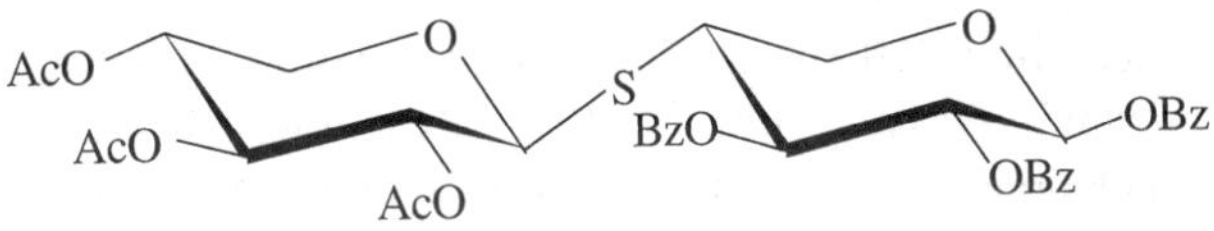

AcO, AcO, AcO, O, S, BzO, O, OBz, OBz

10.53

10.4 Enzymic methods (David *et al.* 1991; Ichikawa *et al.* 1992; Gijsen *et al.* 1996)

10.4.1 Reaction of galactosyltransferase

The uninformed organic chemist risks thinking that enzymes are not only products whose isolation requires immense training, but are very fragile in pure state and quite costly, serving essentially to demonstrate metabolic pathways on a micro or nanomolar scale. These viewpoints have now been changed. Techniques of immobilization allow them to be used several times—and under the best conditions—and cloning opens the way to massive production. Organic chemists who do not follow attentively the development of this new class of reagents might see the results of their efforts reduced to zero by brillant synthetic shortcuts. At any rate, it is very important to note that, contrary to what can be imagined, when enzymic reactions are usable, they allow the preparation of oligosaccharides at a considerably higher scale than do purely chemical coupling methods.

We will first take a look at one particular example, the synthesis of *N*-acetyllactosamine by coupling with galactose to *N*-acetylglucosamine. The activated form of galactose is a 'nucleotide sugar', uridine and galactose pyrophosphate, UDPGal (**10.54**). Coupling (10.9) is catalysed by the galactosyl transferase enzyme present in cow colostrum.

(10.9) UDPGal + GlcNAc → Gal-β-(1→4)-GlcNAc + UDP

In this form, this reaction has no preparative value because it consumes 1 eq of UDPGal, a very expensive product. It is thus necessary to combine it with the regeneration of UDPGal, which is done in three steps.

(10.10) $$\mathrm{UDP} + \mathrm{CH_2{=}C(OPO_3H_2){-}CO_2H} \rightarrow \mathrm{UTP} + \mathrm{CH_3\,COCOOH}$$

(10.11) UTP + α-D-glucopyranosyl phosphate ⇌ UDPGlc + H_2PO_3–O–PO_3H_2

(10.12) UDPGlc → UDPGal

(10.13) H_2O_3P–O–PO_3H_2 + H_2O → 2 H_3PO_4

Reaction (10.10) is the phosphorylation of phosphate, catalysed by the enzyme pyruvate kinase (PK). Reaction (10.11) is the synthesis of the 'nucleotide glucose' UDPGlc, uridine diphosphate glucose (**10.55**) from triphosphate, catalysed by the UDP-pyrophosphorylase enzyme (UP). Reaction (10.12) is the epimerization of the nucleotide glucose to the nucleotide galactose. It is noteworthy that nature prefers to manufacture UDPGlc and epimerize it rather than directly manufacture UDPGal. The enzyme is an epimerase (E). As shown, reaction (10.11) is reversible; the equilibrium is displaced to the right by a fifth enzyme, the inorganic pyrophosphatase which eliminates pyrophosphate from the medium by catalysing its hydrolysis to phosphate. The sum of reactions (10.9) to (10.13) gives the results of the operation in reaction (10.14).

(10.14) α-D-glucopyranosyl phosphate + *N*-acetylglucosamine + enolpyruvate phosphate → *N*-acetyllactosamine + pyruvate + 2 H_3PO_4

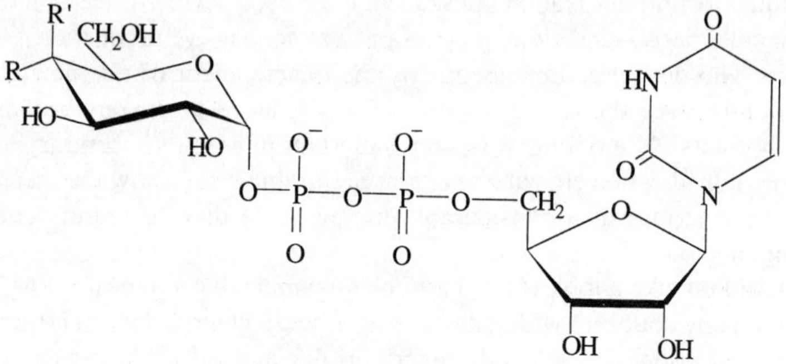

R' CH_2OH R O HO HO O^- O^- O–P–O–P–O–CH_2 O O O HN O N O OH OH

10.54 R = H, R'= OH

10.55 R = OH, R'= H

The essential energy source is the enolpyruvate phosphate, a compound easily accessible in great quantities by chemical synthesis. Likewise, the source of the galactopyranosyl residue, the α-D-glucopyranosyl phosphate **10.56**, is also accessible without problems. Nucleotides only play a catalytic role. In fact all the enzymes involved are active at pH 8; thus substrate and enzymes can be mixed in the same vessel and a cycle, to which the only nucleotide added is a catalytic amount of UDP Glc (2%, mol/mol), can be achieved. This cycle will manufacture *N*-acetyllactosamine following equation (10.14) until the substrates are all used (Fig. 10.4) (Wong *et al.* 1982).

CH_2OH, HO, HO, HO, O, OPO_3H_2

10.56

None of the enzymes in this system are difficult to obtain. However, it is advantageous to immobilize them on an insoluble support. For this purpose, agarose, a natural polysaccharide from D-galactose and 3,6-anhydro-L-galactose, is particularly well suited to this work on a laboratory scale. The support is first activated with cyanogen bromide. Figure 10.5 shows the activation mechanism of a *cis*-diol system and its coupling with the enzyme symbolized by E–NH_2 via

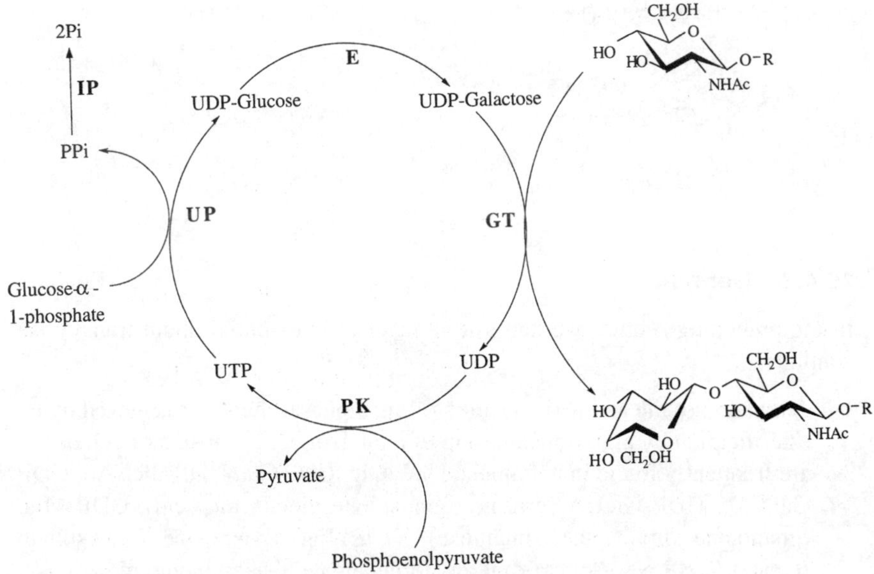

Fig. 10.4 Enzymic galactosylation cycle.

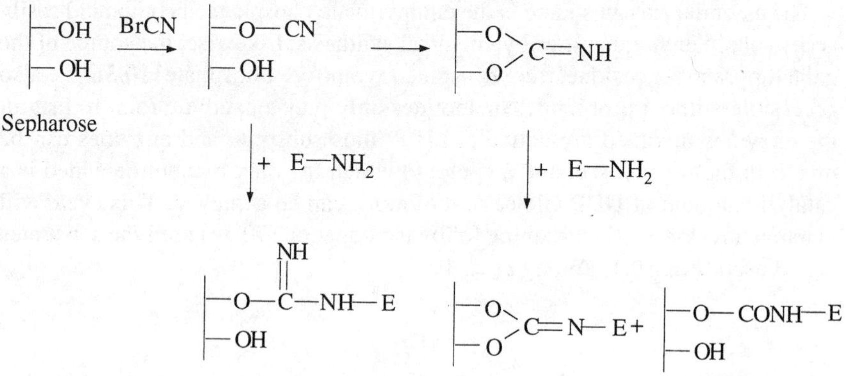

Fig. 10.5 Proposed mechanism for immobilization of an enzyme E-NH_2 on agarose using cyanogen bromide.

the intermediate of its amine functions. The couplings probably involve the formation of isoureas or imidocarbonates. From now on the enzymic activity is bound to an insoluble gel. The five gels are mixed in a reactor in which they must be kept in suspension by stirring and once finished, the mixture of enzymic gels is recovered by filtration, usable for a new galactosylation step.

This system has allowed the galactosylation of a great number of derivatives of *N*-acetylglucosamine. When the substrate is the branched tetrasaccharide **10.57** which has two galactosylable *N*-acetylglucosaminyl residues, coupling takes place exclusively with one of the branches to give **10.58**.

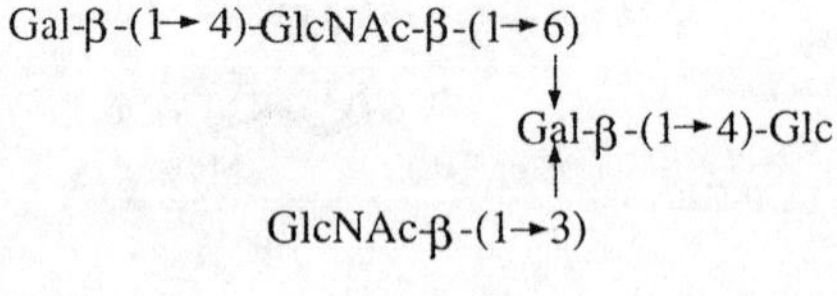

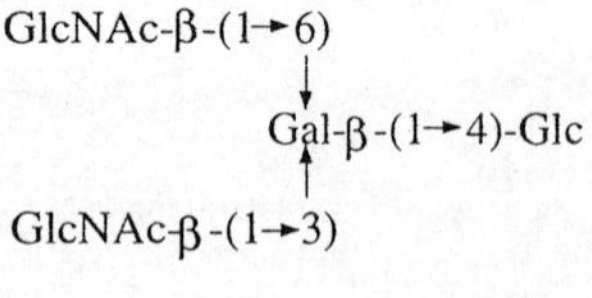

10.4.2 General

In the preceding section we had a look at what is essential about transferase couplings.

1. Activation of the anomeric position of the monosaccharide is achieved by its esterification by a phosphate group of a nucleotide. Eight sugar nucleotides are frequently found in mammals: UDP Glc, UDP Gal, UDP GlcNAc, UDP GalNAc, UDP GlcUA (uridine diphosphate glucuronic acid), GDP Man (guanosine diphosphate mannose), GDP Fuc (guanosine diphosphate fucose), CMP NeuAc (cytidine monophosphate *N*-acetylneuraminic acid). Sugar nucleotides are very expensive and optimization of a regeneration cycle is necessary.

2. The reaction is catalysed by a glycosyl transferase, which is doubly specific. It functions with a particular sugar nucleotide and transfers it to a specific position of a particular sugar.
3. For this reason, *enzymic coupling is a coupling reaction without protection.* The operations are so simple that it is possible to work at a much higher scale than with methods without enzymes.
4. In the case where coupling is executed on a non-reducing end unit of an oligosaccharide chain, the nature of this chain, even at a distance, can modify the efficiency of the enzyme. We have seen that the glycosidation rate is nil on one of the branches of tetrasaccharide **10.57**. Conversely, it is five times higher with chitobiose, GlcNAc-β-(1→4)-GlcNAc, than with GlcNAc.

According to a recent review (Gijsen *et al.* 1996), 31 transferases have been cloned and can therefore be obtained from culture cells. The cost of these enzymes will undoubtedly remain rather high for a while, taking into account the time it takes to set up the cloning procedure. In fact, efficient syntheses can be achieved with partially purified enzymic preparations extracted from mammal organs (liver, kidney, brain) purchased at the butcher's shop.

Fucosyl transferases

Fucosyl transferases use GDP Fuc **10.59**, which may be prepared on a gram scale from L-fucose (Adelhorst and Whitesides 1993; Veeneman *et al.* 1991). Hermann *et al.* (1993) proposed fucosylation with regeneration in a similar manner to one of the biosynthetic routes of fucose according to the reactions (10.15), whose similarity to reactions (10.9) to (10.12) will not escape the reader. The α-D-mannopyranosyl phosphate can be prepared rapidly by a non-enzymic route.

(10.15)

$$\begin{gathered}\text{GDP Fuc} + \text{R} - \text{OH} \rightarrow \text{fucoside} + \text{GDP}\\ \text{GDP} + \text{phosphoenolpyruvate} \rightarrow \text{GTP} + \text{pyruvate}\\ \text{Mannosyl phosphate} + \text{GTP} \rightleftharpoons \text{GDP Man} + P_2O_7H_6\\ \text{GDP Man} + \text{NADPH} \rightarrow \text{GDP Fuc} + \text{NADP}\end{gathered}$$

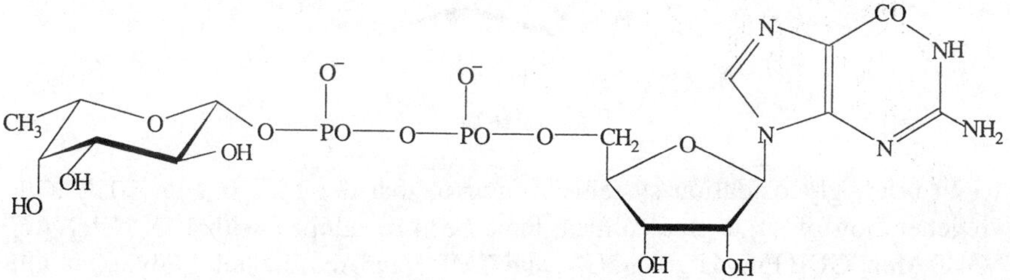

10.59

As in galactosylation, it is necessary to plan the hydrolysis of the pyrophosphate. One novel fact is that the mannose → fucose conversion implies a reduction which takes place with the universal cellular reducing agent, NADPH, nicotinamide adenine diphosphate, in the reduced form. Thus an extra cycle must be added for its regeneration, which leads to a total number of six mixed enzymes.

Two fucosyl transferases are involved in the biosynthesis of blood group substances (see Chapter 16). An α-1,2-fucosyltransferase introduces an α-L-fucopyranosyl unit to position 2 of the non-reducing β-D-galactopyranosyl unit end. Another, the α-1,3/4-fucosyltransferase, has been cloned. It introduces an α-L-fucopyranosyl unit at either position 4 or 3 of the galactose in the non-reducing end units, Gal-β-(1→3)-GlcNAc and Gal-β-(1→4)-GlcNAc, respectively. The enzyme α-1,3-fucosyltransferase introduces a L-fucose unit at position 3 of GlcNAc in lactosamine and derivatives. An α-1,4-fucosyltransferase is also known.

N-Acetylglucosaminyl transferases

These enzymes catalyse the addition of β-GlcNAc residues from the donor UDP GlcNAc to the terminal non-reducing mannose units of the core pentasaccharide of glycosaminide proteins (Section 13.2.3).

α-1,2-Mannosyl transferases

These enzymes catalyse the transfer of a mannose unit from GDP Man to position 2 of various α-mannosides to produce the Man-α-(1→2)-α-Man structural unit.

Glucuronyl transferases

β-D-Glycosides of glucuronic acid **10.60** are manufactured by higher organisms during the process called 'detoxification' to facilitate the elimination of a foreign aglycon. The sugar nucleotide is still UDP GlcUA. Enzymic glycosylation liberates UDP as a by-product. The latter is transformed to UDP Glc as in the galactosylation cycle, but at this stage, instead of epimerization to UDP Gal, enzymic oxidation of the primary alcohol function to a carboxyl group is produced with NADP as oxidant. The latter is reduced and must be regenerated.

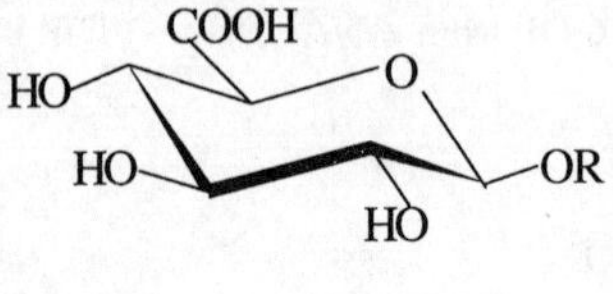

COOH
HO O
HO OR
HO

10.60

In brief, glycosidation systems similar to that depicted in Fig. 10.3, with regeneration of sugar nucleotides, have been developed with UDP GlcNAc, GDP Man, GDP Fuc, UDP GlcUA, and CMP Neu5Ac. Enzymic sialylation will be described in detail in Section 12.4.

Non-specific reactions of galactosyl transferase

The bovine milk enzyme is still active in the transfer of hexoses not too different from D-galactose to appropriate substrates. The donor is then the corresponding uridine diphosphate hexose nucleotide (for a review see Nishida *et al.* 1993).

Conversely, it is also active in the transfer of D-galactose from UDP Gal to modified substrates. Among D-glucose derivatives, the best substrate was 3-acetamido-3-deoxy-D-glucose, with a relative initial rate at 20 mM about 3% of the glucose reaction. Surprisingly, the product was the non-reducing disaccharide **10.61** (26%). This is rationalized by a comparison of the two substrate structures, **10.62** and **10.63** drawn with the conventions of alicyclic chemistry. This shows that the stereochemistry of the β-anomer along C-1 to C-4 is superimposable over that of β-D-GlcNAc along C-4 to C-1 (Nishida *et al.* 1993).

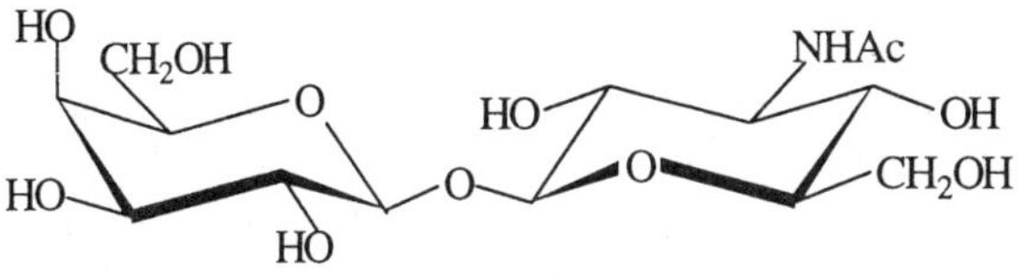

HO CH$_2$OH O HO HO O HO O NHAc OH CH$_2$OH

10.61

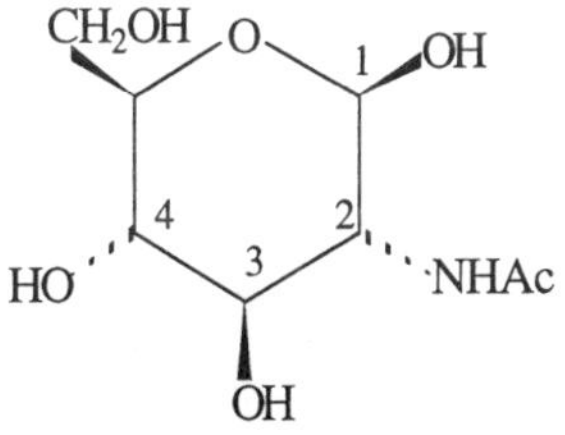

CH$_2$OH O 1 OH 4 2 HO 3 NHAc OH

10.62

CH$_2$OH O OH 4 1 3 HO 2 NHAc OH

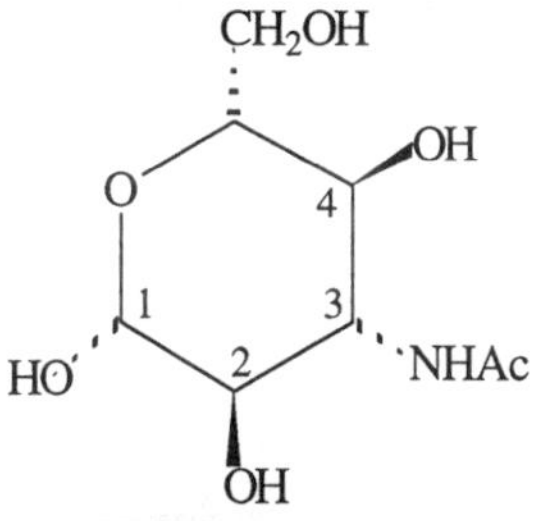

10.63

10.4.3 Glycosidases

The use of glycosidases in the preparation of alkyl glycosides was described in Section 3.5.2. The enzyme capable of hydrolysing a glycoside can also transfer the glycosyl unit to a hydroxyl group of another sugar to give a disaccharide having the same anomeric configuration. That this is so is already considerably advantageous given the generally low price of glycosidases and the absence of protection. Thus in the presence of α-galactosidase, reaction (10.16) takes place.

(10.16)
p-nitrophenyl α-D-galactopyranoside +
methyl α-D-galactopyranoside $\rightarrow$
Gal-α-(1$\rightarrow$3)-Gal-α-OMe + $NO_2C_6H_4OH$

The yield is 28% with respect to the glycosyl donor which is remarkable because this important disaccharide is not easily accessible by other routes. Nonetheless, this requires using the acceptor in great excess and separation of the product is extremely tedious. The yields are generally very low in other conversions of this type.

A β-galactosidase of *Bacillus circulans* can transfer a β-D-galactopyranosyl residue of lactose to *N*-acetylglucosamine according to reaction (10.17).

(10.17) Gal-β-(1→4)-Glc + GlcNAc → Gal-β-(1→4)-GlcNAc + Glc

Recently this very stable and inexpensive enzyme was used in the continuous synthesis of *N*-acetyllactosamine in an 'enzyme membrane reactor'. In these reactors, one of the walls of the reaction chamber is semi-permeable and retains the enzyme while allowing the products to pass through (Fig. 10.6). The mixture of reagents in aqueous solution are fed into the reactor by *a*, remain for residence time τ in contact with the enzyme, then are pumped through the reactor towards the outlet *d* by the arrival of a new charge. One important problem of this technique is to determine the optimum value of τ, in the present case, in order to limit interfering hydrolyses. In the experiment described, a chamber of 10 mL containing 30 mg of enzyme is pumped through in 100 h with 2.6 L of a lactose solution (120 mM) and *N*-acetylglucosamine (300 mM) with τ being equal to 0.25 h or 0.5 h. *N*-Acetyllactosamine is obtained (11.3 g) (Hermann *et al.* 1993). The scale can be increased without diminishing the yield.

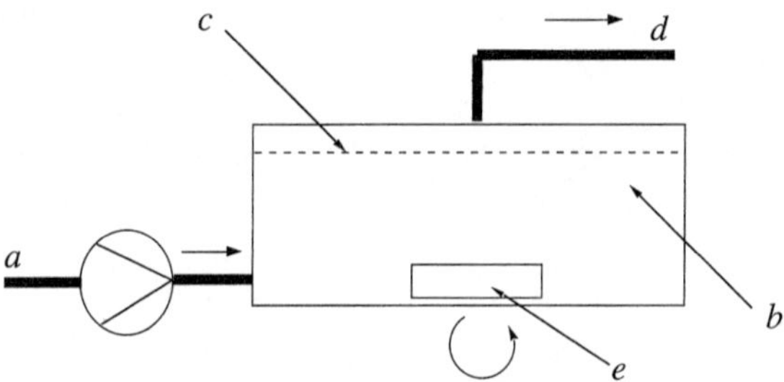

Fig. 10.6 Enzyme membrane reactor. *a*, input; *b*, reaction chamber; *c*, semi-permeable membrane; *d*, outlet; *e*, stirring rod.

10.5 Fluorohydrolysis (Defaye *et al.* 1994)

Treatment of chitine **10.64** [β-D-GlcNAc*p*-(1→4)]*n*-β-D-GlcAc (*n* is very high) of the carapace of shellfish with pure liquid hydrogen fluoride at 0°C gives quantitatively a collection of oligomers **10.65** [βD-GlcNAc*p*-(1→4)]*n*-α-D-GlcNAc*p*-(1→F) (*n* = 1–11). The reaction is preparative. Amalogous oligomers play a role in recognition interactions between higher plants and their hosts, be they symbiotic or parasitic.

References

Adelhorst, K. and Whitesides, G. M. (1993), *Carbohydr. Res.*, **242**, 69–76.
Alais, J. and Veyrières, A. (1981), *Carbohydr. Res.*, **93**, 164–165.
Alais, J., Maranduba, A., and Veyrières, A. (1983), *Tetrahedron Lett.*, **24**, 2383–2386.

Augé, C. and Veyrières, A. (1976), *Carbohydr. Res.*, **46**, 293–298.
Boons, G.-J. (1996), *Tetrahedron*, **52**, 1095–1121.
Castro-Palomino, J. C. and Schmidt, R. R. (1995), *Tetrahedron Lett.*, **36**, 6871–6874.
Czernecki, S. and Ayadi, E. (1995), *Can. J. Chem.*, **73**, 343–350.
Danishefsky, S. J., Behar, V., Randolph, J. T., and Lloyd, K. O. (1995), *J. Am. Chem. Soc.*, **117**, 5701–5711.
David, S., Malleron, A., and Dini, C. (1989), *Carbohydr. Res.*, **188**, 193–200.
David, S., Augé, C., and Gautheron, C. (1991), *Adv. Carbohydr. Chem. Biochem.*, **49**, 175–237.
Defaye, J., Guillot, J. M., Biely, P., and Vršanká, M. (1992), *Carbohydr. Res.*, **228**, 47–64.
Defaye, J., Gadelle, A., and Petersen, C. (1994), *Carbohydr. Res.*, **261**, 267–277.
Garegg, P. J., Henrichson, C., Norberg, T., and Ossovski, P. (1983), *Carbohydr. Res.*, **119**, 95–100.
Garregg, P. J. and Lindberg, A. A. (1988), *Carbohydrate chemistry* (ed. J. F. Kennedy), pp. 500–526, Claredon Press, Oxford.
Gijsen, H. J. M., Qiao, L., Fitz, W., and Wong, C. H. (1996), *Chem. Rev.*, **96**, 443–473.
Hanessian, S. and Banoub, J. (1977), *Carbohydr. Res.*, **53**, C-13–C-16.
Hanessian, S. and Vatèle, J. M. (1981), *Tetrahedron Lett.*, **22**, 3579–3582.
Herrmann, G. F., Kragl, U., and Wandrey, C. (1993), *Angew. Chem., Int. Ed. Engl.*, **32**, 1342–1343.
Hough, L., Richardson, A. C., and Thelwall, L. A. W. (1979), *Carbohydr. Res.*, **75**, C11–C12.
Ichikawa, Y., Look, G. C., and Wong, C.-H. (1992), *Anal. Biochem.*, **202**, 215–238.
Kahne, D., Walker, S., Cheng, Y., and Van Engen, D. (1989), *J. Am. Chem. Soc.*, **111**, 6881–6882.
Kiso, M. and Anderson, L. (1985), *Carbohydr. Res.*, **136**, 309–323.
Kuhn, R. and Kirschenlohr, W. (1956), *Justus Liebigs Ann. Chem.*, **600**, 135–143.
Lafont, D. and Descotes, G. (1988), *Carbohydr. Res.*, **175**, 35–48.
Lemieux, R. U. and Ratcliffe, R. M. (1979), *Can. J. Chem.*, **57**, 1244–1251.
Lubineau, A. and Malleron, A. (1985), *Tetrahedron Lett.*, **26**, 1713–1716.
Lubineau, A., Le Gallic, J., and Malleron, A. (1987), *Tetrahedron Letters*, **28**, 5041–5044.
Mootoo, D. R., Konradsson, P., Udodong, U., and Fraser-Reid, B. (1988), *J. Am. Chem. Soc.*, **110**, 5583–5584.
Nilsson, U., Ray, A. K., and Magnusson, G. (1990), *Carbohydr. Res.*, **208**, 260–263.
Nishida, Y., Wiemann, T., Sinnwell, V., and Thiem, J. (1993), *J. Am. Chem. Soc.*, **115**, 2536–2537.
Paulsen, H. (1982), *Angew. Chem., Int. Ed. Engl.*, **21**, 155–173.
Pougny, J.-R., Nassr, M. A. M., Naulet, N., and Sinaÿ, P. (1978), *Nouv. J. Chem.*, **2**, 389–395.
Sarkar, A. K. and Matta, K. L. (1992), *Carbohydr. Res.*, **233**, 245–250.
Schmidt, R. R. (1986), *Angew. Chem., Int. Ed. Engl.*, **25**, 212–235.
Schmidt, R. R. and Kinzy, W. (1994), *Adv. Carbohydr. Chem., Biochem.*, **50**, 21–123.
Spijker, N. M. and van Boeckel, C. A. A. (1991), *Angew. Chem., Int. Ed. Engl.*, **30**, 180–183.
Stevens, J. D. (1972), *Carbohydr. Res.*, **21**, 490–492.
Veeneman, G. H., Broxterman, H. J. G., van der Marel, G. A., and van Boom, J. H. (1991), *Tetrahedron Lett.*, **32**, 6175–6178.
Wong, C. H., Haynie, S. L., and Whitesides, G. M. (1982), *J. Org. Chem.*, **47**, 5416–5418.

11 Associations with anions, cations, and inorganic molecules

11.1 Associations with metal cations

11.1.1 Introduction

In a mixture of sugars and salts in solution it is quite obvious that there will always be a loose association between hydroxyl groups and metal cations. Likewise, when a sugar is crystallized from a concentrated solution of salts, X-ray analysis will show associations between the sugar, cations, and water molecules. This is not actually very important from our point of view. For example, there is a crystalline addition compound corresponding to the formula saccharose $\cdot NaBr \cdot 2H_2O$, whereas there is no visible association in solution between saccharose and sodium ions. In this work only associations which involve at least three hydroxyl groups of the same sugar molecule will be considered (Angyal 1989). We will not deal with associations in an alkaline medium which are, in fact, alkoxides. Examining solid structures (a limited number) will provide the basis for our discussion. Simple techniques such as paper electrophoresis, thin-layer cation-exchange chromatography, or NMR spectroscopy, give information on complexation in solution.

11.1.2 Structures in the solid state

Unfortunately, complexes whose solid structures could be determined were generally not derived from ordinary sugar molecules whose complexation in solution is easy. The calcium complex of methyl D-*glycero*-α-D-*gulo*-heptopyranoside, $C_8H_{16}O_7 \cdot Ca^{2+} \cdot (Cl^-)_2 \cdot H_2O$, **11.1**, indicates the disposition characteristic of a tridentate complex on a pyranoid ring. The cation is coordinated to oxygen atoms O-1, O-2, and O-3 of one molecule and to O-4′, O-6′, and O-7′ of another. Besides, there is a water molecule and a chloride ion which induce an eight-fold coordination of the calcium. It is interesting to note the axial–equatorial–axial, *aea*, arrangement of the O-1, O-2, and O-3 ligands in structure **11.1**. This is the most favourable disposition for complexing a pyranoid ring. The side chain adopts a conformation which allows the O-4, O-6, and O-7 hydroxyl groups to have the same geometrical relationship as O-1, O-2, and O-3. All of the hydroxyl groups of the sugar are involved in complex formation.

The complex β-D-mannofuranose $\cdot Ca^{2+} \cdot (Cl^-)_2 \cdot 4H_2O$ (**11.2**) is crystallized from a very concentrated aqueous solution of mannose in the presence of an excess of $CaCl_2$. It is remarkable because the proportion of β-D-mannofuranose in pure water is only 0.3%. Complexation thus displaces noticeably the tautomeric equilibrium in aqueous solution. The cation is coordinated to O-1, O-2, and O-3 of one molecule, and to O-5′ and O-6′ of another. Three water molecules bring the coordination of Ca^{2+} to eight. It is impossible to describe coordination with reference to a classical polyhedron. This type of complexation (*cis–cis*, vicinal triol) is characteristic of furanose complexes.

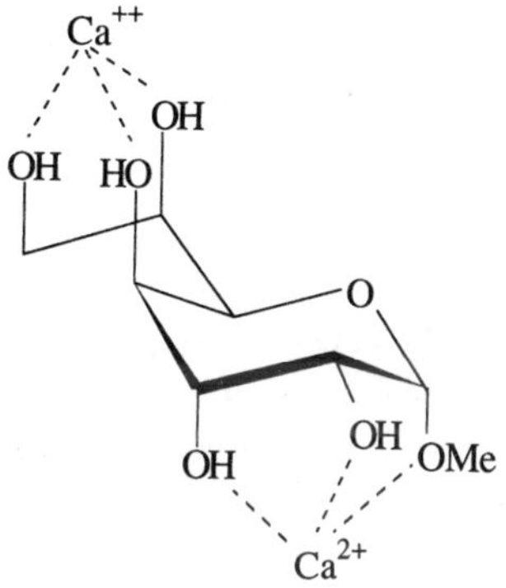

11.1

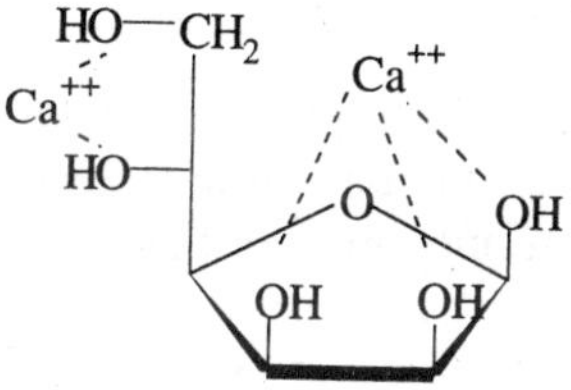

11.2

A monosaccharide cannot provide more than three oxygen atoms for coordination to a cation but with a disaccharide, coordination can be observed at four or five sites. Thus the non-reducing disaccharide, α-D-allopyranosyl-α-D-allopyranoside, presents the *aea* arrangement on each of its monosaccharide units. In the complex **11.3** with $CaCl_2$ and 5 H_2O, Ca^{2+} is coordinated to O-1, O-2, O-3, O-2′, and O-3′. Four water molecules bring the coordination to nine, which is rather rare. The di-β-D-fructopyranose 1,2:1′,2′-dianhydride **11.4** readily complexes with $CaCl_2$, $SrCl_2$, $BaCl_2$, and $LaCl_3$. The hydroxyl groups involved are O-1, O-3, O-1′, and O-3′.

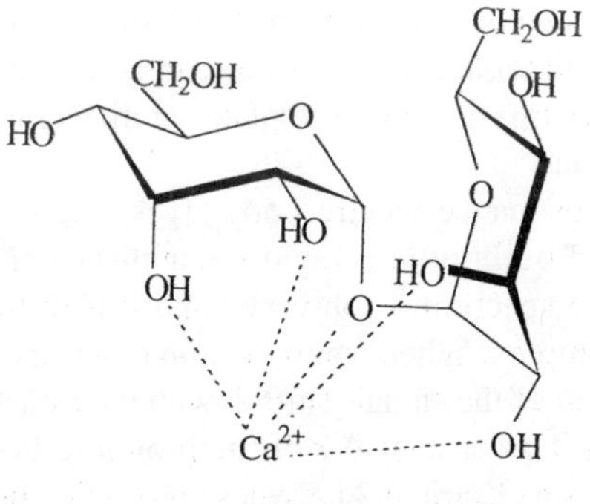

11.3

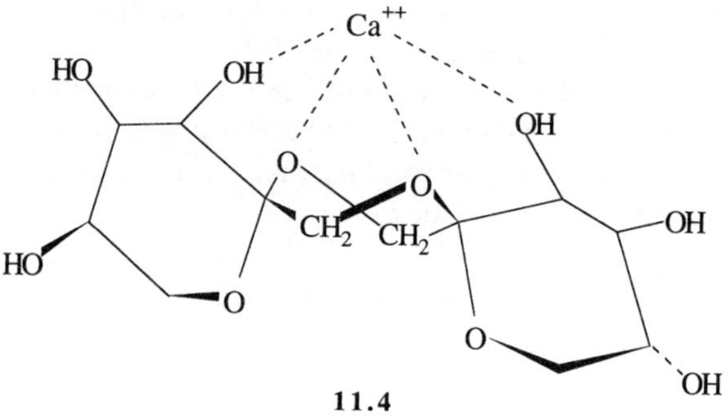

11.4

11.1.3 Complexes in solution

Paper electrophoresis, in the presence of a supporting electrolyte, shows that all sugars migrate towards the cathode, which indicates that at least a fraction of the molecules are coordinated to cations. The cations inducing the greatest mobility are Ca^{2+}, Sr^{2+}, and Ba^{2+}. The rate of migration indicates the extent of coordination. The fastest molecule is the *cis*-inositol **11.5**. Coordination involves three axial hydroxyl groups. This is the best possible orientation for complexation, but this molecule does not belong to the sugar family and, of course, pyranose derivatives with three *cis*-axial hydroxyl groups do not exist. Glucose scarcely migrates.

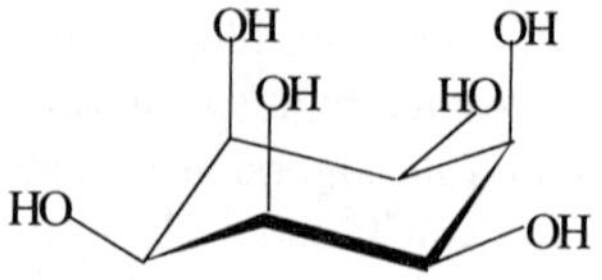

11.5

Coating thin-layer plates with a cation-exchange resin is a well-known method. Sodium ions can be readily exchanged for other desired cations (Cu^{2+}, Ca^{2+}, La^{3+}, etc.) by immersing them in the corresponding salt solution. With water as eluent, sugars migrate on these plates at different rates following their degree of complexation.

Nuclear magnetic resonance spectroscopy gives precise information on complexation in solution. Equilibrium is rapidly established on an NMR time scale, hence only an average spectrum is observed and it is difficult to determine the spectrum of a pure complex. When complexation of a sugar or polyol with a diamagnetic ion occurs, all of the signals shift downfield. Equation (11.1) allows the variation of the shielding constant $\Delta\sigma$ of the proton to be calculated when the nucleus is subjected to an electric field E whose projection on the C–H bond is E_x.

$$\Delta\sigma = -2 \times 10^{-12} E_x - 10^{-18} E^2 \qquad (11.1)$$

By referring to an *aea*-type complex (e.g. **11.1**), we can note that the E_x value is at its maximum along the central axial C–H bond and its lowest value on the equatorial C–H bonds. This allows the complexing site to be identified. For example, the addition of calcium ions modifies the spectrum of methyl α-D-allopyranoside in aqueous solution and we observe that it is the H-2 signal which shows the greatest shift, followed by those of H-1 and H-3. From this structure **11.6** is deduced. In these experiments, the shifts are in the order of 0.2 ppm.

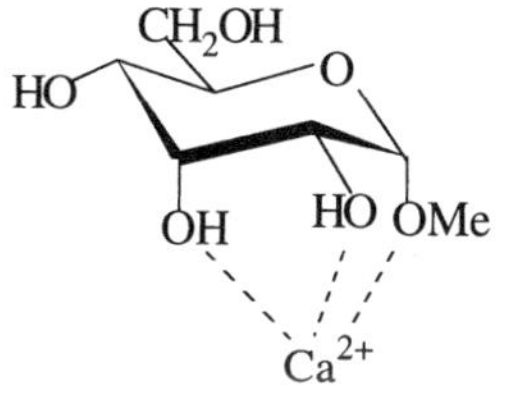

CH$_2$OH HO O OH HO OMe Ca^{2+}

11.6

There may be some traps in these deductions. Complexation can bring about configurational inversion of the ring which, in turn, leads to a variation of the signal positions, much greater than complexation. Owing to fact that **11.7** is in conformational equilibrium with **11.8**, this cyclohexane polyol can present two efficient complexation sites, one being *aea* and the other triaxial, which do not exist in its most stable conformation.

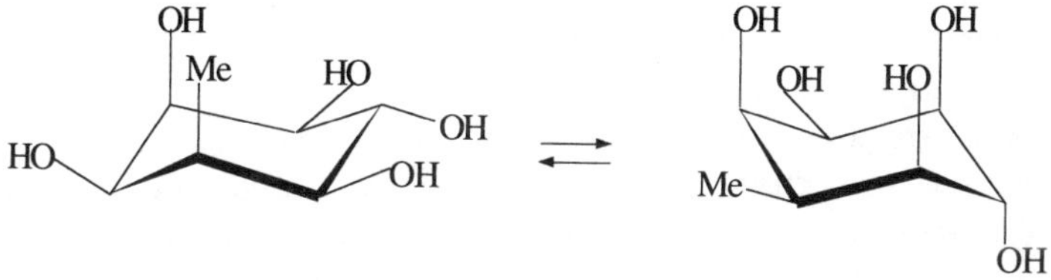

OH Me HO HO OH OH ⇌ OH OH HO OH Me OH

11.7 **11.8**

Signal shifts in the presence of paramagnetic cations are observed not only with sugars but with nearly all the organic families as well. They are much greater and their interpretation is much more complicated. We feel that these phenomena are outside the scope of this chapter.

Even before any crystal structure was determined, examination of electrophoretic mobility had already shown that sugars which can present the *aea* sequence in one of their conformations complex well. This is the most favourable arrangement; the 1,3,5-triaxial arrangement on a cyclohexane is even better but not possible with a pyranose. For a furanose, this *aea* sequence corresponds to the *cis* arrangement of three hydroxyl groups, as in the β-D-mannofuranose discussed above. The flexible furanose can thus adopt a conformation which places the hydroxyl groups in a situation very closely related to the *aea* arrangement.

In general, the stability of the complexes increases with the cation valencies, in the order univalent < divalent < trivalent; but there is another factor, the ionic radius of the cation, whose optimal value is between 100 and 110 pm, as in the case of Na^+, Ca^{2+}, and La^{3+}. The complexes of Li^+, whose radius is 68 pm, are very weakly stable. It is predicted that Cu^{2+}, whose radius is 72 pm, must form mediocre complexes and this has been generally verified. However, there are good indications that strong complexation exists in copper acetate solutions. The true complexing cation would thus be $[Cu_2(OH)_2]^{2+}$. Also, complexes with almost all the trivalent lanthanide cations were examined by the TLC method described above. Ploting the R_f of ribose as a function of the radii of the cations gave a smooth curve with a minimum for Sm^{3+} (Israëli *et al.* 1994). The best ionic radius was found to be 95–98 pm. In the crystalline complex of galactitol with the praseodymium cation, galactitol $\cdot 2\ PrCl_3 \cdot 14H_2O$, the hexitol is in the planar, zig-zag form (Section 2.10) with two cations attached to O-1, O-2, O-3 and O-4, O-5, O-6 (Angyal and Craig 1993).

The complexation equilibrium (11.2) corresponds to an equilibrium constant K given by formula (11.3); this measure of complex-formation strength is called the *stability constant.*

$$\text{sugar} + X^{n+} \rightleftharpoons (\text{sugar} \cdot X^{n+}) \tag{11.2}$$

$$K = \frac{[(\text{sugar} \cdot X^{n+})]}{[\text{sugar}][X^{n+}]} \tag{11.3}$$

With sugars, the measurement is not very accurate as the K value is relatively low. Since solutions are sometimes very concentrated, activities must be introduced which, for these complex cations, are not easily accessible. Several complexation reactions can take place with different stoichiometries. These sources of error are disregarded. The most direct method for calculating the K value is based on a potentiometric determination of the non-complexed cation in solution, in the presence of varying quantities of sugar. When complexation involves a conformational change of the sugar which greatly modifies its NMR spectrum, the fraction of the complexed sugar can be deduced from the shifting of the signals of the spectrum. With Ca^{2+}, stability constants for a 'good complexation' are close to 5 M^{-1} (α-D-allopyranose, 5.1–6.5; α-D-ribopyranose, 4.6–5.5) which decreases to 0.1 for a 'poor' one as with methyl α-D-xylofuranoside. Stability constants are much higher in methanol or ethanol. This explains the dissolving power of alcohol solutions of certain salts vis-à-vis sugars, as for example in concentrated alcohol solutions of calcium chloride.

Among the applications mentioned in Chapter 1, we talked about the separation of sugars on a cation-exchanger column in calcium form, in the context of HPLC analysis. Let us point out that glucose can be separated from fructose on a kilogram scale by this method. In its preparative form, the use of these columns is strongly recommended as they have a very large capacity, they can be used several times without regeneration, and the eluent is water. There are also synthetic applications. The addition of $CaCl_2$ to an aqueous solution of sugar increases the proportion of the pyranose tautomers presenting an *aea* arrangement in one of their conformations and the furanose tautomers having an adjacent *cis,cis*-triol arrangement. In this way the course of the Fischer glycosidation can be radically changed (see Section 3.3) and oriented towards the major production of furanosides, generally not very accessible using other methods. Thus the yield of methyl α-D-ribofuranoside increases from 4 to 69%.

11.2 Structure of liquid water

11.2.1 Introduction

Knowing the structure of water is especially important for understanding its interaction with sugars which, for reasons which will be explained in due time, will be examined at the end of this chapter. However, since the concept of hydrophobic interactions appears several times in Sections 11.3 and 11.4, it would be better for the reader to be familiar with it at once.

11.2.2 Structure of water (Franck 1984)

In ice, each water molecule is the acceptor of two hydrogen bonds involving the hydrogen protons of two neighbouring water molecules. The configuration around each oxygen atom is that of a regular tetrahedron, **11.9**. These tetrahedrons are assembled to give a tridymite-type crystal. The latter contains notable cavities because of the length of the oxygen–oxygen bonds through hydrogen which constitutes the edge of the polyhedrons.

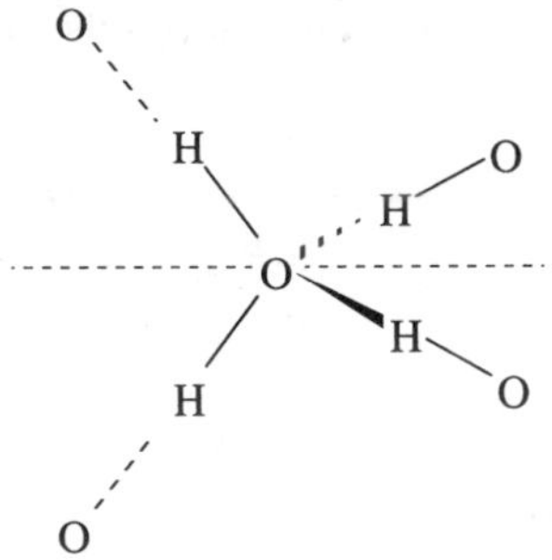

11.9

In liquid water, due to molecular agitation, the assignment of a fixed position to the atoms cannot be made, nor their precise determination by X-ray or neutron diffraction. But when the same methods are used with thin layers of liquid, it is possible to calculate from diffraction images *a function of the radial distribution g (r)*, a measure of the probability of finding an oxygen atom *J* at distance *r* from an oxygen atom *i*. This function, represented in Fig. 11.1, shows

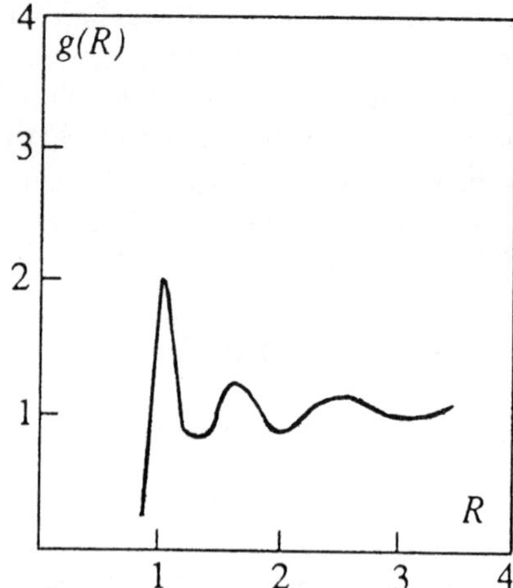

Fig. 11.1 The radial distribution function in liquid water. Ordinates: $g(r)$; abscissae: distance to a particular oxygen, in units of the van der Waals diameter of the water molecule.

that there is a strong probability of finding, in liquid water, a second oxygen atom at distance r close to the van der Waals diameter of the water molecule, 282 pm, and a weaker probability of finding a third one at distance $1.6r$. Then the function quickly subsides which indicates that the oxygen atom i in consideration no longer controls the order at a great distance. The two peaks observed testify to the persistence in liquid water of associations according to model **11.10** with the tetrahedral configuration. By integrating the distribution function from 0 to r, the number $n(r)$ of the immediate neighbours can be calculated according to equation (11.4) whereby ρ indicates the density.

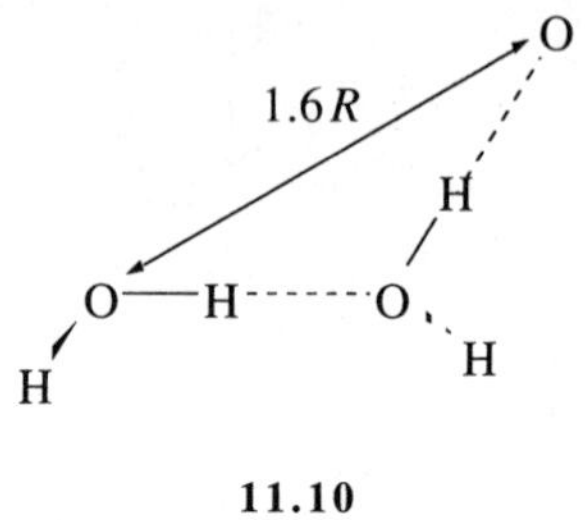

11.10

$$n(r) = \rho \int_0^r 4\pi \, g(r) \, r^2 \, \mathrm{d}r \quad (11.4)$$

We find that an oxygen atom is surrounded by 4.4 directly neighbouring oxygen atoms. Hence there is a certain degree of organization.

In liquid water, there is certainly a greater variety of association modes, but it is believed that the cavities are larger than in ice. When aqueous solutions of an inert gas or hydrocarbons are cooled (obviously very diluted!), something called a *clathrate* settles. In this structure, the solute, when its dimensions permit it, is imprisoned in a regular dodecahedron composed of water molecules. It is thought that this type of cage and others larger pre-exist in pure water. Naturally, they are disrupted and reformed constantly, since the life span of a hydrogen bond in water is in the order of 10^{-10} seconds. The total volume, water plus future solute, diminishes at the time of dissolving which can be interpreted by supposing that the solute simply fills the originally empty volumes. The thermodynamic changes associated with the transfer of a methane molecule from an organic solvent to water, for example, is in agreement with this interpretation. The free enthalpy is positive (water is a very poor solvent for methane!), but in fact the bound enthalpy is negative (exothermic reaction) and the tendency is reversed by a strong decrease in entropy. This reveals an increase in the order of the system. We imagine the solute, when it is far from saturation, as being surrounded by a sphere of water molecules which direct their O–H bonds tangentially when possible, such as to move away from the solute molecule, **11.11**. The general opinion is that this implies a certain rigidification of the water structure. If the concentration is increased, two molecules A and B approach each other and end up being reunited in the same but larger cavity, **11.12**. This is a favourable process because the water–hydrocarbon interface is lessened. These phenomena characterize *hydrophobic hydration*.

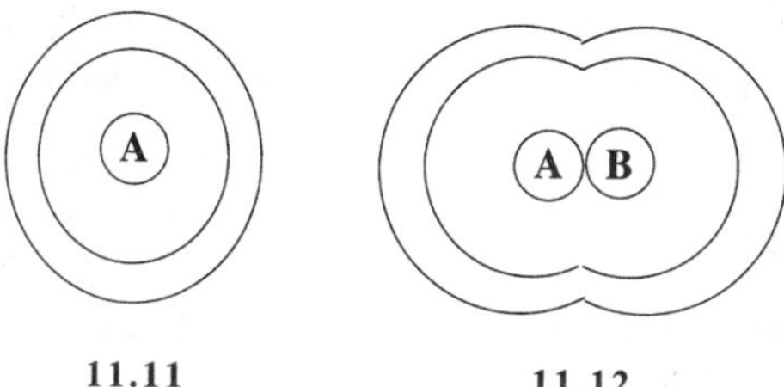

11.11 **11.12**

11.3 Cyclodextrins (Clarke *et al.* 1988)

We will now begin with the study of a family of complexing molecules which are, on paper, the polycondensation products of α-D-glycopyranose matching the general formula **11.13**.

R–O–[CH_2OH, HO, HO, O]$_n$–R'

11.13

In this family there are linear polysaccharides (R = H, R′ = OH) but we will begin with cyclic oligosaccharides, the cyclodextrins. Indeed their properties are often known with precision and they can serve as models for understanding the behaviour of amylose (see Section 11.4) despite an unquestionable difference in structure. These are the 1→4′-linked cyclic oligomers constructed from α-D-glucopyranosyl units. The three most important ones, α-, β-, and γ-cyclodextrins, consist of 6, 7, and 8 α-D-glucopyranosyl residues having the usual D-$^4C_1$ conformation. The rings are cone-shaped with primary alcohol functions at the small base and the secondary alcohol functions at the large base. Structure **11.14** is that of β-cyclodextrin, presently the most utilized. The interior cavity is covered with hydrogens linked to C-3 and C-5 and glycosidic oxygens. It is essentially hydrophobic. The conformation is stabilized by hydrogen bonds between two hydroxyl groups belonging to two adjacent α-D-glucopyranose residues. The cavity's dimensions are in the order of 470–520 pm for α-cyclodextrin, 600–640 pm for β-cyclodextrin, and 750–830 pm for γ-cyclodextrin. It is remarkable that β-cyclodextrin has relatively little solubility in water (18.5 g/L), whereas the α- and γ-cyclodextrins are eight and twelve times more soluble, respectively.

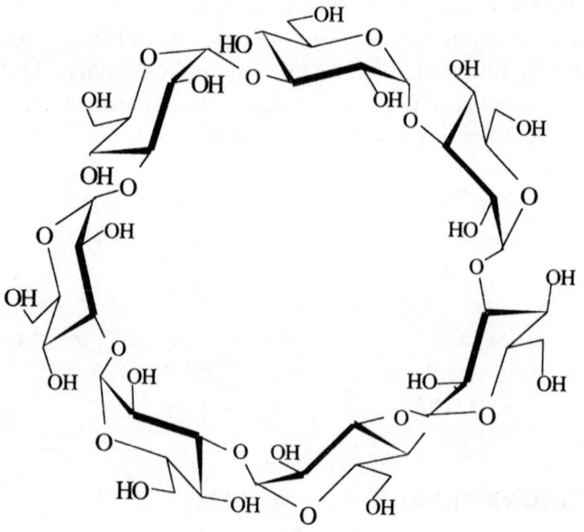

11.14

Cyclodextrins are degradation products from starch by the bacterium *Bacillus macerans*. They are separated by precipitation with complexing agents and purified by recrystallization. They could probably be produced industrially by the tonne. The formation of cyclodextrins is interpreted as a *trans* glycosidation of amylose under enzymic control. The latter, in aqueous solution, adopts at least partially the shape of a helix with a period of six glucopyranose units. The bacterial enzyme catalyses the junction of two glucopyranose residues separated by four, five, or six residues on the chain, but spatially close because of the helical structure **11.15**.

11.15

The most characteristic property of cyclodextrins is their ability to form inclusion complexes in aqueous solution with molecules smaller than their cavities. This is shown, among other ways, in NMR studies. Thus, addition of the α-cyclodextrin to the *p*-nitrophenolate anion in aqueous solution (pD $\approx$ 11) induces a shift of 14–35 Hz of the aromatic proton signals examined at 100 Hz. A study of the chemical shift as a function of concentration allows the calculation of a dissociation constant of the complex, $K_D = 3.7 \times 10^{-4}$ M. Values of the same order are found upon analysing the evolution of other physical properties (Bergeron *et al.* 1977). The complexation site can be determined on a cyclodextrin by using the proton nuclear Overhauser effect (Yamamoto *et al.* 1987). Let us recall that cross-relaxation of two protons depends on their distance *r* according to the expression r^{-6}. The spatial proximity of two protons therefore is expressed by an increase in the signal's intensity. One experiment described is based on this principle but uses a more sophisticated 2D NMR technique. The solution examined was 40 mM in α-cyclodextrin and 80 mM in *p*-nitrophenoxide in D_2O (pD 10), and at least 95% of the cyclodextrin was complexed. It was shown that H-3 of cyclodextrin was in close proximity to the *ortho*-(*o*) and *meta*-(*m*) proton resonances of the *p*-nitrophenoxide, and H-5 to only the *meta* proton resonance. This indicates that phenolate is incorporated into the cavity having the disposition outlined in **11.16**.

H(5) NO_2 (5)H
H *m* *m* H
H(3) (3)H
H *o* *o* H
$O^{\ominus}$

11.16

There are a number of other methods used to demonstrate complexation. For example, it is well known that protons located close to the C_6 axis of an aromatic nucleus give NMR signals shifted towards higher fields, sometimes up to negative δ values. The inclusion of an aromatic molecule in a cyclodextrin produces this effect on the H-3 and H-5 signals, but it is relatively weak (–0.2 ppm) (Demarco and Thakkar 1970). There are several other possible means of investigation: modifications of electronic spectra; increasing the fluorescence of the host which, in the cavity, is protected against the impact of exterior molecules; changes in circular dichroism. An increase in the solubility of substances poorly soluble in water is also observed. On the list of complexed molecules are found alcohols and aliphatic carboxylic acids having low molecular weight, cyclohexane- and adamantane-carboxylic acids, benzenes substituted by various functional groups, heterocycles (pyridines, pyrimidines, indoles), and finally inorganic salts.

It appears that the final explanation for the inclusion mechanism has yet to be found. In the expression of the free enthalpy of complexation the bound enthalpy term is negative, and rather variable, in the order of 20 kJ mol^{-1}, but its contribution is diminished by an entropic term which is nearly always positive. This regular behaviour of host molecules of extremely varying structures otherwise suggests a common mechanism of complexation. Above we saw that two molecules of a hydrophobic solute, A and B, **11.11** and **11.12**, have a tendency to be joined together by driving away the water molecules that separate them. In this case, the hydrophobic interior of the cyclodextrin's cavity plays the role of molecule A, and the host B comes in contact by driving out all or a part of the water molecules that normally occupy this cavity. It seems quite probable that complexation is also due to the participation of van der Waals forces, to varying degrees depending on the configuration of the host.

Currently the exploration of the properties of cyclodextrins is being actively pursued. Because of their lack of toxicity and their ability to complex their hosts in water, it is possible to envision their applications in the pharmaceutical industry for the administration of insoluble drugs. They can also play a catalytic role. Thus β-cyclodextrin accelerates the cycloaddition of cyclopentadiene and acrylonitrile by simultaneous inclusion of the two molecules. Various functional groups can be introduced by derivatization of the hydroxyl groups in order to modify complexing properties. By introducing a weakly basic functional group close to the host cavity, an artificial hydrolase has been constructed. Finally, because cyclodextrins are chiral molecules, we can expect to observe a more or less efficient resolution of racemic hosts. This is what has sometimes been observed and chiral columns have even been built with immobilized cyclodextrins. However, it does not appear that cyclodextrins have become popular as resolution reagents.

11.4 Amylose

Starch, a food-reserve substance from plant grains, is a mixture of two polysaccharides, amylose and amylopectin (Kennedy and White 1988). Amylose is linear with, as a repeating unit, α-D-glucopyranose linked as in compound **11.13** (R = H, R′ = OH), *n* being frequently in the order of 6000. On the other hand, amylopectin is a branched polysaccharide in which similar chains, but much shorter—from 17 to 26 units–are linked to each other by 1→6′ bonds. These two constituents are separated by treatment of an aqueous dispersion of starch by hydroxylated organic compounds as thymol and *n*-butanol which form insoluble complexes with amylose. Amylopectin remains in solution and we will no longer discuss this. Concerning amylose, several crystallized species can be prepared from an aqueous solution, depending on growth conditions. The basic structure of A- and B-amyloses is a double helix (Wu and Sarko 1978), constituted by the winding around each other of two *parallel* chains, each one having the shape of a right-handed helix, as in **11.17**. One complete turn corresponds to six glucopyranose residues for each strand. The A- and B-amyloses differ by their arrangement in the crystalline unit. In B-amylose, the helical conformation repeats after 2080 pm. The V-amyloses, which precipitate in the presence of a complexing agent such as dimethyl sulfoxide or *n*-butanol, are single helices, also with six α-D-glucopyranosyl units per turn, perhaps a little less regular. The V_h-amylose is a left-handed helix (Rappenecker and Zugenmaier 1981). In solution, amylose seems to have a single or double helical conformation, probably rather flexible or a random coil. The helical conformation reproduces to a certain extent the hydrophobic surfaces characteristic of cyclodextrins. We can imagine that this avoids the formation of the hydrophobic aqueous layer which normally surrounds the non-polar molecules in water. The expression 'autocomplexation' could be used. However, the helix presents polar groups towards the solvent bulk.

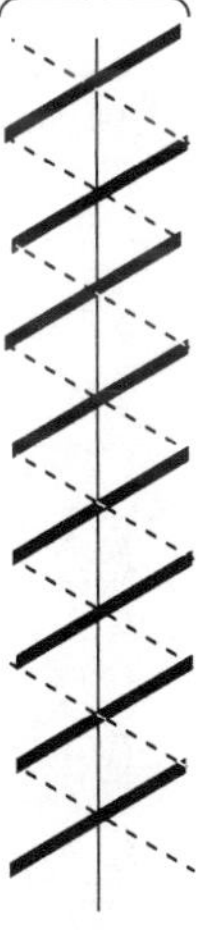

11.17

11.5 Iodine complexes

11.5.1 Introduction

The blue colour of amidon with iodine, well-known to high school students, was observed for the first time nearly 200 years ago. This is due to the amylose fraction. Complexation with smaller homologues can be observed and, since it is easier to characterize these complexes, we will deal with them first.

11.5.2 Complexes with α-cyclodextrin

When an aqueous solution of α-cyclodextrin is placed in contact with an ethereal iodine solution, reddish-brown crystals are formed at the surface having the composition $(C_6H_{10}O_5)_6{\cdot}I_2{\cdot}4H_2O$. X-ray diffraction analysis indicates that the iodine molecule is colinear with the C_6 axis of the cyclodextrin. The closest contacts in the perpendicular direction to this axis take place between one of the iodine atoms and the C-5 and C-6 atoms, and the other iodine atom and the O-4 oxygen atom. The environment is thus hydrophobic for the first, and hydrophilic for the second. The stacking on top of each other of the molecules in the crystal is such that the two apertures of the cavities of each cyclodextrin molecule are clogged by neighbouring molecules. This creates a *cage*-type structure (McMullan *et al.* 1973).

When complexes are prepared in the presence of iodide, four different complex types can be obtained, depending on the conditions. Here we will only discuss the blackish-brown complex (α-cyclodextrin)$_2{\cdot}Cd_{0.5}{\cdot}I_5{\cdot}27H_2O$. On the C_6 axis of the molecule are found four of the five iodine atoms of I_5^-, the central fifth one being disordered. The disordered iodine atom is located between the two α-cyclodextrin molecules which face each at their wide bases, **11.18**. Structures of **11.18** are packed one on top of the other in a way as to create a continuous channel filled with iodine atoms. This is a *channel*-type complex (Noltemeyer and Saenger 1980). With guest molecules other than iodine, either one or the other of the two types of crystalline structures discussed here, i.e. *cage* or *channel*, can be observed.

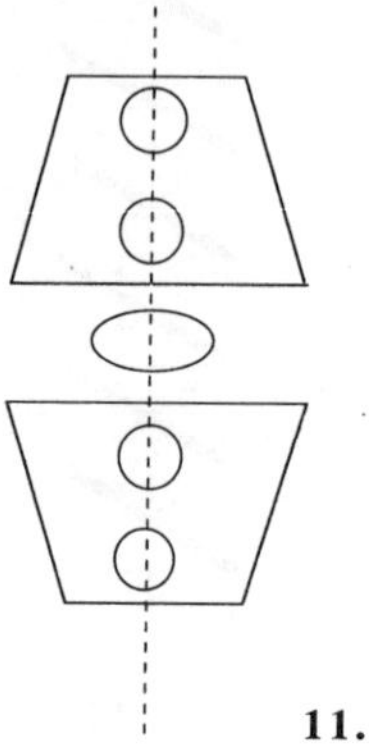

11.18

11.5.3 (*p*-Nitrophenyl α-maltohexaoside)$_2$·Ba(I$_3$)$_2$·27H$_2$O complex

The prefix *malto* is used to designate more precisely a family of linear oligosaccharides having the general formula **11.13** (R = H, R′ = OH) with n being even. For this reason, they can be seen as derivatives of the α-anomer of maltose **11.13** (R = H, R′ = OH, n = 2). Maltohexaose (n = 6) can be considered as a small fragment of amylose. In this study (Hinrichs and Saenger 1990), the α-glycoside of p-nitrophenol, which we will designate as M, was used to avoid problems with possible mutarotation. The evaporation of decimolar aqueous solutions of M and barium triiodide leads to brownish needle-like crystals $M_2 \cdot Ba(I_3)_2 \cdot 27H_2O$. This complex is therefore not an intense blue but rather the colour of iodine in aqueous solution. We have oversimplified the description of this molecule, which is rather complicated in detail, in order to concentrate our attention on points which we feel are more interesting.

The lattice unit contains four anions I_3^- which form an extended broken line, **11.19**. The distances between the triiodides are somewhat shorter than van der Waals distances. Each group of two adjacent triiodides is in contact with two sugar hydroxyl groups and two water molecules. Therefore, the environment is essentially not hydrophobic and this is perhaps why the colour of the complex is close to that of iodine in water or alcohol rather than the blue of the starch–iodine complex. Two molecules M are wrapped around each triiodide, each as a left-handed helix, but antiparallel to each other. They are represented in formula **11.20** in a V-shaped line ending with a hexagon which makes up the aglycon, while the iodine column is drawn linear. A stacking interaction takes place between the aromatic rings of the aglycons of the two antiparallel molecules M which probably reinforce the system's cohesion. The two interdependent chains around a triiodide anion are depicted as stretched out in **11.21**. The dotted lines represent the direct hydrogen bonds between these chains. The system's cohesiveness is largely provided by the Ba^{2+} cation which presents a ten-fold coordination. It is linked to four α-D-glucopyranosyl residues from four different M chains, twice to the O-5 and O-6 and twice to the O-2 and O-3 oxygen atoms, and to two water molecules. The antiparallel double helix is highly hydrated. The hydrogen bond donors or acceptors are all involved, including most of the ring oxygens. The remarkable predominance of chelations involving O-5 and O-6 as well as O-2 and O-3 are observed, leading to five-membered rings, **11.22**. The superposition of the lattice units as shown in **11.20** gives an infinite cylindrical cavity in the crystal, defined by the M helices and filled with triiodide anions in a zig-zag conformation.

11.5.4 Iodine–amylose complex

Complexation in an iodine solution by amylose proceeds until the polysaccharide has absorbed nearly 20% of its weight in iodine. The exact amount as well as the wavelength of the maximum absorption of the blue colour, varying from 606 to 642 nm, depends on experimental conditions and the origin of the

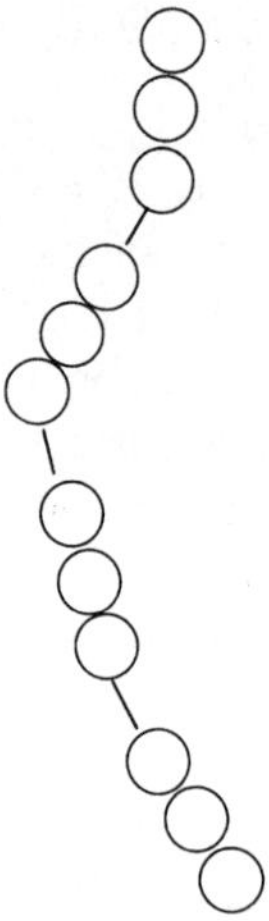

11.19

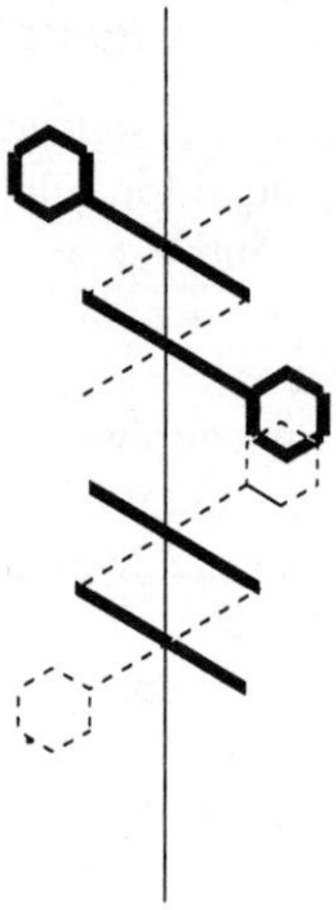

11.20

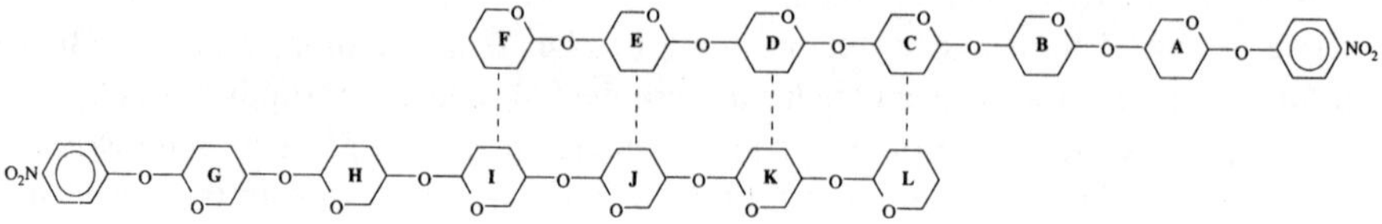

11.21

11.22

amylose. The colouring can also be obtained by treating V-amylose with iodine vapour. It is generally believed that iodine occupies the central cavity of the helix (Rundle 1947). This could explain the absence of the reaction with A- and B-amyloses in a compact double helix which lack sufficient available space on their axis. A structure in the solid state of the V_h-amylose–iodine complex has been made (Bluhm and Zugenmaier 1981). The complex with dimethyl sulfoxide was decomposed with boiling methanol, then exposed to iodine vapour in a chamber saturated with humidity. At that moment, the complex took on the characteristic intense blue colour. The structure can be described as a linear or nearly linear structure on which are aligned iodine atoms, nearly equidistantly separated by 295 ± 4 pm. The amylose coils around to form a left-handed helix with six α-D-glucopyranose residues per turn. There are three iodine atoms for each helical turn. The iodine atoms are in alignment; there is no contact with the oxygen atoms and the colour is blue, whereas in the *p*-nitrophenyl α-D-maltohexaoside with I_3^-, the iodine chain is in a zig-zag conformation, there are contacts with the oxygen atoms, and the colour is brown.

According to a recent article, complexation of strictly anhydrous amylose (which avoids the formation of an iodide anion) would give a complex of molecular iodine (Murdoch 1992).

11.6 Interactions of sugars with liquid water

11.6.1 Their importance

Approaching this question at this point in the chapter should not imply that it is of secondary importance. It has been put off until now because of the uncertainties surrounding the various interpretations. We will essentially explain hypotheses and experimental attempts to obtain answers. Let us recall one obvious fact: glucose and the majority of free sugars are very soluble in water. There remains some truth in the old term carbohydrate for it is the closest organic family to water. Also, sugars attached to peripheral living cells in aqueous medium maintain a space of continuity between a hydrophobic membrane in its deepest parts and the aqueous medium. The omnipresence of glucose in the living world suggests a very ancient source. Does it owe its emergence to some particular adaptation to the water of the 'primeval ocean'? This is pure speculation but there are also problems in the present world. A good number of interactions between a cell and its environment are based on the recognition of a membrane-bound oligosaccharide bathing in an aqueous medium by a lipid or protein receptor. The thermodynamic parameters of this reaction can generally be measured but their concrete interpretation requires knowing the type of bond involved in the initial state. By stretching the meaning of these words a little, we could say that liquid water is the primordial and universal receptor of all sugars and consider the sugar–protein receptor interactions in an aqueous medium as the passage of one receptor to another.

11.6.2 Model of stereospecific hydration of hexopyranoses

Elucidating the details of sugar and water interactions amounts to differentiating between the behaviour of a thin layer in contact with the solute from that of the bulk of water molecules surrounding it. We can observe variations with concentration of certain solvent properties concerning their oxygen or hydrogen atoms, or of the entire molecule, and they are interpreted as an effect of perturbations affecting the fraction of bound water compared to the bulk. Differences between the hexoses are observed which force us to rule out at once the idea that each hydroxyl group can be both donor and acceptor in an indistinguishable fashion. By comparing the effect of a collection of sugars on water, we try to construct a *specific hydration model*, that is to say, to set forth the rules which define the build-up of hydrates as a function of the configuration and conformation of the dissolved sugar. Immediately we notice that exploiting the experiments will be a difficult task since each sugar in aqueous solution is a mixture, in the worst case, of two pyranoses and two furanoses. Another source of difficulty, apparently neglected by many authors, is the conformational instability which causes a mixture of shapes, even for a well-defined tautomer (see Chapter 2). Of course, hexopyranoses are rigid, with the exception of idose, but with pentopyranoses there is a risk of conformational change in the experimental conditions. Finally, there is the problem of the flexibility of furanoses.

Another source of difficulty is that the schools of thought concerned with these problems seem to have a different conception about 'adaptation'. According to one of these schools, an elevated disturbance of the properties of the water solution compared to those of pure water reveals that there are a great many water molecules associated with the sugar, hence this hydration layer is relatively stable. The authors therefore consider that this indicates a good integration into the water network. Thus, we reckon that β-D-glucopyranose would bring with it four water molecules to form a structure whose life span, in the order of μs, would be a thousand times greater than that of hydrogen bonds in the solvent. This optimal hydration would be a result of the favourable configuration of its hydroxyl groups. The distance between the equatorial pairs O-1–O-3 and O-2–O-4 is very close to that between an oxygen atom and its *second* neighbour in pure water. This would allow the formation of **11.23** which integrates well into the water network. As we go further away from this optimal configuration, hydration diminishes. α-D-Glucopyranose would only bring with it three water molecules. The domination of the β-anomer of D-glucopyranose in aqueous solution, contrary to the anomeric effect, is interpreted as a result of its better adaptation to the water network (Franck 1987). As the primary hydroxyl group is also involved in the network of hydrogen bonds, certain authors have formulated the paradoxal statement: all polar groups of β-D-glucose participate in hydrogen bonding, consequently it behaves as a hydrophobic solute in the sense that a foreign molecule approaching it only sees C–H bonds. In conclusion, according to these principles, we see that as the calculated hydration is an increasing function of disturbance, the best adaptation to the water network corresponds to the maximal disturbance of the properties of pure water.

Another school of thought defends the opposite point of view (Galena *et al.* 1992). If a sugar were perfectly integrated into the network of water, the water molecules surrounding it could not be distinguished one from another, and there would be no disturbance of the physical properties of pure water. In practice, the most integrated sugar is the one which disturbs the least, α-D-talose **11.24**; the least integrated, the one which disturbs the most, α-D-galactose **11.25**.

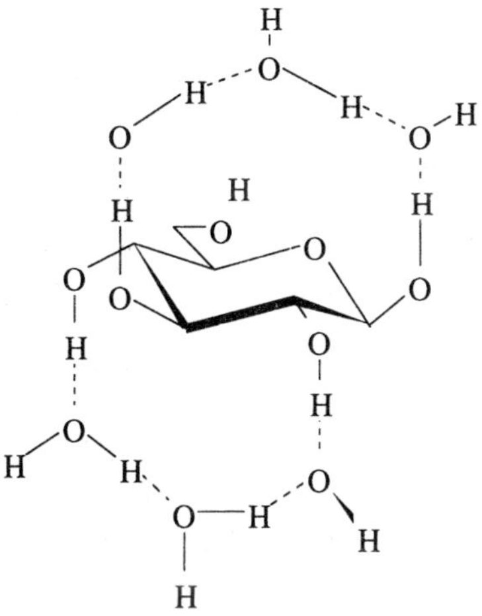

11.23

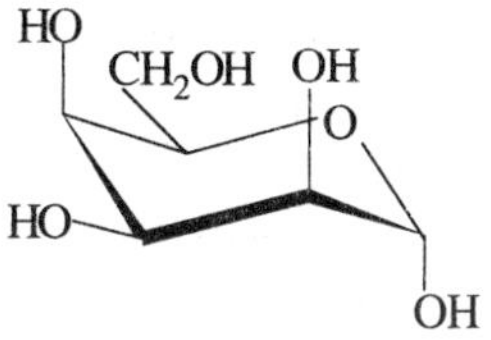

11.24

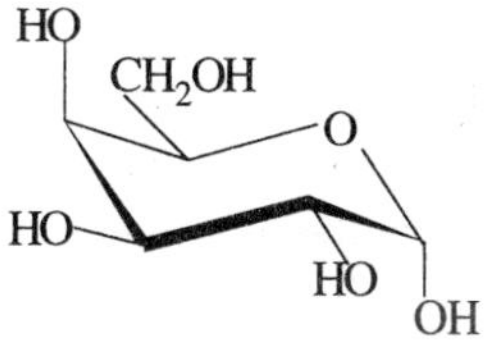

11.25

With sugar **11.24**, the O-2, O-4, and O-6 oxygen atoms form a nearly equilateral triangle whose side is approximately equal to the length between an oxygen atom and its *nearest* neighbour in the water network. There would be three distances integrated into the network of water, out of one for α-D-galactopyranose **11.25**.

11.6.3 Principles of measure

Relaxation

Let us take the example of the relaxation observed in NMR studies (Uedaira *et al.* 1989). Under the effect of molecular agitation, magnetization relaxes following a supposed exponential rule $M = M_o\,(1 - e^{-t}/T_1)$; T_1, the *longitudinal relaxation time*, is a measure of mobility. In pure water, T_1 measured with the naturally abundant ^{17}O isotope is 7.3 ms at 25°C. It is multipled by approximately 1.6 in hexose solutions with a molality of 2.0. This can be explained by considering that water bound at least temporarily to the sugar as a layer of solvation is less mobile than that of the solvent bulk. Disturbance increases in the order mannose, galactose, and glucose.

Measure of compressibility

An acoustic procedure allows the measuring of partial molar isentropic compressiblities. According to the authors, if a solute is perfectly integrated into water, the same value for pure water should be observed, 8.17×10^{-4} mL mol^{-1} bar^{-1}. In the ideal case, associated water molecules are imperceptible. If there is poor integration, there should be an intermediary layer less compressible than pure water. Indeed, negative partial molar compressibilities are found: talose –11.9, mannose –16.0, glucose –17.6, galactose -20.8×10^{-4} mL mol^{-1} bar^{-1} at 25°C.

Kinetic effect of the medium

Here the kinetic effect has to do with the influence of a foreign solute on the reaction rate in solution. Reaction (11.5) is part of the hydrolysis of benzamide in water, independent of the pH. It proceeds through a polar transition state containing two water molecules. The reaction is slowed down by all hexoses. This is interpreted by supposing that there are common water molecules in the hydration layers of benzamide, from the transition state and the sugar, the result of which modifies in a non-parallel fashion the free enthalpies of the starting product and the transition state. If the sugar behaves in a truly hydrophobic manner, we can imagine that there will be a stabilizing interaction with the slightly polar benzamide. The slowing down increases in the sense of galactose, glucose, mannose, and talose. The authors believe that this means there is an increase in the hydrophobic character and confirms the results of the measure of compressibility. We must point out, however, that of the series studied, talose is the only hexose which contains 31% of furanose in aqueous solution.

(11.5) PhCO — N(triazole, Ph) → [PhCO — N(triazole, Ph) ··· H—O—H···O(H)—H]‡ → PhCOOH + HN(triazole, Ph)

Cycloaddition in aqueous phase

The cycloaddition of a diene to a dienophile has the reputation of not being very sensitive to the polarity of the solvent. However, it is considerably accelerated in water (Rideout and Breslow 1980). Thus reaction (11.6) is a thousand times faster in water than in isooctane. It has been suggested that this is a result of the *hydrophobic effect*. The increase in rate is even greater in solutions of LiCl, a salt which diminishes the solubility of lipophilic molecules in water by increasing the hydrophobic effect. In Section 11.3 we have already discussed a similar effect of β-cyclodextrin. Diene and dienophile can be complexed simultaneously in the adequately sized hydrophobic cavity. α-Cyclodextrin, smaller in size, can only complex one of the reactants and slows down the reaction. When the rate of a reaction depends that much on the nature of the solvent, it is tempting to employ this as a chemical tool for studying this solvent. Its parameters of activation must be a priori very sensitive to the addition of compounds which can modify their structure. They have been compared in methanol and water (Lubineau *et al.* 1994). Table 11.1 (lines 1 and 2) show that the increase in rate going from a methanolic solution to an aqueous solution comes from an increase in the entropy of activation. This could be the result of a decrease in entropy when going from methanol to water, more important for the initial state than the final one.

(11.6)

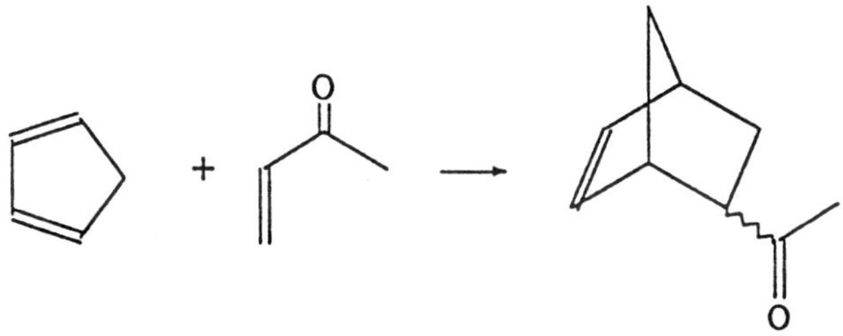

The ideal chemical probe for the structure of a solvent should bring into play reactants that are less volatile and more soluble in water, this being apparently incompatible with a lipophilic character. The use of a diene linked to a free sugar, according to reaction (11.7), approaches this ideal. A comparison of lines 3 and 4 of Table 11.1 shows that this reaction is also accelerated in pure water and that this is also the consequence of an increase in the entropy activation. Thus, the hydrophobic effect can take place between the diene and dienophile close to the sugar part.

(11.7)

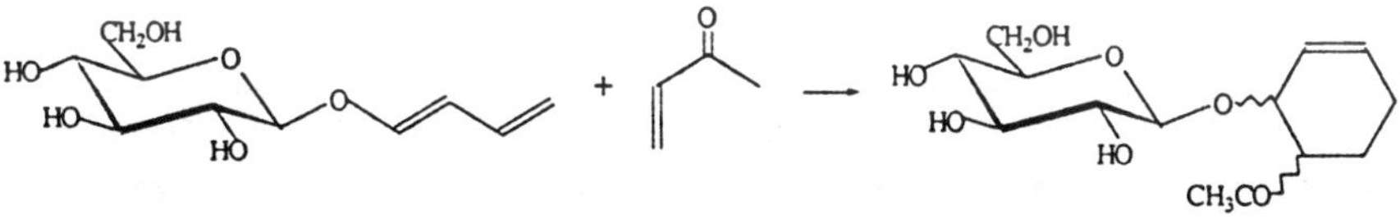

Reaction (11.7) is very sensitive to the addition of sugars to the medium. Table 11.2 gives some precise figures. These sugars, not active in weak concentrations, accelerate the reaction in a linear fashion from 0.2 M. At a 2 M concentration, glucose accelerates the reaction more than a saturated solution of β-cyclodextrin. Addition of glucose, mannose, galactose, sorbitol, ribose, etc. have comparable accelerating effects, but this perhaps overshadows very different mechanisms. By way of example, while both glucose and ribose having the

Table 11.1 Thermodynamic activation parameters in cycloadditions at 25°C. (from Lubineau *et al.* 1994) (reproduced with kind permission from Gauthier-Villars, © 1994)

Reaction	Solvent	$10^4 k_2$ ($M^{-1}\,s^{-1}$)	$\Delta H^{\ddagger}$ (kJ mol^{-1})	$\Delta S^{\ddagger}$ (J mol^{-1} T^{-1})	$\Delta(\Delta H^{\ddagger})$* (kJ mol^{-1})	$T\Delta(\Delta S)$* (kJ mol^{-1})
(11.6)	water	543	38.0 ± 1.7	−140.9 ± 5.0	0	9.83
(11.6)	methanol	10.4	38.0 ± 1.0	−173.9 ± 3.4		
(11.7)	water	2.85	40.0 ± 0.6	−178.8 ± 2.1	6.4	9.63
(11.7)	methanol–water (1:1, v/v)	0.85	33.6 ± 0.8	−211.1 ± 2.6		

* increase from the methanolic solvent to water

Table 11.2 Influence of additives on the second-order rate constants of reaction (11.7) at 25°C in water. (from Lubineau *et al.* 1994) (reproduced with kind permission from Gauthier-Villars, © 1994)

Additive	None	Methanol	Glucose			Saccharose		β-Cyclodextrin
Molarity	–	(50%)	1 M	2 M	3 M	1 M	2 M	(saturated)
$10^5 k_2$ ($M^{-1} s^{-1}$)	28.5	8.5	34.0	45.0	61.3	44.9	74.9	40.2

same molality (2.6 M) accelerate the reaction, ribose accelerates the reaction because of a lowering in the enthalpy of activation whereas glucose behaves as lithium chloride and accelerates the reaction by acting favourably in both entropic and enthalpic terms, contrary to the usual rule of compensation. It has been shown that this acceleration is not a result of an increase in the viscosity of the medium which, on the contrary, proves to be an unfavourable factor.

This study shows the structuring character of glucose (it increases the hydrophophic effect) compared to that of ribose and offers new support to the specific hydration model of sugars. Interacting in a privileged way to the tetracoordinated network of liquid water, glucose maintains this structure by cooperative hydrogen bonding and favours the interactions between lipophilic molecules, thus limiting unfavourable contacts with structured water. This could explain the favourable behaviour of glucose (compared to that of ribose) in the stabilization of proteins such as ovalbumin against thermic denaturation.

References

Angyal, S. J. (1989), *Adv. Carbohydr. Chem. Biochem.*, **47**, 1–43.

Angyal, S. J. and Craig, D. C. (1993), *Carbohydr. Res.*, **241**, 1–8.

Bergeron, R. J., Channing, M. A., Gibeily, G. J., and Pillor, D. M. (1977), *J. Am. Chem. Soc.*, **99**, 5146–5151.

Bluhm, T. L. and Zugenmaier, P. (1981), *Carbohydr. Res.*, **89**, 1–10.

Clarke, R. J., Coates, J. H., and Lincoln, S. F. (1988), *Adv. Carbohydr. Chem. Biochem.*, **46**, 205–249.

Demarco, P. V. and Thakkar, A. L. (1970), *J. Chem. Soc., Chem. Commun.*, 2–4.

Franks, F. (1987), *Pure Appl. Chem.*, **59**, 1189–1202.

Franks, F. (1984), *Water*, The Royal Society of Chemistry, London.

Galema, S. A., Blandamer, M. J., and Engberts, J. B. F. N. (1992), *J. Org. Chem.*, **57**, 1995–2001.

Hinrichs, W. and Saenger, W. (1990), *J. Am. Chem. Soc.*, **112**, 2789–2796.

Israëli, Y., Morel, J.-P., and Morel-Desrosiers, N. (1994), *Carbohydr. Res.*, **263**, 25–33.

Kennedy, J. F. and White, C. A. (1988), The plant, algal and microbial polysaccharides. In *Carbohydrate chemistry* (ed. J. F. Kennedy), pp. 220–262, Oxford University Press, Oxford.

Lubineau, A., Bienaymé, H., Queneau, Y. and Scherrmann, M.-C. (1994), *New. J. Chem.*, **18**, 279–285.

McMullan, R. K., Saenger, W., Fayos, J. and Mootz, D. (1973), *Carbohydr. Res.*, **31**, 211–227.

Murdoch, K. A. (1992), *Carbohydr. Res.*, **233**, 161–174.

Noltemeyer, M. and Saenger, W. (1980), *J. Am. Chem. Soc.*, **102**, 2710–2722.

Rappenecker, G. and Zugenmaier, P. (1981), *Carbohydr. Res.*, **89**, 11–19.

Rideout, D. C. and Breslow, R. (1980), *J. Am. Chem. Soc.*, **102**, 7816–7817.

Rundle, R. E. (1947), *J. Am. Chem. Soc.*, **69**, 1769–1772.

Uedaira, H., Ikura, M. and Uedaira, H. (1989), *Bull. Chem. Soc. Jpn.*, **62**, 1–4.

Wu, H.-C. H. and Sarko, A. (1978), *Carbohydr. Res.*, **61**, 7–25 and 27–40.

Yamamoto, Y., Onda, M., Kitagawa, M., Inoue, Y. and Chûjô, R. (1987), *Carbohydr. Res.*, **167**, C11–C16.

12 Sialic acids and sialylated oligosaccharides

12.1 Natural state

Natural sialic acids (Schauer 1982; 1991) are derivatives of 5-amino-3,5-dideoxy- D-*glycero*-D-*galacto*-nonulosonic acid **12.1**. This awkward name has been replaced by 'neuraminic acid'. The most common derivative is *N*-acetyl-neuraminic acid **12.2** whose configuration is easy to memorize because, in the Fischer representation, **12.3**, it is presented as an aldolic condensation product of *N*-acetylmannosamine (2-acetamido-2-deoxy-D-mannose) and pyruvic acid. When the expression 'sialic acid' is used without any other precision, it is in reference to derivative **12.2**. It exists in the free state or glycosidated in the D-$^2C_5$ conformation, which allows an equatorial disposition of the three-carbon side chain. Structure **12.2** represents the stable β-anomer of the free sugar with an axial anomeric hydroxyl group and all-equatorial non-anomeric substituents. An X-ray spectrum of this crystallized β- anomer confirms this conformation and reveals, moreover, that the side chain has the zig-zag conformation with two

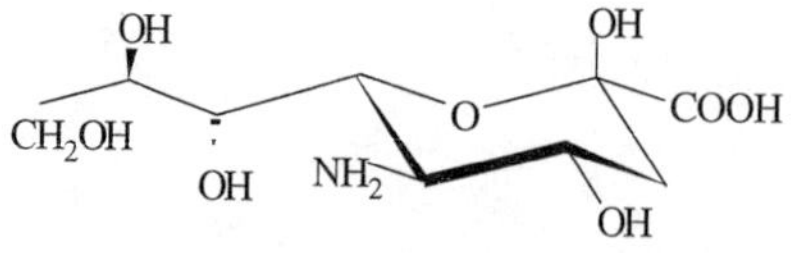

12.1

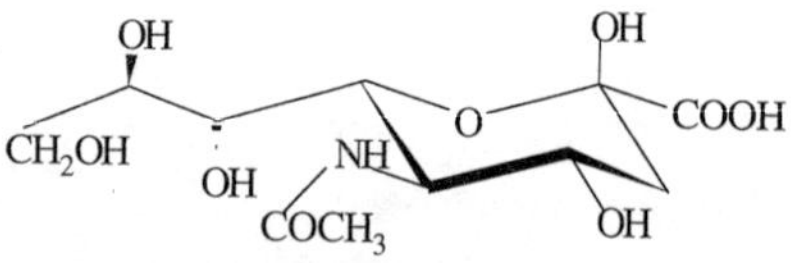

12.2

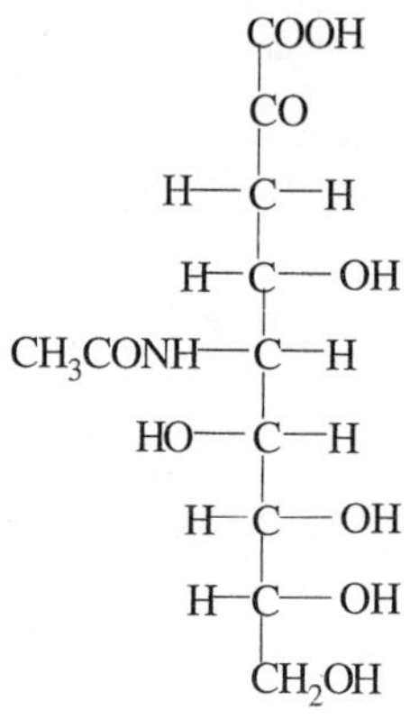

12.3

anti-disposed hydroxyl groups at positions 7 and 8. This D-$^2C_5$ conformation is also indicated in aqueous solution by a proton NMR spectrum which depends slightly on the pH because of an ionizable function. In aqueous solution at pH 7.0, the signals of certain protons of α- and β-anomers are well separated (Table 12.1) and the ratio of the β/α concentration is 93:7. Natural glycosides of sialic acid all have the α-anomeric configuration, whereas the natural activation form (see Section 12.4) has the β-configuration.

In the living world, two types of sialyl derivatives are encountered: α-sialopyranoside residues, located in the vast majority of cases at the non-reducing ends of oligosaccharide chains, and polycondensed products called 'polysialosides'. The latter will be treated separately at the end of this chapter. In sialylated oligosaccharides of cell membranes, the peripheral position of sialic acid leads to its participation in a number of recognition phenomena, the attachment of enzymes, hormones, toxins, lectins, bacteria, and viruses, and participation in the development of the nervous system. It plays a major role in the negative charge of cells.

One characteristic of the neuraminic acid configuration is that, until now, at least 27 functional derivatives have already been found in glycoconjugates (Table 12.2). On the basic structure are found various substituents such as acetyl (Ac), glycolyl (Gc), lactyl (Lac), acetoxyacetyl, methyl (Me), sulfate (S), and phosphate (P). Neuraminic acid **12.1** is abbreviated to Neu followed by a number to indicate the position of the substituent on the oxygen or nitrogen atom, and the symbol for the subsituent. For example, 9-*O*-acetyl-*N*- glycolylneuraminic acid is written Neu9Ac5Gc. The last acid (bottom right of Table 12.2) Neu2en5Ac is an elimination product of the anomeric hydroxyl group, **12.4**. This is found in trace amounts in human fluids and, in a derivative form, it is a nondesirable by-product in certain chemical reactions. This diversity of natural derivatives is particular to sialic acids, while other sugars in glycoconjugates are much less prone to occur in derivatized form.

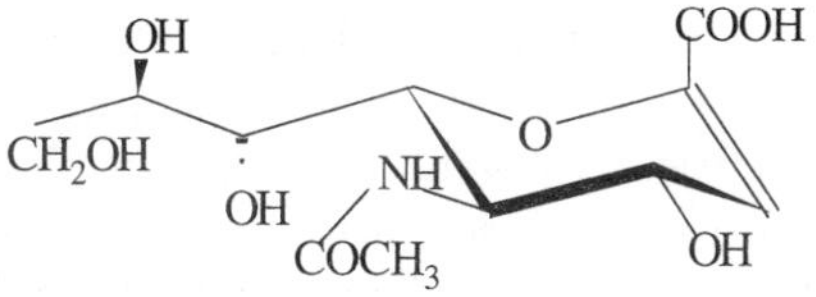

OH COOH
CH_2OH O NH
OH $COCH_3$ OH

12.4

Table 12.1 NMR signals of certain Neu5Ac protons in aqueous solution (pH 7.0) (from Haverkamp *et al.* 1982).

Anomer	H-3 ax	H-3 eq	NAc
β	1.827	2.208	2.050
α	1.621	2.730	2.030

Table 12.2 Natural derivatives of neuraminic acid.

Neu 5 Ac (**12.2**)	Neu 5 Gc
Neu 4,5 Ac_2	Neu 4 Ac 5 Gc
Neu 5,8 Ac_2	Neu 7 Ac 5 Gc
Neu 5,9 Ac_2	Neu 9 Ac 5 Gc
Neu 4, 5, 9 Ac_3	Neu 7,9 Ac_2 5 Gc
Neu 5, 7, 9 Ac_3	Neu 8,9 Ac_2 5 Gc
Neu 5, 8, 9 Ac_3	Neu 7, 8, 9 Ac_3 5 Gc
Neu 5, 7, 8, 9 Ac_4	Neu 5 Gc 8 Me
Neu 5 Ac 9 Lac	Neu 9 Ac 5 Gc 8 Me
Neu 4, 5 Ac_2 9 Lac	Neu 5 Gc 8S
Neu 5 Ac 8 Me	Neu 5 ($COCH_2OAc$)
Neu 5,9 Ac_2 8 Me	
Neu 5 Ac 9 P	Neu 2 en 5 Ac (**12.4**)

Sialic acids also differ from other monosaccharides of glycoconjugates by their total absence in the plant kingdom. They are typical products of the animal kingdom, although they are found in a few bacteria which are pathogenic to man, and protozoa. They appear in echinoderms, starfish, and sea urchins, where we find the methyl ether Neu5Ac8Me and sulfate Neu5Gc8S of the alcohol function at C-8. Among the vertebrates, the greatest variety is found in mammals and there are clear differences between the derivatives that are found in humans and, for example, in cattle. The concentration in normal human saliva is 25 μM. A case has been reported of a mentally retarded child whose urine levels were 10 g a day, or 10^4 times the normal. Glycoproteins of normal secretions are often rich in sialic acid as in submaxillary gland mucins, far-eastern swallow nests (9–36%), and the jelly enveloping sea urchins (up to 70%).

Sialic acid is unstable in an acidic medium (see Section 10.4.1), thus the protocol for separation by acid hydrolysis is a compromize between insufficient hydrolysis and excessive degradation. We have already given, in Section 3.5.1, hydrolysis conditions for NeuAc and even milder conditions which allow the separation of its *O*-acetates without total deacetylation. A great variety of enzymes exist, the neuraminidases, which hydrolyse the glycosidic bond of sialosides. Preparations from bacterial sources which are practical for use after immobilization on agarose can be found in catalogues. Their activation requires a slightly acidic medium, sometimes with mineral ions (e.g. Ca^{2+}, 4 mM) and the hydrolysis rate depends on the neuraminic acid substituents and the position of the bond on the penultimate residue. Activity could be nearly zero. The primary reaction product of the neuraminidase of *Clostridium perfringens* with an α-sialoside is the α-anomer which slowly isomerizes to the β-anomer.

Finally, it is important to mention a sugar closely related to sialic acid by its structure and biological origin, namely, the 3-deoxy-D-*glycero*-D-*galacto*-nonulosonic acid **12.5**, or KDN. This acid has been isolated on a microgram scale from the polysialoglycoprotein of rainbow trout eggs. It is located exclusively at the non-reducing end of polysialyl chains and could play a role in the activation of eggs from the salmonid family by protecting them against sialidases. It appears as a hydrolytically deaminated sialic acid.

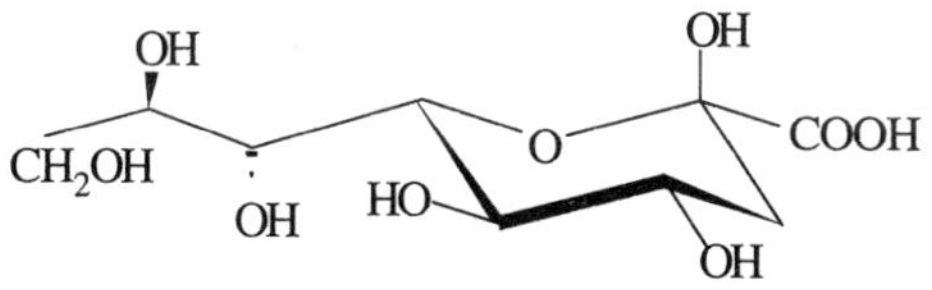

12.5

12.2 Preparations of sialic acids

Chinese grocery shops carry packaged products for making swallow's nest soup. Ten percent of their weight contains bound sialic acid which can be recovered by hydrolysis with warm diluted sulfuric acid, followed by separation on an anion-exchange column. The process is simple, but the raw material is relatively expensive. Since there is currently a strong demand for sialic acids by researchers of human biology, enzymic-type procedures have been proposed which do not require exotic ingredients (David *et al.* 1991).

Mammals produce sialic acid by aldolic condensation of phosphoenolpyruvate and *N*-acetylmannosamine 6-phosphate (reaction 12.1). A *kinase* enzyme catalyses the phosphorylation of *N*-acetylmannosamine and a phosphatase catalyses the hydrolysis of the phosphate of sialic acid. These phosphorylation and dephosphorylation steps are irreversible, such that the synthesis can be total even with low concentrations of the substrate. A variation of reaction (12.1), observed with the bacterium *Neisseria meningitidis*, uses non-phosphated *N*-acetylmannosamine. However, these were not the enzymes used in the preparative synthesis, which used instead a microbial aldolase which catalyses equilibrium (12.2). This enzyme probably plays a catabolic role in these organisms, but it functions in the synthetic sense in the presence of an excess of pyruvate.

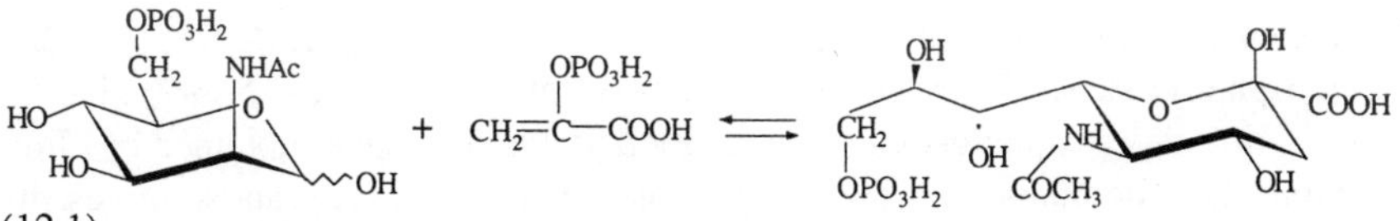

(12.1)

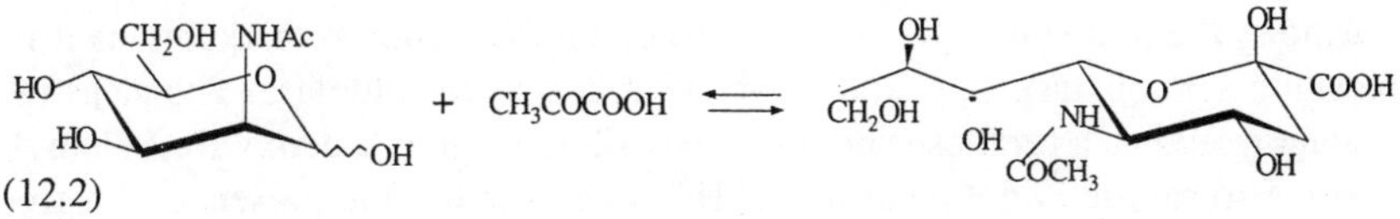

(12.2)

The microbial aldolase is commercially available at a reasonable price and its cloning has been patented. The immobilized enzyme on agarose conserves 80% of its activity after being used four consecutive times and allows *N*-acetylneuraminic acid to be synthesized on a very large scale. Due to the high price of pure *N*-acetylmannosamine, a mixture of *N*-acetylmannosamine and *N*-acetylglucosamine is used as starting products, obtained by alkaline epimerization of the latter sugar (the price of which is negligible). Because the D-*gluco* epimer is strictly not a substrate, the enzyme only uses *N*-acetylmannosamine (Augé *et al.* 1984). The immobilized aldolase has also allowed the preparation, in appreciable quantities, of a certain number of sialic acids found in Table 12.2. The starting product is the 2-azido-2-deoxy-β-D-*manno* derivative which, by manipulating protecting groups, makes it possible to obtain mannosamines diversely substituted at positions 2, 4, and 6. *N*-Acetylated acids, Neu5,9Ac_2 and Neu5Ac9Lac, *N*-glycolylated acids, Neu5Gc and Neu9Ac5Gc, and an acetate of Neu5Gc, Neu5($COCH_2OAc$) were prepared in this way (Augé *et al.* 1988). These acids are widespread in animals, but cannot be purified from natural sources in quantities sufficient enough for a complete study. Aldolase thus accepts substrates other than ManNAc, to which we will come back later. Another adaptation of enzymic synthesis of sialic acid uses *N*- acetylglucosamine as substrate, and a system of two enzymes which combines an *epimerase* with the sialyl aldolase. The latter brings into equilibrium the *N*-acetylglucosamine and its *manno* epimer in the reactor at a pH (7.0–8.0) compatible with the functioning of aldolase. These two enzymes are not immobilized, but are in aqueous solution, and are retained in the reactor by a semi-permeable membrane which allows the substrates and product to pass through (Kragl *et al.* 1991).

Now we are going to take a look at the problem of the enzyme's specificity. It is instructive to go back to KDN, **12.5**. The sialyl aldolase brings it into equilibrium with its two aldolic cleavage products, D-mannose and pyruvate. Conversion can be oriented in the synthetic direction and KDN can be prepared most efficiently with this enzyme from these two common products on any scale (Augé and Gautheron 1987). In fact, D-mannose is a substrate which is at least as good as ManNAc. This suggests that the nature of the substituent at position 2 is not very important provided that it is axial (Augé *et al.* 1989; Augé *et al.* 1990*a*). Indeed, this assumption has allowed the preparation of nonulosonic acids with varying substituents at position 5 such as 5-deoxy **12.6**, 5-azido **12.7**, and 5-phenyl **12.8**. 'Deoxygenated mannoses' at positions 2, 4, 5, or 6 give 5-, 7-, 8-, or 9-deoxy nonulosonic acids in good yields. The sugar obtained by eliminating the side chain of D-mannopyranose, that is D-lyxopyranose, gives the octulosonic acid **12.11**. Its deoxygenation product at C-2, 2-deoxy-D-*threo*-pentose is the simplest substrate known to this day; it gives acid **12.12** in a preparative fashion.

Among the non-substrates we have noted GlcNAc, but, remarkably as it is, the sialic acid of starfish, Neu5Gc8Me **12.9**, resists the aldolase. The cleavage product would be a furanose and, moreover, 2-acetamido-2-deoxy-5-*O*-methyl-D-*manno*-furanose is not a substrate. However, neither the presence of a free hydroxyl group at position 8 in sialic acid nor a pyranose configuration of the cleavage hexose are necessary because 8-deoxy acid **12.10** is a substrate (David *et al.* 1992).

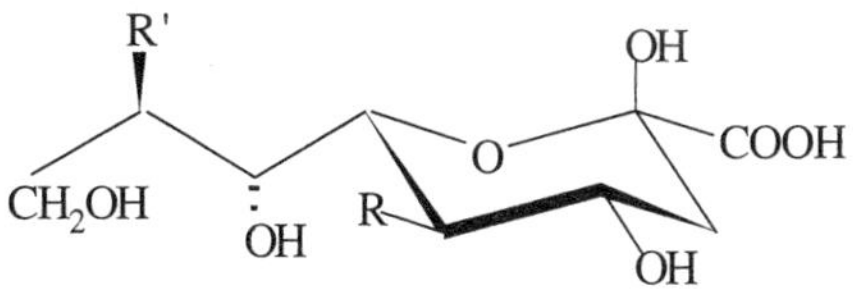

12.6 R= H, R'= OH
12.7 R= N_3, R'= OH
12.8 R= Ph, R'= OH
12.9 R= $NHCOCH_2OH$, R'= OMe
12.10 R= NHAc, R'= H

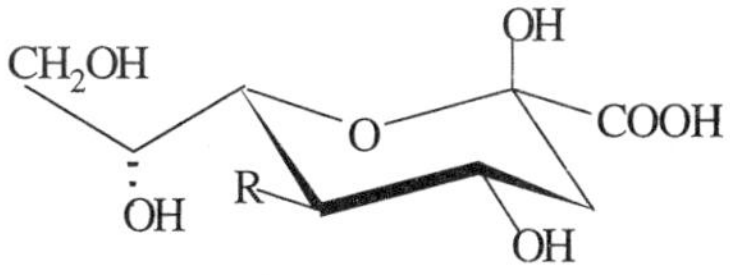

12.11 R= OH
12.12 R= H

On the other hand, reactions are observed in good yields with the L-series such as L-mannose which gives the nonulosonic acid **12.13** (Gautheron-Le Narvor *et al.* 1991) and L-gulose (Lin *et al.* 1992). Acid **12.13** is the enantiomer of the KDN. This multiplicity of observations shows that it is premature to define the specificity of the enzyme, but that it already appears a remarkable tool in organic chemistry.

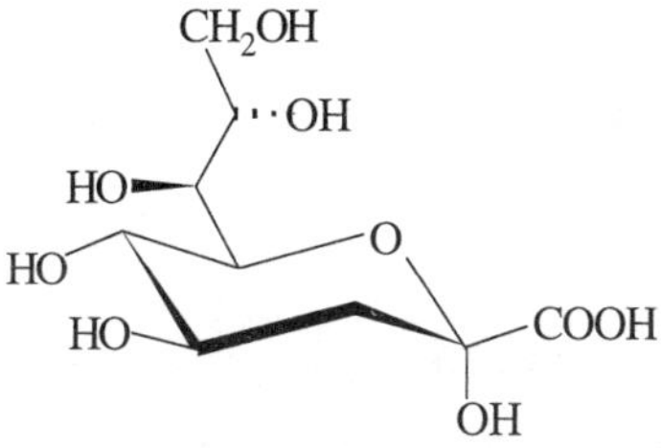

12.13

12.3 Chemical coupling

The chemical coupling of sialic acid, the 'sialylation' of a sugar, presents special problems (Okamoto and Goto 1990). Since it is a ketose, coupling means the construction of a quaternary carbon. There is no substituent at C-3, hence no participation to depend on in that respect. Finally, the coupling reaction conditions favour dehydration which gives a protected version of acid **12.4**, particularly when the hydroxyl group to be coupled is not very nucleophilic. For this reason, the classic peracetylated halides **12.14** (X = F, Cl, Br) rarely give satisfactory results with protected sugars and more often give anomeric mixtures in less than average yield. As a solution, the introduction of a temporary equatorial substituent at C-3 has been proposed. It can function either by participation, or by steric blocking of the β-face. It is with this in mind that a bromide similar to **12.15** was first prepared as a sialylating reagent, but with protection of the alcohol functions by benzyl groups. This reagent has allowed the sialylation of secondary alcohols in good yields (Okamota and Goto 1990). The simplest access to a reagent of this type seems to be the addition of phenylsulfenyl chloride, PhSCl, to the peracetate of the methyl ester of the alkenic acid **12.4** which gives reagent **12.15**. The latter condenses at the 3′ position of a partially protected lactoside to give **12.16** in 70% yield. Compound **12.17** is obtained by desulfuration with tributylstannane (Ito and Ogawa 1987; Ito and Ogawa 1988; Ito *et al.* 1989).

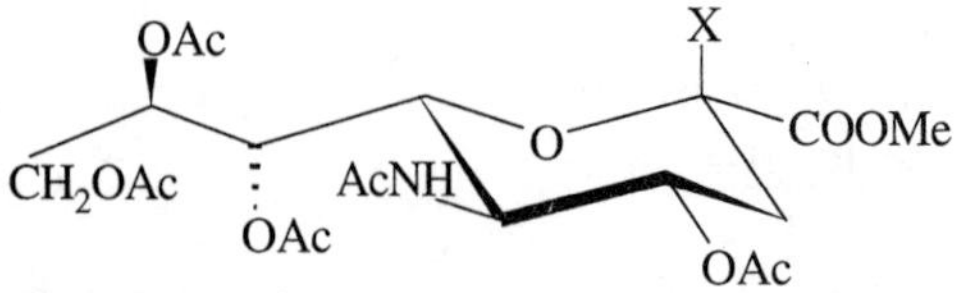

OAc X COOMe O CH_2OAc AcNH OAc OAc

12.14

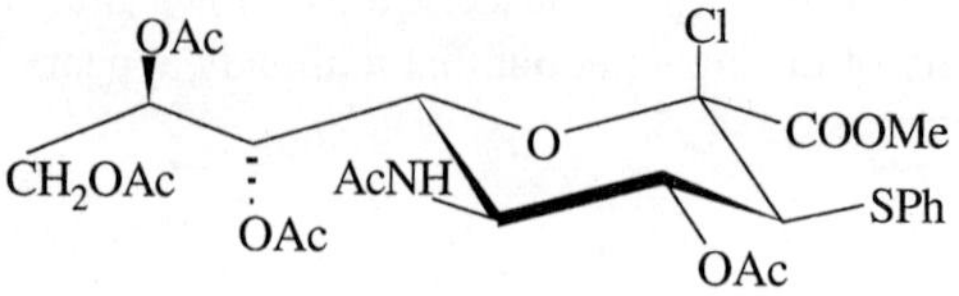

OAc Cl COOMe O CH_2OAc AcNH SPh OAc OAc

12.15

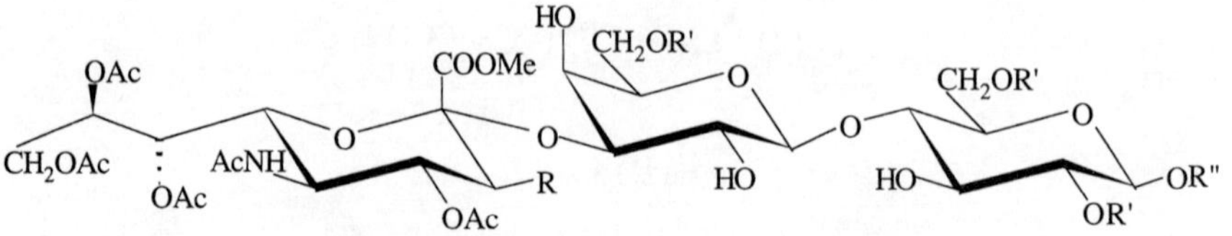

HO CH_2OR' OAc COOMe O CH_2OR' O O O O CH_2OAc AcNH R HO HO OR" OAc OAc OR'

12.16 R= SPh, R'= $COCMe_3$, R"= $CH_2CH_2SiMe_3$
12.17 R= H, R'= $COCMe_3$, R"= $CH_2CH_2SiMe_3$
12.18 R= H, R'= Bz, R"= CH_2CH_2Br

In fact, it was recently shown that direct activation of thioglycosides of sialic acid gave satisfactory yields with secondary hydroxyl groups. Thus the anomeric mixture of protected methyl thioglycosides of sialic acid, in the presence of the *N*-iodosuccinimide–CF_3SO_3H promoter, condenses at O-3 of the galactose (α = 59%; β = 10%) (Hasegawa *et al.* 1991). Ethyl xanthate was finally proposed (Marra and Sinaÿ 1990), prepared by following reaction (12.3). This sialylating reagent is activated by successive additions of silver triflate and MeSBr, precursors of the promoter CF_3SO_3SMe. With 2 eq of xanthate and a lactoside free at positions 2′, 3′, and 4′, **12.18** is obtained selectively and in high yield (β = 1%). In these couplings (Hasegawa *et al.* 1991; Lönn and Stenvall 1992), the solvent always contains a high proportion of acetonitrile (pure if possible) and it is believed that the intermediate formation of an axial nitrilium **12.19** favours the α-introduction of the oxygen.

(12.3) Br, COOMe, O, OBz + KSCSOEt → COOMe, S-CS-OEt, O, OBz

Me, C, N +, O, COOMe, O, H, R

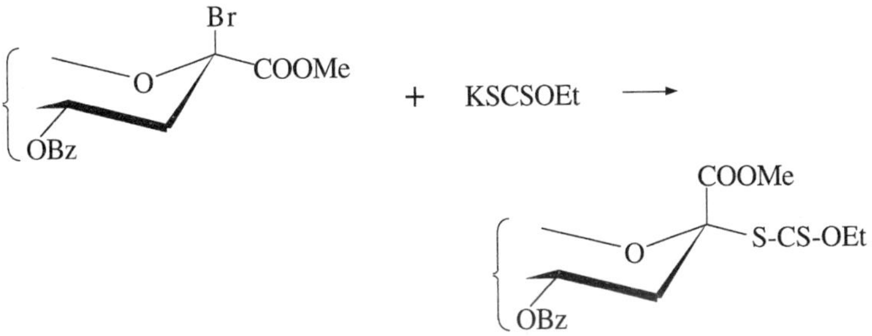

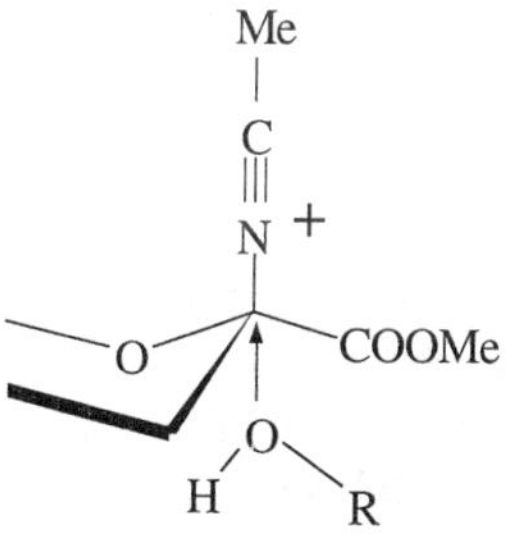

12.19

12.4 Enzymic coupling

The natural activated form of sialic acid is the 'nucleotide sugar' CMPNeu5Ac **12.20** that the cells produce from cytidine triphosphate CTP in the presence of a *synthetase* (cytidine 5′-monophosphosialic acid synthetase) according to reaction (12.4). Among the activation reactions of sugars, the latter is remarkable from several viewpoints; the substrate of the enzyme is the non-phosphorylated sugar, the nucleotide sugar is a phosphodiester instead of a pyrophosphate (see Section 10.4), the reaction is irreversible, and the bond has the β-orientation. The synthetase enzyme is isolable from calf brain purchased at the butchers (David *et al.* 1991). The enzyme of *E. coli* has been cloned and overexpressed in a strain of *E. coli* for which it represents 10% of soluble proteins (Shames *et al.* 1991; Liu *et al.* 1992). As substrate it accepts modified neuraminic acids such as Neu5Gc and Neu5,9Ac$_2$, and less efficiently, KDN. However, in cells, CMPNeu5Gc is synthesized by enzymic oxidation of CMPNeu5Ac. Acetates and methyl ethers result from transformations, mediated by enzymes, of glycoconjugate-bound sialic acids. *S*-Adenosylmethionine is the methyl donor.

(12.4) CTP + [sialic acid] →

[CMP-sialic acid] + $P_2O_7H_4$

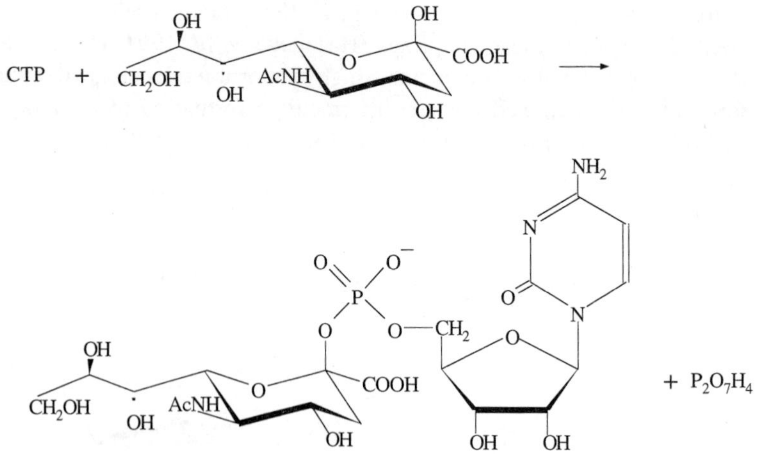

12.20

Reaction (12.4) uses a stoichiometric quantity of CTP, a rare and costly product. Thus the common nucleotide phosphate, cytidine monophosphate, is used as the starting material, and is phosphorylated twice following reactions (12.5) and (12.6). The first phosphorylation, with ATP, is catalysed by the nucleoside monophosphate kinase enzyme. Because of the price of ATP, it must be coupled with a regeneration reaction of the latter which is a traditional biochemical reaction (reaction 12.7), catalysed by the pyruvate kinase enzyme. Phosphorylation of CDP to CTP is possible directly with phosphoenolpyruvate and pyruvate kinase. Besides the two kinases and synthetase, an inorganic pyrophosphatase is used to hydrolyse the pyrophosphate formed in reaction (12.4) and makes it irreversible. The four enzymes are immobilized separately on agarose and the gels are mixed and suspended in an aqueous solution at pH 7.5 in the reactor. This system, supplied with cytidine monosphosphate, sialic acid, and phosphoenolpyruvate in the presence of catalytic amounts of ATP, synthesizes the nucleotide sugar that can be extracted from the medium after the gels are separated by filtration. Figure 12.1 outlines the enzymic reactions coupled in the reactor. The following yields were obtained (David *et al.* 1991): CMPNeu5Ac, 60%; CMPNeu5Gc, 80%; CMPNeu5,9Ac, 52%; CMP-KDN, 26%.

(12.5) CMP + ATP → ADP + CDP

(12.6) CDP + phosphoenolpyruvate → CTP + pyruvate

(12.7) ADP + phosphoenolpyruvate → ATP + pyruvate

Coupling with the acceptor sugar which takes place with simultaneous liberation of CMP, is catalysed by a *sialytransferase*, as shown in the biochemical reaction in Fig. 12.2. Sialytransferases introduce a sialyl unit in the form of α-D-pyranosyl with a sugar at the non-reducing end of an oligosaccharide. A recent review (Harduin-Lepers *et al.* 1995) lists 19 sialytransferases (ST). The definition of their specificity takes into account the configuration of the acceptor, the sialylation position, and the sequence structure which carries the acceptor. Transferases ST6 catalyses sialylation at position 6 of galactose, GlcNAc, and GalNAc. Transferases ST3 introduce a sialyl residue at position 3 of a non-reducing galactose unit end. Transferases ST8 catalyse sialylation at position 8 of a non-reducing sialyl unit end to give the sequence Neu5Ac-α-(2→8)-Neu5Ac. Seventeen enzymes of this family have been cloned from birds or mammals.

The most common sialyltransferase introduces the α-D-sialopyranosyl residue at the primary position of galactose in methyl β-D-galactopyranoside, methyl β-lactoside, and *N*-acetyllactosamine, where the galactose is free or at the non-reducing end of an oligosaccharide. It is abundant in pork liver, found at the butchers. In immobilized form, it has allowed the preparation of tetrasaccharide glycoside **12.21** whose sequence is often found in glycoproteins (Augé *et al.* 1990*b*). The three glycosidic couplings of **12.21** were carried out in three different ways: the first with mannose by an organic chemical process (with CF_3SO_3Ag), the second with the galactosyltransferase, and the third with the sialyltransferase (46%). This enzyme also accepts the nucleotide CMPNeu5,9Ac_2 as substrate. Transfer to the lactosamine gives trisaccharide **12.22** (115 mg, 57%) which is a part of the influenza C virus receptor in red blood cells (Augé *et al.* 1990*b*). It would have been very complicated to prepare by common chemical methods a trisaccharide specifically acetylated on only one of its eleven hydroxyl groups.

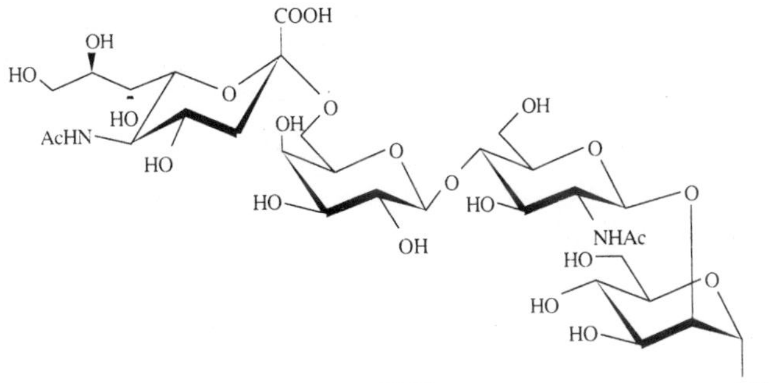

12.21

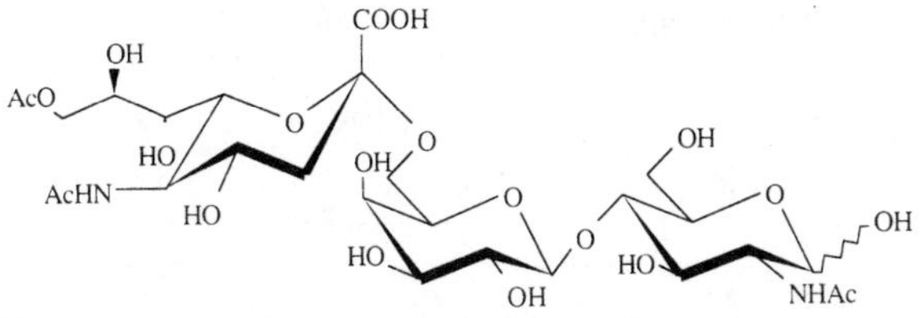

12.22

Nonetheless, it seems that the majority of the sialylated oligosaccharides involved in recognition phenomena present the glycosidic bond Neu5Ac-α-(2→3)-Gal. Without a great deal of difficulty, a transferase can be isolated from pork liver usable in this coupling whose natural substrate is the sequence β-D-Gal-(1→3)-D-GalNAc. This transferase also catalyses the coupling on β-D-Gal-(1→3)-D-GlcNAc and allows the preparation of trisaccharide **12.23** (140 mg), the epitope 'CA 50' of an antigen abundantly found in association with certain tumours (Lubineau *et al.* 1992).

12.23

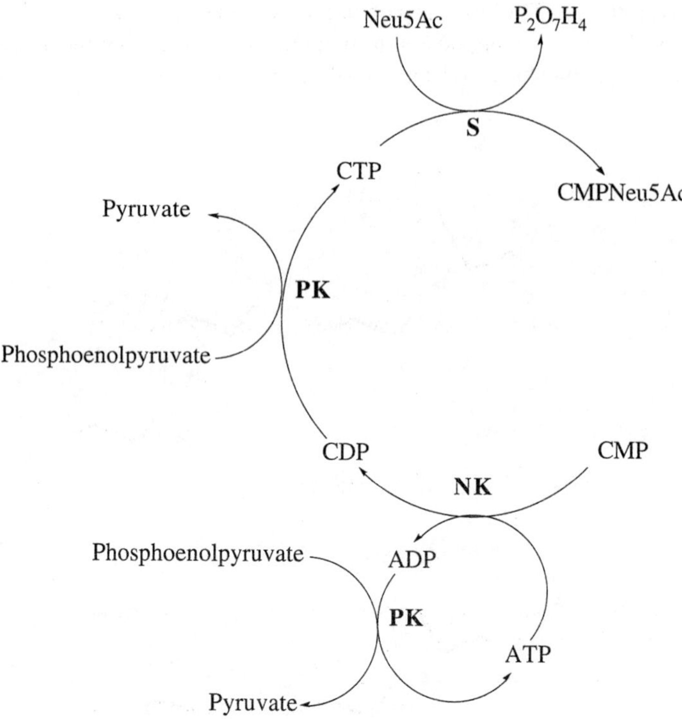

Fig. 12.1 Enzymic synthesis of CMPNeu5Ac with regeneration of cofactors: **NK** = nucleotide monophosphate kinase, **PK** = pyruvate kinase, **S** = synthetase.

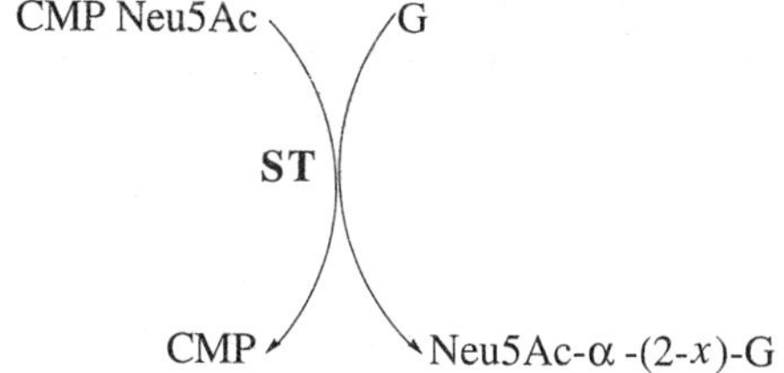

Fig. 12.2 Enzymic sialylation with sialytransferase **ST**.

We can observe that linking the reaction schemes in Fig. 12.1 and 12.2 forms a complete sialylation cycle which, in principle, should function with only catalytic amounts of CMP. However, an undesirable enzymic activity of the synthetase isolated from calf brain, impossible to eliminate economically, prevents the functioning of this cycle, which was carried out instead with the cloned synthetase. The two transferases, α-(2→6) and α-(2→3) were also cloned (Ichikawa *et al.* 1991; Ichikawa *et al.* 1992; Tsuji 1996).

The UDP-Gal (described in Section 10.4.1) and CMPNeu5Ac regeneration systems have been combined in a one-pot reaction. This methodology has been used in the synthesis of the so-called 'sialyl Le^x' tetrasaccharide **12.24**. Four monosaccharides were coupled using β-1,4-galactosyltransferase, α-2,3-sialyltransferase, and α-1,3-fucosyltransferase in sequence. (Section 10.4.1). It has been reported that kilogram quantities of the sialyl Le^x tetrasaccharide were prepared in this way (Gijsen *et al.* 1996).

NeuAc-α-(2→3)-Gal-β-(1→4)-GlcNAc-β-(1→3)-Gal

↑

Fuc-α-(1→3)

12.24

12.5 Polysialic acids (Roth *et al.* 1993)

12.5.1 Introduction

This section supposes that the reader is already familiar with a few of the ideas about glycoproteins and immunochemistry that will be developed later in Chapters 13 and 15, which may be referred to if necessary. Instead of following a more logical order, we prefer to condense the greatest number of experiments concerning the remarkable character of sialic acids in a single chapter.

The configurations of the glycosidic bond in natural polycondensed polymers of sialic acid are α-(2→8) and α-(2→9). Exceptionally, in the polysialic acid of starfish, the interglycosidic bond involves the primary alcohol function of the glycollyl residue in Neu5Gc8Me (Roth *et al.* 1993). Oligomers **12.25** ($n = 2, \ldots 5$) have been isolated and characterized from a ganglioside present in starfish.

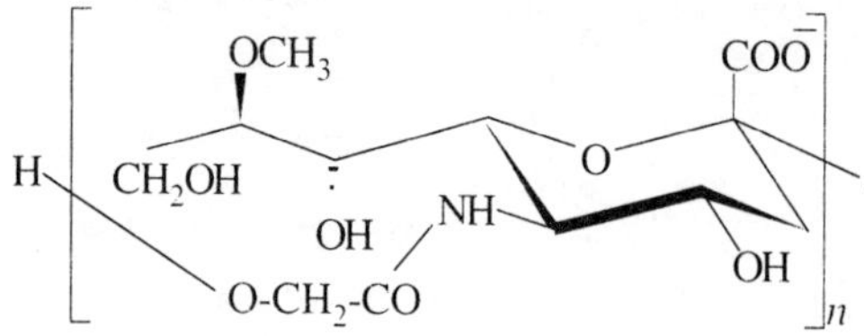

12.25

12.5.2 Microbial polysialic acids

The meningococcus, *Neisseria meningitidis*, is divided into 10 distinct serotypes. Groups B and C are responsible for 80% of the meningitis cases reported, one of the major causes of death in children and adults. These bacteria contain linear polysialic acids. In Group C, the bond between the sialyl residues, partially or non-acetylated, is α-(2→9). This structure, **12.26**, is a powerful immunogen; the polysaccharide of Group C is an official vaccine. However, there is no response in children less than two years old, perhaps because at this age, a similar polysialyl structure is found in their tissues.

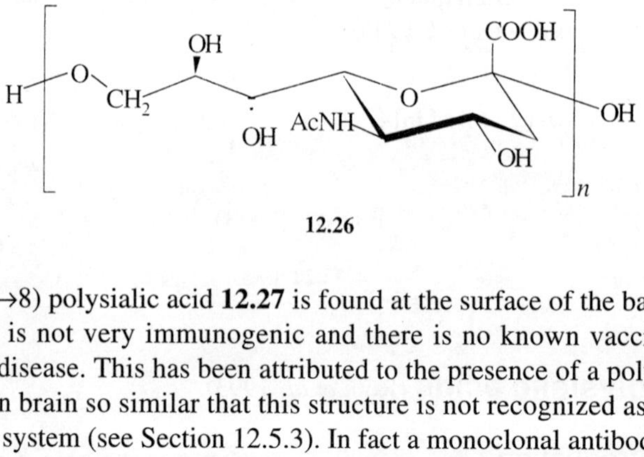

12.26

The α-(2→8) polysialic acid **12.27** is found at the surface of the bacteria from Group B. It is not very immunogenic and there is no known vaccine for this form of the disease. This has been attributed to the presence of a polysialic acid in the human brain so similar that this structure is not recognized as foreign by the immune system (see Section 12.5.3). In fact a monoclonal antibody common to these two polysialic acids (cross reaction) is known. The bacterium *Escherichia coli* KI, also the cause of meningitis, produces an α-(2→8) polysialic acid, possibly acetylated. Finally, this same polymer is observed on the surface of certain tumour cells.

The α-(2→8) polysialic acid has been examined with the hope of obtaining a good immunogen through modification. It reacts with an antibody from horse serum, and in a traditional fashion, research was done to recognize the epitope by trying to inhibit the combination by oligosaccharides. Normally, an oligosaccharide epitope corresponds to, at most, five to six monosaccharide residues;

12.27

however, oligomer **12.27** ($n = 5$) is not active, and n should be equal to at least ten in order to observe inhibition. This appears to be a *conformational epitope*, meaning that it is not only the oligosaccharide sequence that counts but also the general shape. That shape of the epitope of the α-(2→8) polysialic acid would be a very elongated helix with eight to twelve residues per turn.

In a bacteriophage, an α-(2→8) endosialidase enzyme is found which hydrolyses the internal bonds of this polysialic acid to a high degree of specificity. The cleavage site must be flanked by at least five sialoside residues on the side of the non-reducing end and by three on the other. The recognition site is exceptionally large, as was already observed in the immunochemical reaction.

12.5.3 Neural cell adhesion molecule, N-CAM

Cell adhesion plays a major role in the embryonic development of the nervous system. Rather than a great number of adhesion molecules being involved, it is now believed that there are only a small group of different specificities and that the variety of phenomena observed is based on the coordination of their expression with cytoplasmic events. The N-CAM molecules are the best known (Cunningham *et al.* 1987).

Figure 12.3 shows a schematic representation (Regan 1991) of one of these molecules, N-CAM$_{180}$, the index 180 indicating the molecular weight, close to 180 000, of the polypeptide backbone. This is a transmembrane protein consisting of a single polypeptide chain of 1072 amino acids. Going from the NH_2 end, which bathes in the external medium, there is first an area of about 400 amino acid residues. In this region there is an oligosaccharide side chain, probably

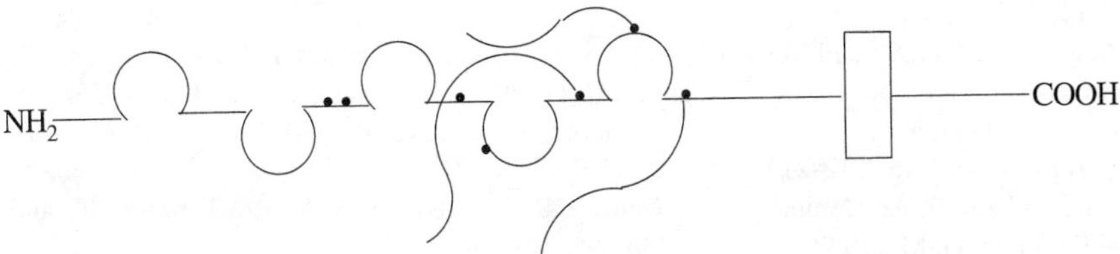

Fig. 12.3 Schematic representation of the N-CAM$_{180}$ molecule. The black dots correspond to the glycosidation site (from Regan 1991) (reproduced with kind permission from Elsevier Science Ltd, Kidlington, UK).

linked to asparagine 203, which carries the sulfated tetrasaccharide epitope called L2/HNK-1, **12.28**, which is not sialylated. The domains involved in cell adhesion are also found in this region, according to a mechanism called 'homophilic bonding'. This means that a molecule of N-CAM of one cell adheres to a molecule of N-CAM of another cell. Thus this adhesion does not involve sialic acid. Following the chain towards the carboxyl end, new glycosylated chains are encountered on asparagines 404, 430, and 459. The latter terminate with polysialyl chains having the α-(2→8) bond. At around residue 692, the polypeptide backbone begins to cross the cell membrane from which it exits towards residue 710 and a long root of around 300 residues extends into the cytoplasm.

HSO_3-3-GlcUA-β-(1→3)-Gal-β-(1→4)-GlcNAc-β-(1→3)-Gal

12.28

The total amount of sialic acid varies during embryonic development, between <10% and 30%. Observations have shown that these polysialic acid chains inhibit adhesion. Thus their biosynthesis seems to be a mechanism of regulating the adhesive properties of the polypeptide area ending with NH_2. Generally speaking, it is conceivable that the correct development of the nervous system sometimes involves adhesion, sometimes repulsion. The exact mechanism is not yet known with certainty. There may be a simple mechanical effect. The growth of the polysialyl chain greatly increases the volume of the molecule N-CAM and this could keep the surface of the neighbouring cell too far away for an interaction to take place. This volume is due, in part, to strong hydration. There could be repulsion by negative charges of carboxylates or encapsulation of the adhesion sites by polysialyl chains, or conformational modifications.

References

Augé, C. and Gautheron, C. (1987), *J. Chem. Soc., Chem. Commun.*, 859–860.

Augé, C., David, S., and Gautheron, C. (1984), *Tetrahedron Lett.*, **25**, 4663–4664.

Augé, C., David, S., Gautheron, C., Malleron, A., and Cavayé, B. (1988), *New. J. Chem.*, **12**, 733–744.

Augé, C., David, S., and Malleron, A. (1989), *Carbohydr. Res.*, **188**, 201–205.

Augé, C., Gautheron, C., and Fernandez, R. (1990*b*), *Carbohydr. Res.*, **200**, 257–268.

Augé, C., Gautheron, C., David, S., Malleron, A., Cavayé, B., and Bouxom, B. (1990*a*), *Tetrahedron*, **46**, 201–214.

Cunningham, B. A., Hemperly, J. J., Murray, B. A., Prediger, E. A., Brackenbury, R., and Edelman, G. M. (1987), *Science*, **236**, 799–806.

David, S., Augé, C., and Gautheron, C. (1991), *Adv. Carbohydr. Chem. Biochem.*, **49**, 176–237.

David, S., Malleron, A., and Cavayé, B. (1992), *New. J. Chem.*, **16**, 751–755.

Gautheron-Le Narvor, C., Ichikawa, Y., and Wong, C.-H. (1991), *J. Am. Chem. Soc.*, **113**, 7816–7818.

Gijsen, H. J. M., Qiao, L., Fitz, W., and Wong, C.-H. (1996), *Chem. Rev.*, **96**, 443–473.

Harduin-Lepers, A., Recchi, M.-A., and Delannoy, P. (1995), *Glycobiology*, **5**, 741–758.

Hasegawa, A., Nagahama, T., Ohki, H., Hotta, K., Ishida, H., and Kiso, M. (1991), *J. Carbohydr. Chem.*, **10**, 493–498.

Haverkamp, J., van Halbeek, H., Dorland, L., Vliegenthart, J. F. G., Pfeil, R., and Schauer, R. (1982), *Eur. J. Biochem.*, **122**, 305–311.

Ichikawa, Y., Lin, Y.-C., Dumas, C. P., Shen, G.-J., Garcia-Junceda, E., Williams, M. A. *et al.* (1992), *J. Am. Chem. Soc.*, **114**, 9283–9298.

Ichikawa, Y., Shen, G.-J., and Wong, C.-H. (1991), *J. Am. Chem. Soc.*, **113**, 4698–4700.

Ito, Y. and Ogawa, T. (1987), *Tetrahedron Lett.*, **28**, 6221–6224.

Ito, Y. and Ogawa, T. (1988), *Tetrahedron Lett.*, **29**, 3987–3990.

Ito, Y., Numata, M., Sugimoto, M., and Ogawa, T. (1989), *J. Am. Chem. Soc.*, **111**, 8508–8510.

Kragl, U., Gygax, D., Ghisalba, O., and Wandrey, C. (1991), *Angew. Chem. Int. Ed. Eng.*, **30**, 827–828.

Lin, C.-H., Sugai, T., Halcomb, R. L., Ichikawa, Y., and Wong, C.-H. (1992), *J. Am. Chem. Soc.*, **114**, 10138–10145.

Liu, J. L.-C., Shen, G.-J., Ichikawa, Y., Rutan, J. F., Zapata, G., Vann, W. F., Wong, C.-H. (1992). *J. Am. Chem. Soc.*, **114**, 3901–3910.

Lönn, H. and Stenvall, K. (1992), *Tetrahedron Lett.*, **33**, 115–116.

Lubineau, A., Augé, C., and François, P. (1992), *Carbohydr. Res.*, **228**, 137–144.

Marra, A. and Sinaÿ, P. (1990), *Carbohydr. Res.*, **195**, 303–308.

Okamoto, K. and Goto, T. (1990), *Tetrahedron*, **46**, 5835–5857.

Regan, C. M. (1991), *Int. J. Biochem.*, **23**, 513–523.

Roth, J., Rutishauser, U., and Troy, II, F. A. (1993), *Polysialic acid, from microbes to man*, Birkhäuser, Basel.

Schauer, R. (1982), *Adv. Carbohydr. Chem. Biochem.*, **40**, 131–234.

Schauer, R. (1991), *Glycobiology*, **1**, 449–452.

Shames, S. L., Simon, E. S., Christopher, C. W., Schmidt, W., Whitesides, G. M., and Yang, L. L. (1991), *Glycobiology*, **1** 187–191.

Tsuji, S. (1996), *J. Biochem.*, **120**, 1–13.

13 Glycoconjugates

13.1 Glycolipids

13.1.1 Definitions and methods of isolation

A glycolipid is an association of an oligosaccharide with a lipid (Morrison 1988; Li and Li 1982; Stults *et al.* 1989). Our description will be limited to two families which are easily definable. In the *sphingosine type*, the oligosaccharide is linked glycosidically to the primary alcohol function of a long-chain amino alcohol. The most common one is sphingosine **13.1**. In glycolipids, the amino function is involved in an amide bond with an acyl residue of a long-chain fatty acid. There is a great variety of derivatives of this type, which have been given the collective name *ceramide (Cer)*. The other important family of glycolipids belongs to the *glycerol* type. The oligosaccharide is linked by a glycosidic bond to the primary alcohol function of a glycerol molecule esterified by two fatty acid molecules **13.2**.

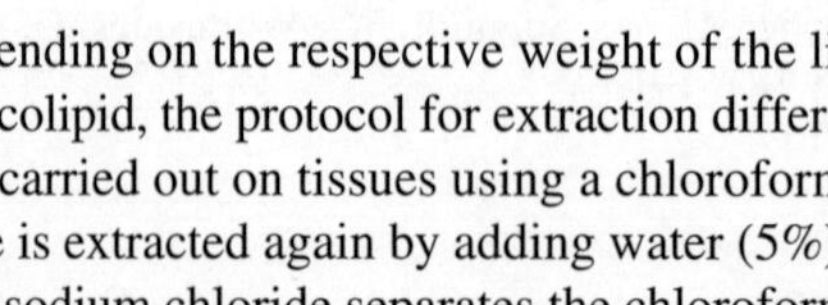

$CH_3(CH_2)_{12}$ — NH_2 — OH — OH

13.1

CH_2OH

CHOCOR

CH_2OCOR'

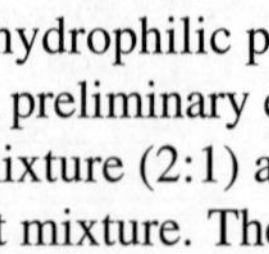

13.2

Depending on the respective weight of the lipophilic and hydrophilic parts of the glycolipid, the protocol for extraction differs. Generally, a preliminary extraction is carried out on tissues using a chloroform–methanol mixture (2:1) and the residue is extracted again by adding water (5%) to the solvent mixture. The addition of sodium chloride separates the chloroform phase containing neutral glycolipids from the aqueous methanol phase containing acid or very hydrophilic glycolipids. The usual techniques for separation are employed such as chromotography on silica gel, dialysis, chromatography on a diethylaminoethylcellulose column, and TLC. Acid methanolysis separates the sphingosine and the fatty acids from the oligosaccharide. Oligosaccharide characterization is carried out using the general techniques described in Chapter 9. The reader should recall that it is advantageous to analyse the glycolipid directly by FAB.

13.1.2 Animal glycolipids

Neutral glycolipids from animal tissue (without sialic acid) are nearly exclusively glycosphingolipids. Families are classified according to the sequence of monosaccharide residues attached to the ceramide. The unit directly linked to the ceramide is in the majority of cases β-D-Glc*p* and exceptionally β-D-Gal*p*. The following residue is always β-D-Gal*p* so that the most numerous families of glycosphingolipids begin with a β-lactosyl unit. In the less frequently encountered *arthro* series, the sequence at the reducing end is Man-β-(1→4)-Glc-β-(1→Cer). Table 13.1 would be difficult to memorize, but it can give the reader an idea of the variety of structures that can be found.

These tetrasaccharide glycolipids, such as they are or with shorter or longer sugar chains, are found in diverse tissues of man and higher animals. *Globo* trihexoside **13.3** is the most important glycolipid in human red blood cells; *lacto* tetrahexoside is extended and carries the epitopes of blood group substances, ABH, Lewis, Ii and Pi; the *muco* family carries epitopes of groups A and H. It is remarkable to observe glycolipids, whose backbones can have as many as 20 repeating units of *N*-acetyllactosamine linked to each other by a β-(1→3′) bond according to formula **13.4**.

Gal-α-(1→4)-Gal-β-(1→4)-Glc-β-(1→Cer)

13.3

→[3)-Gal-β-(1→4)-GlcNAc-β(1]$_n$→3)-Gal-β-(1→4)-Glc-β-(1→Cer)

13.4

13.1.3 Gangliosides

This is the name given to sialylated glycolipids. The sialic acid support is generally a glycosphingolipid from the *ganglio* family. The separation of these glycolipids and their fractionation into mono-, di-, and trisialyl derivatives is based on their acidic character and are carried out on anion-exchange columns (DEAE-cellulose). The sialic acids found are Neu5Ac, Neu5Gc, and their

Table 13.1 Families of neutral glycosphingolipids; all residues are pyranoses from the D-series.

globo	GalNAc-β-(1→3)-Gal-α-(1→4)-Gal-β-(1→4)-Glc-β-(1→Cer)
isoglobo	GalNAc-β-(1→3)-Gal-α-(1→4)-Gal-β-(1→4)-Glc-β-(1→Cer)
lacto	Gal-β-(1→3)-GlcNAc-β-(1→3)-Gal-β-(1→4)-Glc-β-(1→Cer)
neolacto	Gal-β-(1→4)-GlcNAc-β-(1→3)-Gal-β-(1→4)-Glc-β-(1→Cer)
ganglio	Gal-β-(1→3)-GalNAc-β-(1→4)-Gal-β-(1→4)-Glc-β-(1→Cer)
muco	Gal-β-(1→3)-Gal-β-(1→4)-Gal-β-(1→4)-Glc-β-(1-Cer)
gala	GalNAc-α-(1→3)-GlcNAc-β-(1→3)-Gal-α-(1→4)-Gal-α-(1-Cer)

acetates. Hematoside **13.5** is present in the brain and erythrocytes of dogs. It is also found in the erythrocytes of horse and cattle in which Neu5Ac is replaced by Neu5Gc and Neu4Ac5Gc, respectively. The most important ganglioside of the brain, called GM1, is the sialylation product **13.6** of the *ganglio* glycolipid from Table 13.1 with an α-(2→3) glycosidic bond. Polysialylated sequences are also observed with an α-(2→8) intersialosidic bond such as **13.7** and **13.8**, linked at position 3 of the same internal galactose of the *ganglio* glycolipid, and similar sialyl and disialyl extensions starting at position 3 of the non-reducing galactose end. To conclude, in glycosphingolipid as well as in ganglioside families, there is a great variety of oligosaccharide structures built from basic chains. Note the presence of L-fucose on some of them and sulfuric esters.

Neu5Ac-α-(2→3)-Gal-β-(1→4)-Glc-β-(1→Cer)

13.5

Gal-β-(1→3)-GalNAc-β-(1→4)-Gal-β-(1→4)-Glc-β-(1→Cer)
↑
Neu5Ac-α-(2→3)

13.6

Neu5Ac-α-(2→8)-Neu5Ac-α-(2→3)-

13.7

Neu5Ac-α-(2→8)-Neu5Ac-α-(2→8)-Neu5Ac-α-(2→3)-

13.8

For the synthesis of ganglioside, direct coupling of the trichloroacetimidate of tri- and higher oligosaccharides to ceramide was not very efficient, but very good yields could be achieved in the coupling of trichloroacetimidates to a precursor of ceramide. This was a modified sphingosine, with an azido group instead of the amino group, and protection of the secondary alcohol function was carried out by silylation or benzoylation. Coupling of a hexasaccharide with two fucosyl branches was achieved in 75% yield. Glycosidation is followed by reduction of the azido function and attachment of the fatty acid group (Schmidt and Kinzy 1994).

13.1.4 Plant glycolipids

We will only mention galactosylglycerols **13.9** and **13.10**.

```
Gal-β-(1→O-CH2                    Gal-α-(1→6)-Gal-β-(1→O-CH2
          |                                                 |
          CHOCOR                                            CHOCOR
          |                                                 |
          CH2OCOR'                                          CH2OCOR'

          13.9                                              13.10
```

13.2 Glycoproteins (Montreuil 1980)

13.2.1 General

Glycoproteins are proteins covalently linked to oligosaccharides. There is a good deal of proof that these oligosaccharide chains are responsible for important biological properties. The reader will find a number of examples in the following chapters. Glycoproteins, widespread in the living world, are found as soluble compounds, linked to cell membranes, inside cells, or in extracellular fluids. For their extraction or purification, protein chemistry methods are used, but the presence of sugars allows the employment of a very powerful supplementary tool, affinity chromatography on an immobilized *lectin*. Lectins (see Chapter 15) are plant or animal proteins, easily accessible when originating from the plant kingdom, which reversibly bind with monosaccharides or oligosaccharide sequences. A specific sugar ligand corresponds to each one. An immobilized lectin column retains specifically the glycoprotein to which this ligand is attached. Elution takes place with a solution of this specific ligand (a small molecule) which displaces the glycoprotein from recognition sites of the lectin.

Six sugars participate in the oligosaccharide structure of complete animal glycoproteins: galactose, mannose, *N*-acetylglucosamine, *N*-acetylgalactosamine, and sialic acid from the D-series, and fucose from the L-series, all in the pyranoid form. There are two main types of linkages to the polypeptide backbone which can, moreover, coexist in the same glycoprotein. In all cases, the oligosaccharide should be separated with the miminum of degradation and its sequence is analysed using methods discussed in Chapter 9. The adjacent sequence at the junction is called the 'core' of the oligosaccharide. This core concept (Montreuil 1980) is justified by the observation that there is only a very small number of them. Each is presented as an invariant at the heart of a multitude of forms.

13.2.2 Glycoside proteins

The oligosaccharide is involved at its 'reducing' end in a glycosidic bond with one of the hydroxylated side chain residues of the polypeptide. They are the ones derived from L-serine **13.11**, L-threonine **13.12**, L-hydroxylysine **13.13**, and

L-hydroxyproline **13.14**. Formula **13.15** represents a widely occurring type of junction on L-serine (R = H) or L-threonine (R = CH_3). In certain glycoproteins, the β-D-Gal unit is replaced by β-D-GalNAc. From this disaccharide, extensions and branches lead to a very great variety of structures. To determine them, it is necessary to detach the oligosaccharide from the polypeptide. Hydrolysis catalysed by a proteolytic enzyme with high specificity, pronase, can ultimately liberate a *glycopeptide*, that is to say the oligosaccharide still linked to the amino acid of the junction, but sometimes the oligosaccharide acts as a protecting function against enzymic proteolysis. The most characteristic cleaving mode of glycosides from L-serine and L-threonine is through β-elimination in an alkaline medium. This is the expected result of the lability of the proton α to the amide carbonyl which is liberated by base B (Fig. 13.1*a*). Unfortunately, the story does not end here. The sugar whose reducing function has been liberated is in equilibrium with the aldehyde tautomer which also presents an acidic proton in the α-position producing a new β-elimination (Fig. 13.1*b*), and so forth. This is the 'peeling' that we try to avoid when working in the presence of $NaBH_4$ with the hope that reduction of the aldehyde carbonyl will be faster than β-elimination.

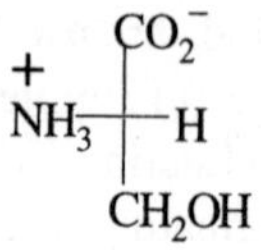

13.11 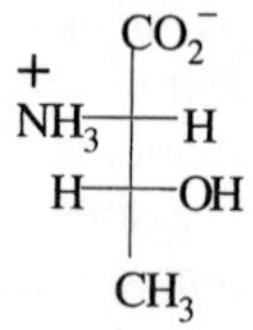 **13.12** 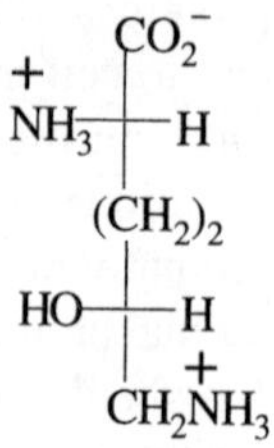 **13.13** 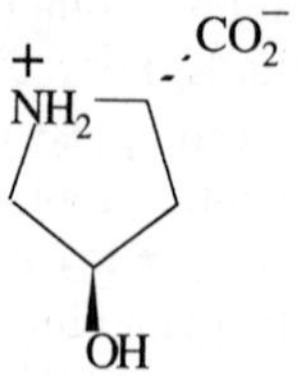 **13.14**

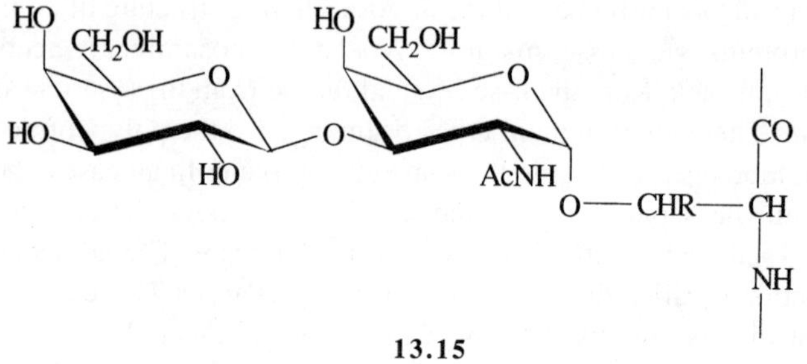

13.15

In fact, the apparent simplicity of the β-elimination reaction is deceiving. Compared to a benzyl, the glycopyranosyl is eliminated abnormally fast, as for example on a model glycopeptide it is eliminated in 10 s at 20°C in sodium carbonate. It is thought that the sugar complexes with the sodium cation and is thus transformed into a good leaving group (Kunz and Rück 1993).

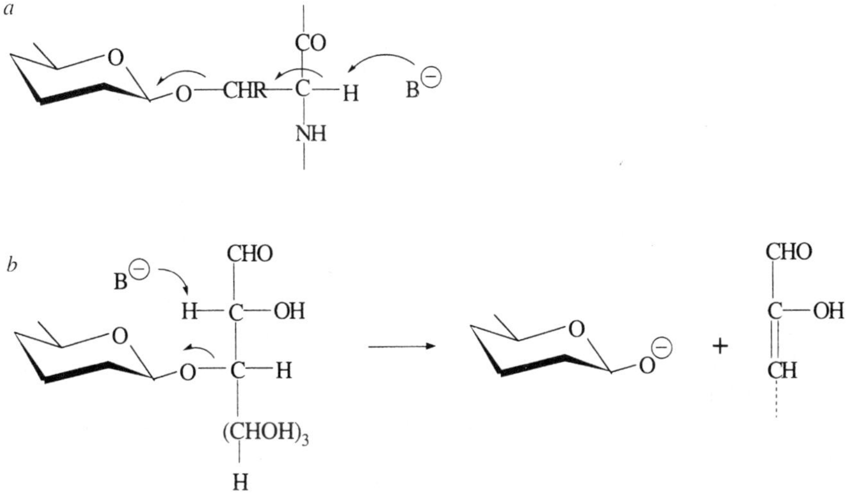

Fig. 13.1 (a) β-Elimination and (b) 'peeling' of glycosides from serine and threonine.

The synthesis of the peptide disaccharide corresponding to **13.15** poses the problem of the 1,2-*cis* glycosidation. The activated disaccharide with a non-participating azido group **13.16** is used, which is condensed to a serine or a threonine protected on the amino and carboxyl groups with the promoter mixture $AgClO_4/Ag_2CO_3$. In order to reduce the azido group to an amino group before *N*-acetylation, H_2S is used. It has been recently reported that the protected galactosamine derivative, 2-acetamido-3-*O*-acetyl-4,6-*O*-benzylidene-2-deoxygalactose, activated as the α-trichloroacetimidate, gave a near 55% yield of the serine α-glycoside (α/β ratio = 20:1) under the usual Schmidt coupling reaction conditions. This represents a very simple access to this structure (Yule *et al.* 1995). The polypeptide side chain can also be extended by the usual methods of peptide synthesis (Paulsen 1990; Meldal and Bock 1990) to obtain **13.17**, for example.

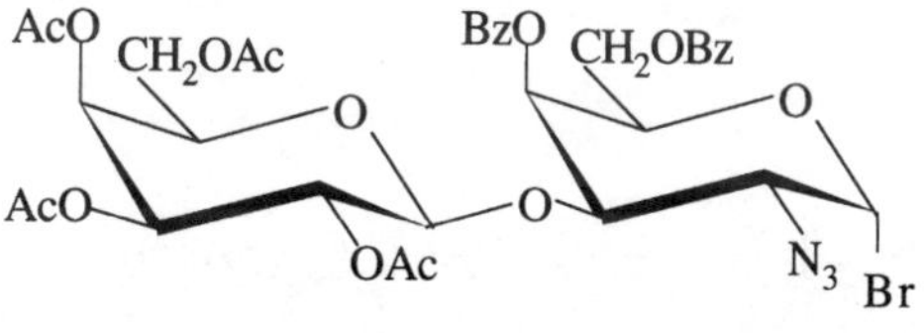

13.16 **13.17**

13.2.3 Glycosaminide proteins

The sugar at the junction is *N*-acetylglucosamine and the partial structure is **13.18**, the sugar being substituted as we shall see further on. Compound **13.18** can be described as being either a glycosylamine amide derived from aspartic acid or, preferably, as a glycosylation product of asparagine. One method for detaching the oligosaccharide without too much degradation is by trifluoroacetolysis (Nilsson and Svensson 1979) which consists in maintaining it for 2 days at 100°C in a mixture of trifluoroacetic anhydride and trifluoroacetic acid, 50:1. The oligosaccharide is perfluoroacetylated and the inductive effect of the fluoroacetylated radicals stabilizes the interglycosidic bonds by destabilizing the protonated intermediate of the hydrolysis, partial formula **13.19**. However, the bond is cleaved in this way with asparagine. Subsequent corrective treatment of the cleavage product is necessary to hydrolyse the ester and amide groups of the trifluoroacetic acid and to reacetylate the amino group. The core of the branching on asparagine is pentasaccharide **13.20**. On the reducing side are found two *N*-acetylglucosamine residues as in chitobiose, then a β-mannose residue. To this β-mannose are attached two α-mannose units at positions 3 and 6. Depending on the structure of the extensions and branching at this core, **13.20**, two major sub-families can be identified. In the first, only α-mannose residues are encountered. In the second, the pentasaccharide is substituted by a diverse number of *N*-acetyllactosamine residues, Gal-β-(1$\rightarrow$4)-GlcNAc, with in addition, fucose residues, sialic acid, etc.

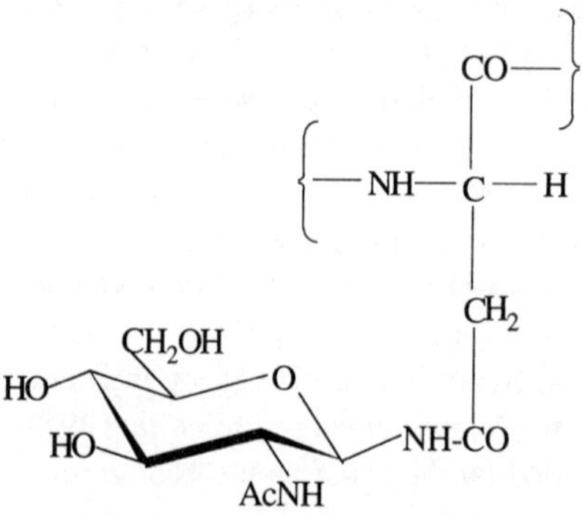

CO—
—NH—C—H
CH_2
CH_2OH
HO
O
HO
NH-CO
AcNH

13.18

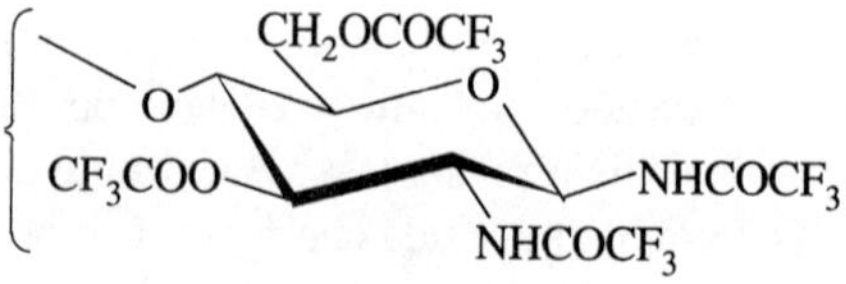

CH_2OCOCF_3
O
O
CF_3COO
$NHCOCF_3$
$NHCOCF_3$

13.19

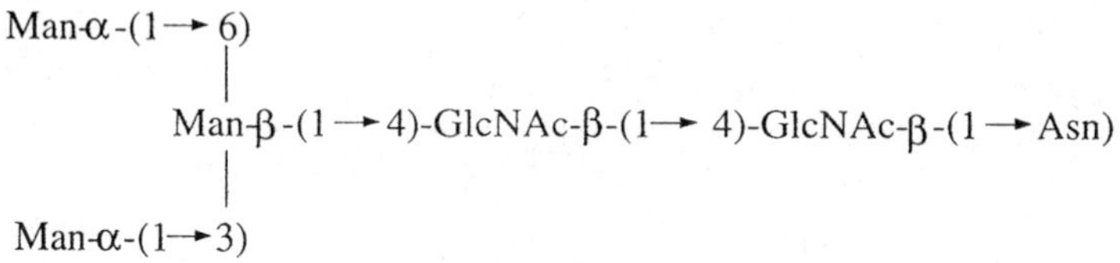

Man-α-(1→6)
Man-β-(1→4)-GlcNAc-β-(1→4)-GlcNAc-β-(1→Asn)
Man-α-(1→3)

13.20

We have already approached the synthetic problems relative to these oligosaccharides. Concerning glycosylamines, we have given the preparation of glucosides of asparagine in Section 3.6.2 and, as a synthetic example using enzymic methods, the synthesis of a peripheral sequence, **12.21**.

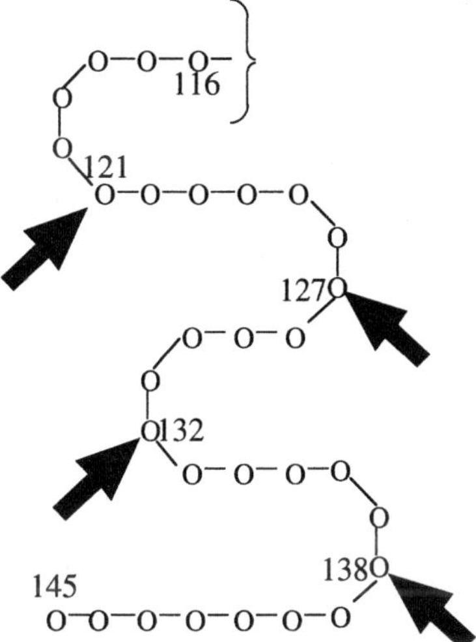

Fig. 13.2 Partial representation of the β sub-unit of hormonocorticogonadotropin, glycosidated at the amino acid residues 121, 127, 132, and 138 (from Montreuil 1980) (reproduced with kind permission from Academic Press).

13.2.4 Conformational problems

In glycosaminide proteins, the oligosaccharide is attached to asparagine from a particular tripeptide residue, Asn-X-Ser(Thr), but not all sequences of this type are necessarily glycosidated. There are other factors, apparently linked to the secondary structure of the polypeptide chain. In general, in proteins, the secondary structure adopts a preferred conformation in which varied geometric forms appear such as helices, folds, hair-pin turns, and loops. The sequences of amino acids in the neighbourhood of glycosylated sites are most often those that are observed in certain turns of protein chains (Fig. 13.2) and others are close to loops. These are, in fact, the most accessible regions of the polypeptide chain, which could explain both the preferential glycosylation and the important role of the oligosaccharide, which is exposed in different circumstances. What we have just said also applies to glycoside proteins.

Concerning the conformation of the oligosaccharide (Paulsen 1990), which indeed is of major importance in recognition problems, a certain number of significant phenomena show up. Results come from NMR analysis using sophisticated techniques, in particular by the nuclear Overhauser effect, modelling calculations, and comparison with the solid state structure of trisaccharide **13.21**. The values of the dihedral angles θ and ϕ which characterize the interglycosidic bond (see Section 9.2) are in keeping with the predictions of the exo-anomeric effect theory. Glycosidation on the primary alcohol function introduces flexibility because of the possibilities of rotation around the C-5–C-6 bond. This is how tetrasaccharide **13.22** (Fig. 13.3), which represents the branching point of the core oligosaccharides linked to asparagine, adopts the conformation outlined, in which we see Man-α-(1→6) (**C**) folded over GlcNAc. In certain oligosaccharides, called 'bisected', there is a supplementary substitution at O-4 of the central mannose. In model **13.23**, the residue **C** is turned towards the new substituent GlcNAc (**E**).

Man-α-(1→3)-Man-β-(1→4)-GlcNAc

13.21

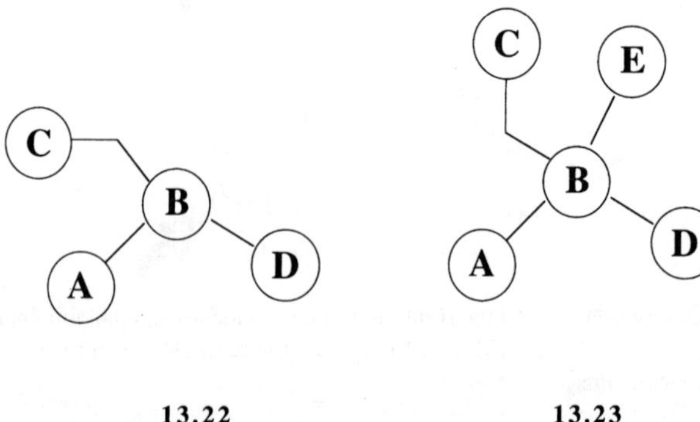

13.22 **13.23**

Fig. 13.3 Conformations at the branch point of the glycosaminides of proteins **A**, GlcNAc; **B**, Man-β-(1→4); **C**, Man-α-(1→6); **D**, Man-α-(1→3); **E**, GlcNAc-β-(1→4).

The complete solid structure of a crystallized immunoglobulin IgG_1 could be determined by X-ray analysis (see Section 15.2). We will only outline the topology of the two heavy chains and two molecules of decasaccharide **13.24** linked to an asparagine residue which each one carries. Figure 13.4 shows that the decasaccharides are found on the inner side of the crescent formed by two heavy chains that face each other, but without contact. Disk **B** represents the branching point, Man-β-(1→4). Disk **A*** corresponds to the chitobiose linked to the asparagine. The considerable interest shown in this conformation is due to the disymmetric role played by two identical trisaccharides linked at positions 3 and 6 of the mannose. The flexible chain linked at position 6 by the intermediary of mannose **C** is in contact with a *hydrophobic* part of the polypeptide (rich in phenylalanine, valine, and tyrosine residues). The other trisaccharide linked via the intermediary of mannose **D**, on the contrary, is bathed in the water of the central cavity and, apparently, behaves as a *hydrophile*. In Chapter 11 the reader has already seen the ambiguities of the hydrophilic–hydrophobic notions when dealing with sugars.

```
                                                            Fuc-α-(1→6)
Gal-β-(1→4)-GlcNAc-β-(1→2)-Man-α-(1→6)                          |
                                    |
                                   Man-β-(1→4)-GlcNAc-β-(1→4)-GlcNAc
                                    |
Gal-β-(1→4)-GlcNAc-β-(1→2)-Man-α-(1→3)
```

13.24

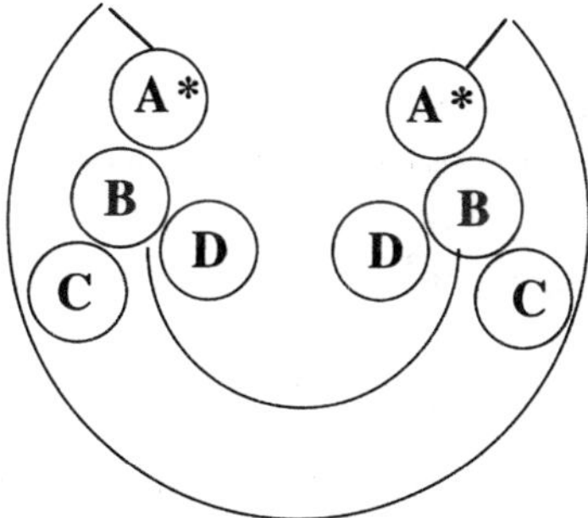

Fig. 13.4 Disposition of decasaccharide **13.24** inside the two chains F_C of the immunoglobulin IgG_1. **A***: chitobiose at the junction; **B**, **C**, and **D** are identical to Fig. 13.3: Man-β-(1→4); Man-α-(1→6); Man-α-(1→3).

13.2.5 The glycosylphosphatidylinositol (GPI) anchor

Many of the cell surface glycoproteins are said to be transmembrane: the polypeptide crosses the lipid layer with the sequence buried in this layer being constituted of hydrophobic amino acids. There is another means of attaching a protein to the surface, that is by way of an 'anchor' built from an oligosaccharide modified at its ends in order to be able to fulfil this anchoring role. More than 100 of these structure types have been recognized, at all levels of evolution from single cell eukaryotes to humans. Analysis of a number of them gave the general structure shown in Fig. 13.5.

Scheme **13.25** shows the detail of the extremity of the molecule attached to the cell wall. The covalently linked glycerolipid is, in fact, part of the membrane. The glucosamine is linked to glyceride via a molecule of myoinositol and a phosphodiester bridge.

CH_2OH
O
O
HO
NH_2
OH
OH
O
OH
CH_2OR'-CHOR-CH_2-O-PO(OH)-O
OH

13.25

Scheme **13.26** gives the detail of the anchoring system to the protein. A phosphodiester bridge links the mannose to an ethanolamine molecule. The carboxylic end of the protein is attached to ethanolamine by an amide bond.

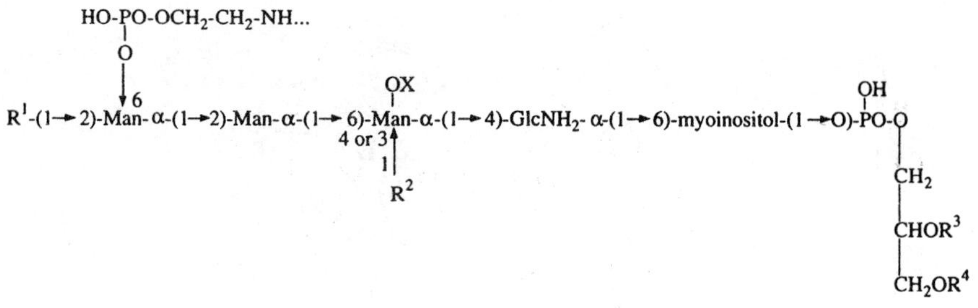

HO-PO-OCH$_2$-CH$_2$-NH...
O
6
OX
OH
R^1-(1→2)-Man-α-(1→2)-Man-α-(1→6)-Man-α-(1→4)-GlcNH$_2$-α-(1→6)-myoinositol-(1→O)-PO-O
4 or 3
1
R^2
CH$_2$
CHOR^3
CH$_2$OR^4

Fig. 13.5 The general structure of the glycosylphosphatidylinositol anchor.

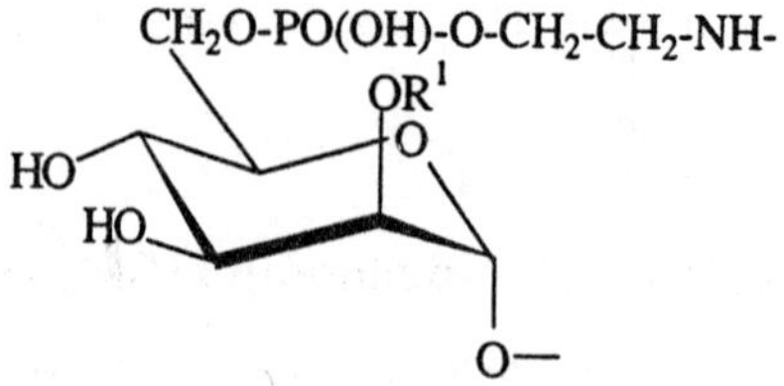

CH$_2$O-PO(OH)-O-CH$_2$-CH$_2$-NH-
OR^1
O
HO
HO
O—

13.26

Substituents R^1 and R^2 are a hydrogen atom, a mono- or an oligosaccharide. The substituent X is the hydrogen atom or a monophosphate of ethanolamine. The GPI anchor has several remarkable characteristics: the end molecule of myoinositol, the non-acetylated glucosamine, the phosphodiester functions with ethanolamine, one of which may be in the middle of the chain, apparently without any obvious utility.

The anchor of a variant surface glycoprotein (VSG) of the parasitic protozoan *Trypanosoma brucei* corresponds to the formula in Fig. 13.5 with R^1 = H, R^2 = Gal-α-(1→6)-Gal-α-(1→3), R^3 = R^4 = COC$_{13}$H$_{27}$, and was obtained by synthesis (Murakata and Ogawa 1992). The anchor Thy-1 of rat brain corresponds to the formula in Fig. 13.5, with R^1 = Man, R^2 = GalNAc-β-(1→4), X = PO(OH)OCH$_2$CH$_2$NH$_2$. This was also synthesized (Stewart Campbell and Fraser-Reid 1995). In the GPI anchors of yeast (*Saccharomyces cerevisiae*) ceramides containing phytosphingosine, such as **13.27**, are found instead of glycerolipids. The rest of the molecule is as in Fig. 13.5, with R^1 = Man-α-(1→2), R^2 = X = H. This anchor has been synthesized (Mayer *et al.* 1994).

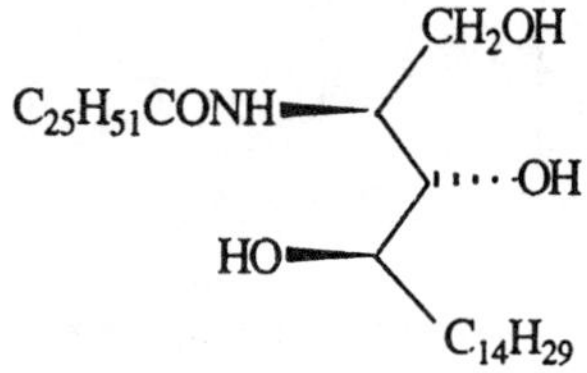

CH$_2$OH
C$_{25}$H$_{51}$CONH
OH
HO
C$_{14}$H$_{29}$

13.27

The construction of oligosaccharides was essentially achieved by the methods described in Section 10.3. Some new problems arose. *Myo*-inositol is achiral while the derivatives present in the GPI molecules are chiral. The preparation of derivative **13.28**, made for the coupling and subsequent phosphorylation, consequently requires a resolution step. Phosphorylation is another problem. Figure 13.6 summarizes the method for the introduction of the ethanolamine phosphate side chain. Condensation of the protected derivative, $PhCH_2OCONHCH_2CH_2OH$ with 2-(cyanoethyl)-*N*,*N*-bis(2-propyl) chlorophosphoramidite in the presence of a tertiary base gives a phosphoramidite which reacts with the hydroxyl derivative ROH to give a phosphite. This is oxidized to a phosphate with peracid. The protecting groups, 2-cyanoethyl and benzyloxycarbonyl, are removed by treatment with base and catalytic hydrogenolysis, respectively.

MeO HO BnO BnO OBn OBn O

13.28

$iPr_2N\text{-}P(Cl)\text{-}OCH_2CH_2CN \xrightarrow{PhCH_2OCONHCH_2CH_2OH} PhCH_2OCONHCH_2CH_2O\text{-}P(NiPr_2)\text{-}OCH_2CH_2CN \xrightarrow{ROH}$

$PhCH_2OCONHCH_2CH_2O\text{-}P(OR)\text{-}OCH_2CH_2CN \xrightarrow{ArCO_3H} PhCH_2OCONHCH_2CH_2O\text{-}PO(OR)\text{-}OCH_2CH_2CN \rightarrow \rightarrow$

$NH_2CH_2CH_2O\text{-}PO(OH)\text{-}OR$

Fig. 13.6 Method for the preparation of the mixed phosphodiesters of the GPI anchor.

13.3 Glycosaminoglycans and proteoglycans

13.3.1 General

The macromolecules treated in this section are essentially extracellular. In association with collagen fibers (a glycoprotein), they contribute to the nature, structure, and rigidity of tissues. These are sugar–protein associations as glycoproteins, but can be distinguished in a rather characteristic way so that they are placed in a special class. The portion of protein is generally small and can go as low as 2% whereas the sugar part represents between 50 and 90%. However, from this point of view, the differences between these two categories of glycoconjugates are fuzzy. What is more characteristic is the long linear chains of proteoglycans, built from a repeating disaccharide motif. Sometimes alterations which appear to be at random destroy the periodicity in the strict sense, but it remains clearly discernable in spite of this (Kennedy and White 1988). 'Glycosaminoglycan' is the name for the polysaccharide and 'proteoglycan' for the entire conjugate. As glycoproteins, proteoglycans are made of either glycosaminide bonds involving L-asparagine, or of glycosidic bonds on the alcohol oxygen atom of L-serine or L-threonine. The repeating or pseudo-repeating chain does not directly attach itself to the protein, but there is a core oligosaccharide, as in glycoproteins.

The bond to L-asparagine takes place via an oligosaccharide rich in mannose **13.20**, which is probably the same as in glycoproteins. On the other hand, the most frequent core structure on L-serine and L-threonine, **13.29**, is entirely different. The reader should note the β-D-xylopyranosyl and β-D-glucuronopyranosyl units. The long chain is linked to O-4 of the β-D-glucoronopyranosyl unit. We will come back to these structures in Chapter 17, while on the subject of heparin, which belongs to this family.

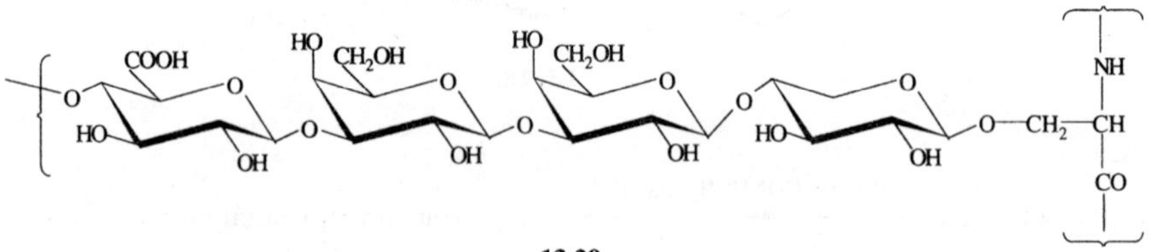

13.29

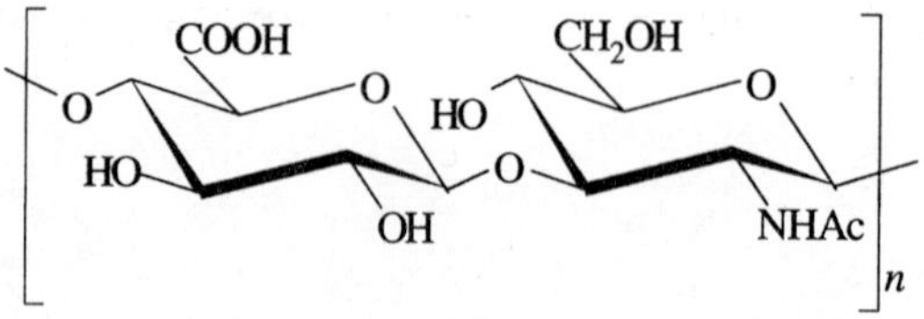

13.30

13.3.2 Periodical or quasi-periodical chains

The fundamental structure of hyaluronic acid is **13.30**, which may be composed of several thousand disaccharide units. In chondroitin **13.31**, *N*-acetylglucosamine is replaced by *N* acetylgalactosamine. There are two natural sulfated derivatives of chondroitin. In one of them, the disaccharide unit is sulfated at O-4 and in the other, at O-6 of the *N*-acetylgalactosamine unit, giving **13.32** and **13.33**. The chains are considerably shorter than in the hyaluronic acid (10 to 60 repeating units). In dermatan sulfate, the basic motif **13.32** is found but a certain number of D-glucuronic residues have undergone inversion of configuration at C-5 which gives an L-iduronic residue and a basic motif, **13.34**. This inversion varies from a few units to 100%. The structure of glycosaminoglucan of heparin is given in Chapter 17. Keratan sulfate is exceptional. The motif **13.35** does not contain a carboxylic acid function, the internal bond is β-(1→4) and the bond with the rest of the chain is β-(1→3). Condensation is low (10 or 30–50 motifs). This polysaccharide is structurally a poly-(*N*-acetyllactosamine) chain, as seen in Sections 13.1 and 13.2, sulfated at O-6 of GlcNAc.

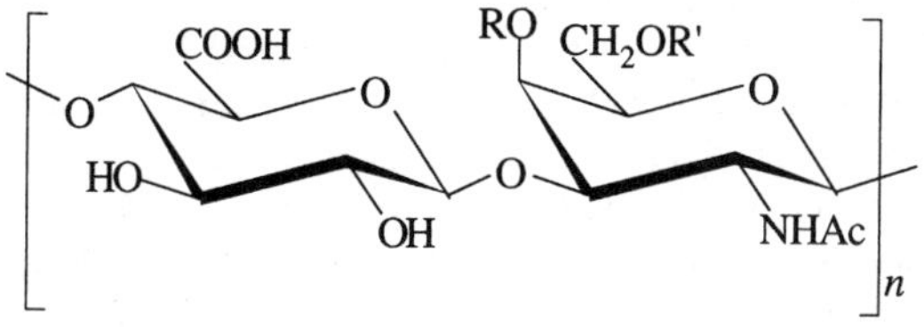

COOH RO CH_2OR' O O O HO O OH NHAc n

13.31 R= R'= H
13.32 R= SO_3H, R'= H
13.33 R= H, R'= SO_3H

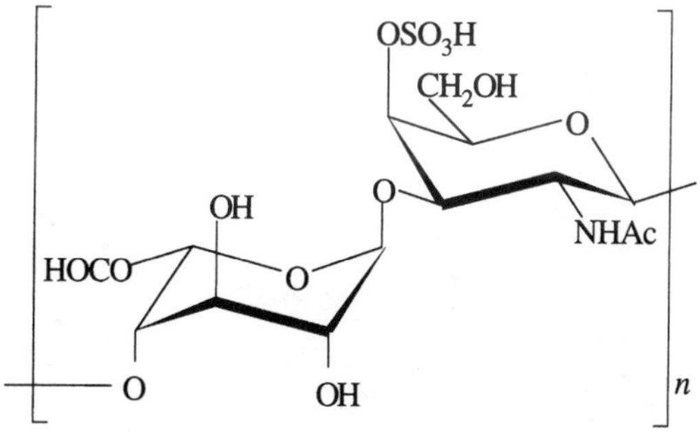

OSO_3H CH_2OH O O OH NHAc HOCO O O OH n

13.34

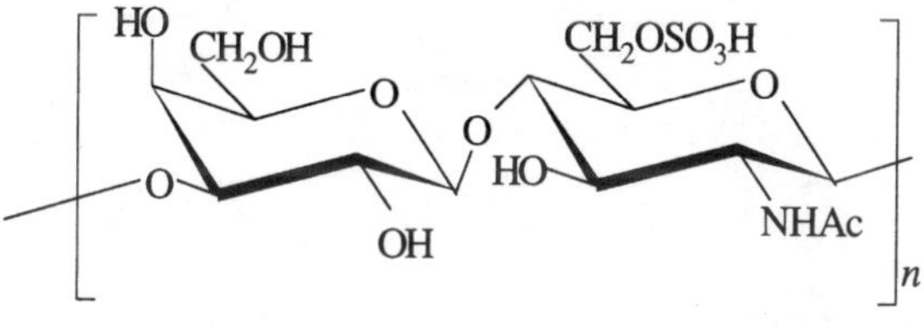

HO CH_2OH CH_2OSO_3H O O O O HO OH NHAc n

13.35

References

Kennedy, J. F. and White, C. A. (1988), The glycosaminoglycans and proteoglycans. In *Carbohydrate chemistry* (ed. J. F. Kennedy), pp. 303–341, Oxford University Press, Oxford.

Kunz, H. and Rück, K. (1993), *Angew. Chem., Int. Ed. Engl.*, **32**, 336–358.

Li, Y.-T. and Li, S.-C. (1982), *Adv. Carbohydr. Chem. Biochem.*, **40**, 235–286.

Mayer, T. G., Kratzer, B., and Schmidt, R. R. (1994), *Angew. Chem., Int. Ed. Engl.*, **33**, 2177–2181.

Meldal, M. and Bock, K. (1990), *Tetrahedron Lett.*, **31**, 6987–6990.

Montreuil, J. (1980), *Adv. Carbohydr. Chem. Biochem.*, **37**, 157–223.

Morrison, I. M. (1988), The glycolipids and gangliosides. In *Carbohydrate chemistry* (ed. J. F. Kennedy), pp. 196–219, Oxford University Press, Oxford.

Murakata, C. and Ogawa, T. (1992), *Carbohydr. Res.*, **235**, 95–114.

Nilsson, B. and Svensson, S. (1979), *Carbohydr. Res.*, **72**, 183–190.

Paulsen, H. (1990), *Angew. Chem., Int. Ed. Engl.*, **29**, 823–839.

Schmidt, R. R. and Kinzy, W. (1994), *Adv. Carbohydr. Chem. Biochem.*, **50**, 21–123.

Stewart Campbell, A. and Fraser-Reid, B. (1995), *J. Am. Chem. Soc.*, **117**, 10387–10388.

Stults, C. L. M., Sweeley, C. C., and Macher, B. A. (1989), *Methods in Enzymology*, **179**, 167–214.

Yule, J. E., Wong, T. C., Gandhi, S. S., Dongxu, Q., Riopel, M. A. and Koganty, R. R. (1995), *Tetrahedron Lett.*, **36**, 6839–6842.

14 The structure of some crystallized sugar–protein complexes

14.1 General. The ABP–L-arabinose complex

14.1.1 Proteins and sugars

In this work we will encounter the following four types of proteins associated with sugars: enzymes concerned with the metabolism of sugars, lectins, specific antibodies, and transport proteins. A good number of these proteins have been crystallized and examined by X-ray analysis for the determination of their tertiary structure. In view of their high molecular weight, analysing their spectra by diffraction analysis is a difficult task. However, knowledge of their primary structure, that is to say the peptide chain, obtained by chemical procedures, can act as a guide. The resolution power of these methods is on average 2.5 Å. It is better than this in the examples we will give. Knowledge of the protein structure only gives hints about the structure of the complex, especially since conformation modifications can occur during complexation. Certain protein–sugar complexes have been obtained as crystals and their structures have been elucidated with comparable degrees of resolution. In general, the receptor site on the protein is an elongated groove, more or less deep, but the rest of the protein seems eminently variable depending on its biological function (Quiocho 1986). Crystalline complexes belonging to the above-mentioned four protein types are known and have been studied. In the area of enzymes, let us point out the complex between poly-*N*-acetylchitiobiose and lysozyme, an enzyme capable of hydrolysing the bacterial cell walls with six receptor sites, and between a maltodextrin and takaamylase, an enzyme which hydrolyses amylose and can associate with a hexasaccharide unit. These studies are extremely important for elucidating the enzymic hydrolysis mechanism. However, since some of these publications are already outdated and the reader who is not a crystallographer will have a few problems in appraising the results, we recommend looking at the discussions concerning them (Quiocho 1986). The idea of this chapter is to describe precisely the immediate environment surrounding the sugar and it is with this in mind that we have chosen our examples. We advise the reader to examine coloured stereoscopic views of original papers.

14.1.2 ABP–L-Arabinose protein complex

An American group (Quiocho 1986; Quiocho 1989) meticulously studied a complex of L-arabinose with a protein called ABP (*arabinose-binding protein*) which they were able to describe at 1.7 Å resolution. The authors believe the binding modes that they observed are of general value. This is why we will begin this chapter by a simplified account of their conclusions, even though L-arabinose does not participate in recognition phenomena in higher animals.

A rather widespread family of proteins, found in the periplasmic space of gram-negative bacteria, complexes certain small molecules and allows them to be transported through the cell wall or activate chemotaxis. Each of these functions involves a consecutive interaction with specific membrane proteins. The molecules transported are amino acids, sulfate, mono- and oligosaccchrides. In this way ABP complexes L-arabinose (K_d 0.98×10^{-7} M), and MBP (*maltodextrin-binding protein*) complexes maltose (K_d 35×10^{-7} M) and maltodextrins. It is in this series that are found the strongest possible bonds between sugars and proteins. The dissociation rate (k_{-1} 1.5 s^{-1}) is indicative of the upper limit of the ionic transport rate.

Hydrogen bonds

Hydrogen bonds play a very important role in the ABP–L-arabinose complex in that all the polar groups are used. Generally speaking, hydroxyls are functional groups characteristic of sugars; with fixed orientations, they can each be involved in three hydrogen bonds, one as donor and the other two as acceptors. The variations of the torsion angle H–C–O–H allow the most favourable structure to be established. Participation in the complexation of the highly directional hydrogen bonds explains the specificity of the interaction. There are three types of hydrogen bonds systems.

In *cooperative hydrogen bonds*, the hydroxyl group of the sugar is simultaneously donor and acceptor according to the following scheme:

$$NH \rightarrow OH \rightarrow O$$

The NH and O atoms are part of the complexation site.

In *bidentate bonds*, two adjacent hydroxyl groups, having an equatorial–equatorial or axial–equatorial disposition, close a ring on two atoms of a planar polar group as in the disposition **14.1**.

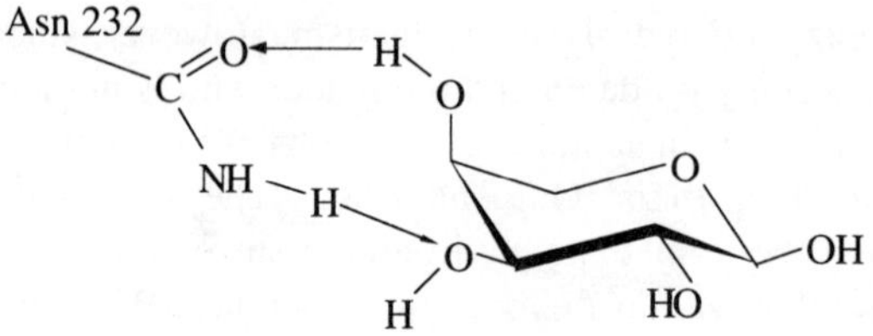

Asn 232
C=O ← H–O
NH–H → O–H
O
OH
HO

14.1

The formation of cooperative and bidentate bonds creates a dense network of hydrogen bonds between the sugar and the main receptors. These bonds are strong. The average distance between the heavy atoms, donor–acceptor, is 2.82 (0.15) Å, and the average angle 164 (9)°. The residues with planar polar groups,

Asn, Asp, Glu, Gln, Arg, and His, are especially present in receptor sites (lysine is only used once). These groups are rigidly bound and their final adaptation to the sugar can only be achieved by a conformational change in the protein.

One common feature of all these periplasmic proteins is their ability to complex α- and β-anomers without a noticeable difference. This can easily be explained by the scheme of hydrogen bonds. The very precise alignment of one of the oxygen atoms of aspartate allows it to accept a hydrogen bond from not only the α-anomer but the β-anomer as well, without perturbing the rest of the network, as shown in **14.2**. Figure 14.1 shows the topological scheme of the hydrogen bond network. The $^4C_1$ conformation of the sugar remained normal in the complex.

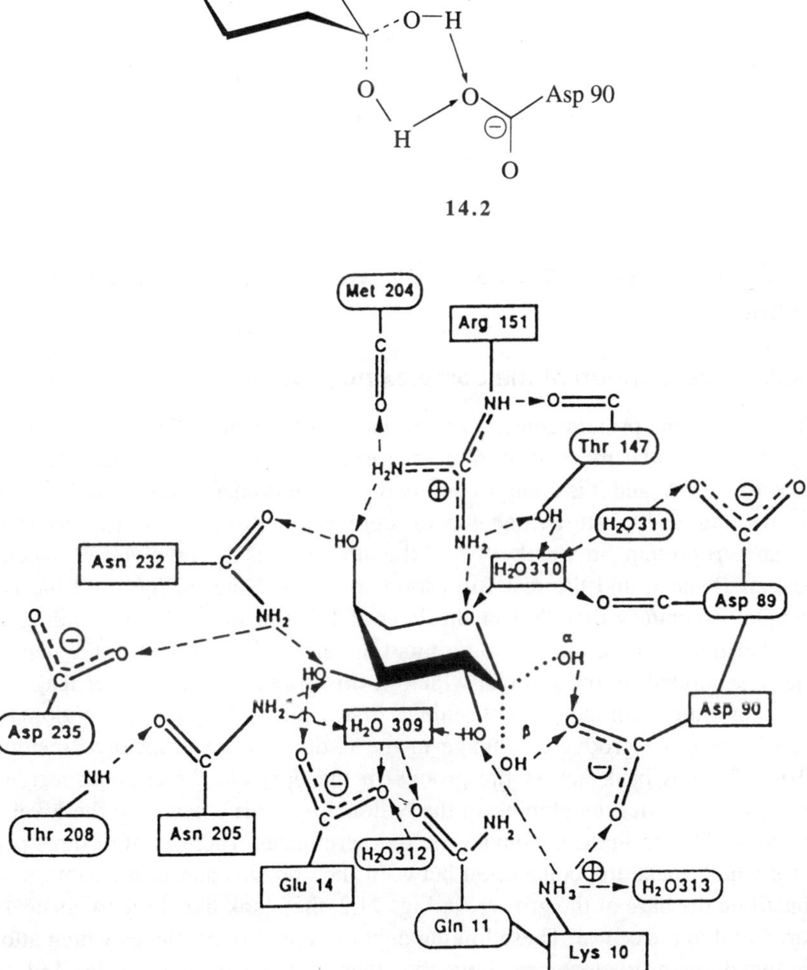

Fig. 14.1 Scheme of the hydrogen bond network in the ABP–L-arabinose complex (from Quiocho 1989) (reproduced with kind permission from the International Union of Pure and Applied Chemistry).

Van der Waals forces

All the heavy atoms of the L-arabinose are in van der Waals contact ($d < 4$ Å) with those of the protein host. There are nearly 54 contacts in all, an unusually high number, due to the dense network of hydrogen bonds. The latter creates an impressive compactness. The 'hydrophobic' α-face of the pyranose, in which the C–H bonds dominate, lies in part on the indolic nucleus of tryptophane 16. In formula **14.3**, this indole is represented by the trace of its horizontal plane of symmetry.

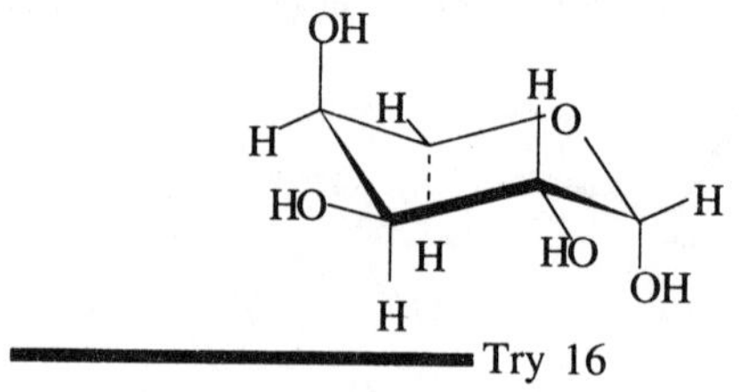

OH H H O H HO H HO OH H Try 16

14.3

14.2 Maltose complexed by the maltodextrin-binding protein

14.2.1 Description of the complexing protein

The complexing protein consists of a sequence of 370 amino acids corresponding to a molecular mass of 40 622. Its form is ellipsoidal with dimensions of $65 \times 40 \times 30$ Å, and it is composed of two globular domains separated by a deep groove. Figure 14.2 shows one way of representing this protein. It is described as 'an α/β protein' in which 40% of the amino acids are involved in α-helix, 20% in β-sheet, and the rest form loops or coils. Note in Fig. 14.2 the two unequal domains called N (on the left) and C (on the right) depending on whether they contain the amino terminal or carboxylic terminal of the peptide chain, separated by the groove, which in this drawing is nearly vertical. The chain, starting from the amino terminal, first winds its way into the N domain, then crosses the groove to engage in the C domain at the level of residues 110–113, goes back across the groove in the opposite direction at residues 268–271 and, after developing in the N domain, returns finally to the level of residues 311–315 in the C domain where it terminates. There are thus three peptides which make the connection between the domains and at the same time, constitute the base of the groove. In Fig. 14.2, they look like three twigs nearly horizontal to the centre. These linking peptides are also the hinges which allow the two domains to close one above the other as the shells of an oyster. Indeed, while the CO–NH system is rigid, there are possibilities of rotation around the carbon located between NH and CO, **14.4**.

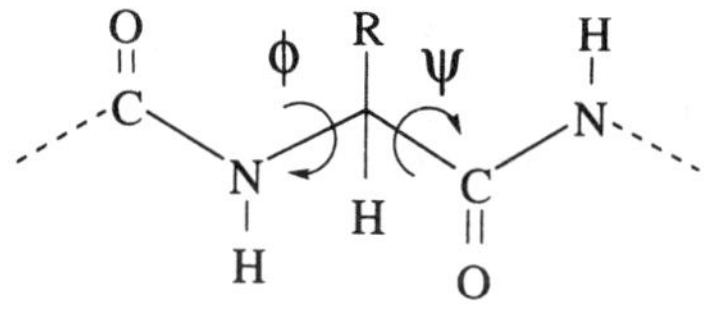

14.4

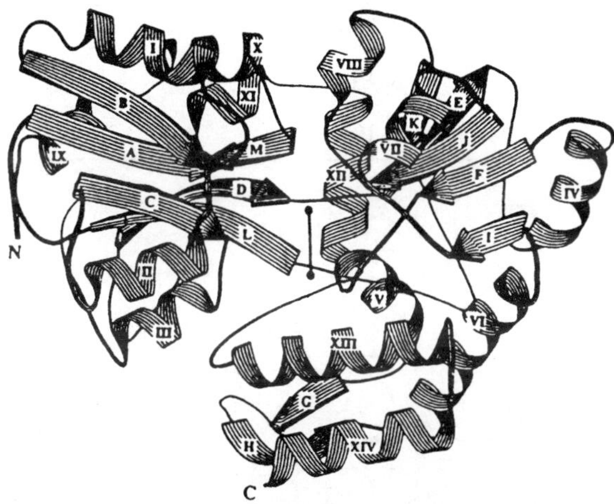

Fig. 14.2 Conventional representation of the transport protein of maltodextrin. The β-folds are represented by arrows. The sequence order of the β-folds in the polypeptide chain is indicated by letters and those of the helices by Roman numerals. The site of the maltose, in the centre, is indicated by two large black dots connected by a vertical bar (Spurlino *et al.* 1991) (reproduced with kind permission from the American Society for Biochemistry and Molecular Biology and the authors).

14.2.2 Complexation of maltodextrins

The transport protein which we have just described complexes an entire family of maltodextrins. The thermodynamic and kinetic parameters are shown in Table 14.1. It is remarkable that there are so few variations from maltose to cyclodextrins. The first three complexes in the table were obtained in the crystalline state.

14.2.3 Complexation mode of maltose

Until now, the conformation of a complexed maltose seemed to be very close to that of free maltose in the crystalline state (Fig. 14.3) (Jeffrey and Sundaralingam 1981). The conformation of each cycle is D-$^4C_1$. The torsion angles are O-5′–C-1′–O-1′–C-4 = +116° and C-1′–O-1′–C-4–C-3 = +122°. The glycosidic angle C-1′–O-1′–C-4′ is 120°. There is an intermolecular hydrogen bond O-3–H–O-2′.

Table 14.1 Maltodextrin complexes with the transport protein of maltose (from Quiocho 1989) (reproduced with kind permission from the International Union of Pure and Applied Chemistry). K_d, dissociation constant, k_1 and k_{-1}, complexation and decomplexation rates.

	$10^7\ K_d$ (M)	$10^{-7}k_1$ ($M^{-1}\ s^{-1}$)	k_{-1} (s^{-1})
Maltose	35	2.3	90
Maltotriose	1.6	2.5	8.4
Maltotetraose	23		
Maltopentaose	50		
Maltohexaose	34		
Maltoheptaose	16		
α-Cyclodextrin*	40	3.6	110
β-Cyclodextrin	18	2.2	4.6

*see Section 11.3.

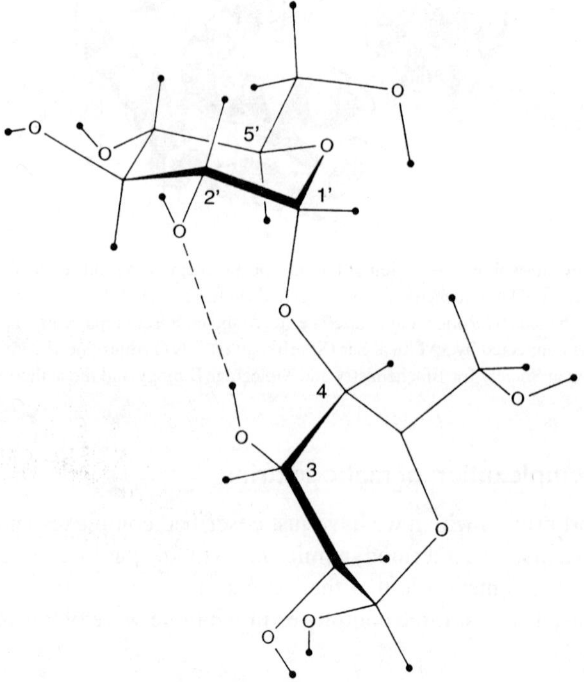

Fig. 14.3 Structure of maltose in solid state (from Jeffrey and Sundaralingam 1981) (reproduced with kind permission from Academic Press). In the text, numbers 3 and 4 correspond to the reducing residue and numbers 1′ and 5′ to the non-reducing residue.

The receptor of maltodextrins resembles a groove or cleft, the walls of which (height 18 Å) are domains N and C and the base (9 × 18 Å^2) consists of three linking peptide segments and helix XIII. The dimensions of this cavity are sufficient enough to accommodate cyclodextrins (see Section 11.3), for example β-cyclodextrin whose diameter is 15.4 Å and height is 7.9 Å (Spurlino *et al.* 1991). When complexation of the maltose takes place at the base of the groove, the two walls approach each other by pivoting on the hinge consisting of the linking peptides (Sharff *et al.* 1992). This movement corresponds to a 35° rotation. The maltose is swallowed up to such a point that only 3.6% of its surface (545.5 Å^2) remains in contact with the solvent. Complexation utilizes residues from each of the walls and the groove base.

Hydrogen bonds

Just as with L-arabinose, attachment is essentially attributed to the hydrogen bonds. These are nearly exclusively linked to polar groups of side chains. This is different from the complexation mode of sulfate, in anionic state, which forms an association mainly with NH of the peptide bonds. The association of maltose takes place by means of 16 hydrogen bonds, 11 of which correspond to a direct linkage to amino acid residues and five probably via water molecules (Fig. 14.4). The ony non-bonded oxygen atoms are ring and interglycosidic oxygens. Nine of the direct hydrogen bonds are established between the *neutral* hydroxyl group of the sugar and *charged* residues: carboxylate, ammonium, and guandidinium. The precise disposition of an aspartate residue allows complexation of not only the α-maltose but also the β-maltose with equal efficiency.

Van des Waals contacts

There are approximately 65 van der Waals contacts (≤ 4 Å). Most of them should be considered as an indirect result of hydrogen bonds which tend to

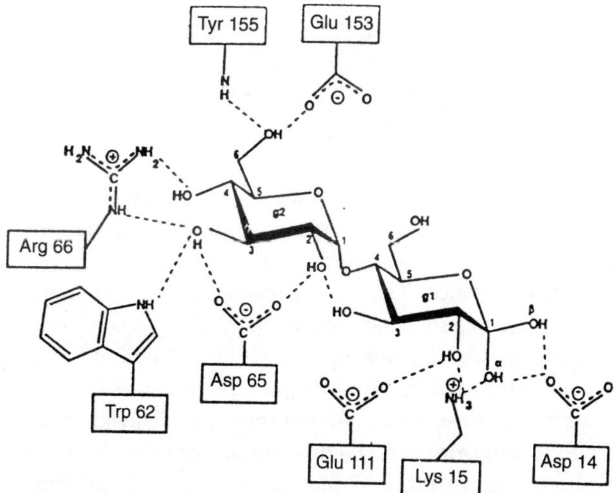

Fig. 14.4 Network of hydrogen bonds of the complexed maltose. The bonds made via water molecules are not shown (from Spurlino *et al.* 1991) (reproduced with kind permission from the American Society for Biochemistry and Molecular Biology and the authors).

create the most compact structure possible. Nearly the entire α-face of the non-reducing terminal unit lies on the indole nucleus of tryptophan 340. The glycosidic bond and a part of the α-face of the reducing unit lies on the phenyl nucleus of tyrosine 155. The α-faces of glucopyranoses forming the maltose, sometimes called the 'hydrophobic faces', present a series of C–H bonds on their surface.

14.3 A lectin–biantennary octasaccharide complex (Bourne *et al.* 1992)

This lectin, 'isolectine I', isolated from the seeds of *Lathyrus ochrus*, has two identical subunits, each composed of a light chain of 52 amino acids and a heavy chain of 181 amino acid residues. Each subunit contains Ca^{2+} and Mn^{2+} ions, necessary for the reaction. The ligands examined are fragments of dodecasaccharide–asparagine **14.5** present in human lactotransferrin: mannose, trisaccharide 234 and octasaccharide 234564′5′6′. The complexes were obtained in the crystalline state. Here we will look at the complex with the octasaccharide, of which a 0.3 mm crystal could be prepared and the structure given at 2.3 Å resolution. We will only give the schematic description of this molecule, topologically correct, stressing the importance of the nature of the bonds. The reader who requires a more detailed picture should refer to the coloured stereoscopic views in the original literature (Bourne *et al.* 1990; 1992).

6 5 4
NeuAc-α-(2→6)-Gal-β-(1→4)-GlcNAc-β-(1→2)-Man-α-(1→6)
2 1
3 Man-β-(1→4)-GlcNAc-β-(1→4)-GlcNAc-β-(1→Asn)
NeuAc-α-(2→6)-Gal-β-(1→4)-GlcNAc-β-(1→2)-Man-α-(1→3) Fuc-α-(1→6)
6' 5' 4'

14.5

Complexation takes place via a multitude of polar bonds (≤ 3 Å). There are 14 direct hydrogen bonds between the oxygen or nitrogen atoms of the sugar and the polar groups of the amino acid residues and seven are achieved by the intermediacy of water molecules. The same sugar may be connected to amino acid residues distant on the peptide chain but spatially close, a common occurrence in complexing by proteins, and a consequence of the folding of this chain. In addition, 14 water molecules are involved in indirect contacts between the sugar and lectin, or between the consecutive units of the sugar. The latter case results in a stabilization of the conformation. To this are contributed 18 hydrogen bonds of which one is direct and the other 17 indirect. Because of the high specificity of lectins, it is expected that complexation involves hydrogen bonds which play a directional role. The important novelty in this structural determination is the massive participation of water molecules in complexation.

The complex is also stabilized by 68 van der Waals contacts (≤ 4 Å), 27 of which with aromatic residues. These contacts are mainly observed between the aromatic rings and the C-5–C-6 atoms of sugars.

14.4 Association between the Le^b-OMe tetrasaccharide glycoside and a lectin from *Griffonia simplicifolia* (Lemieux 1996)

Lectins are proteins which bind sugars (Section 15.4). One lectin from the plant *Griffonia simplicissima*, called GS-IV, binds the methyl glycoside of a tetrasacharide related to blood group substances (Chapter 16), the Le^b tetrasaccharide **16.5**. This association is interesting in several respects: it involves a blood group substance, the complex was obtained in the crystalline state and its structure elucidated by X-ray crystallography, the thermodynamic parameters of the complexation were estimated, and the specificity of the lectin was closely investigated.

Figure 14.5 gives the structure of the complex at 2.8 Å resolution. Of the ten hydroxyl groups in Le^b–OMe, the three at positions 3b, 4b, and 4c are hydrogen-bonded to amino acid residues of the lectin (Asp 89 and Ser 49) as proton donors. The two hydroxyl groups at positions 3c and 2d are hydrogen bonded to the lectin as proton acceptors. The five others remain fully in contact with water. There are good reasons, not detailed here, for believing that these conclusions are still valid in homogeneous aqueous solution.

These observations definitively confirm earlier conclusions which had been suggested by the testing of monodeoxygenated analogues obtained by synthesis. The extent of binding in these experiments was measured by a change in the UV absorption caused by increasing amounts of the ligand up to saturation. The thermodynamic parameters for the association of the Le^b–OMe tetrasaccharide

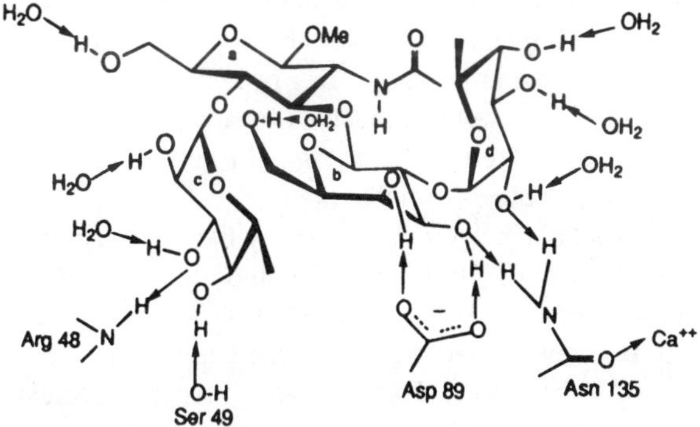

Fig. 14.5 Hydrogen bonds formed between the epitope of Le^b-OMe and the receptor site of GS-IV. Reproduced with permission from *Acc. Chem. Res.*, (1996), **29**, 373–380. Copyright 1996, American Chemical Society.

with the native lectin were $\Delta G^{\ominus}$ –6.3, $\Delta H^{\ominus}$ –13.3, $T\Delta S^{\ominus}$ –7.0 kcal mol^{-1}. Monodeoxygenation at one of positions 3b, 4b, or 4c (Fig. 14.5), that is, removal of one of the 'key polar groups', which act as proton donors in the intact ligand, practically abolishes binding. On the other hand, only small changes in affinity were observed in the binding of Le^{b}–OMe analogues deoxygenated at any one of the other seven positions, as estimated by the measurements of the free enthalpy of binding, $\Delta G^{\ominus}$, which oscillated between –5.5 and –6.4 kcal mol^{-1}. Van't Hoff plots, in the 15–45°C range, allowed the calculation of the corresponding $\Delta H^{\ominus}$ and $\Delta S^{\ominus}$. It was thus discovered that the near insensitiveness of $\Delta G^{\ominus}$ to these deoxygenations concealed rather large but virtually compensating changes for $\Delta H^{\ominus}$ and $\Delta S^{\ominus}$, except at position 6a which is distant from the receptor. In all cases, binding caused a decrease in entropy. Perhaps the interpretation of this (Lemieux 1996) may be better understood by recalling the description of the interaction of sugars with water given in Section 11.6.

In the vicinity of a hydrophobic surface, water molecules are held in a structure, more rigid than that of bulk water. Consequently, when two hydrophobic surfaces come into contact, the escape of water molecules from these layers to bulk water is accompanied by an increase in entropy. On the other hand, in the case of sugars, according to the theory of the specific hydration model, the polar hydroxyls may interact with water in two different ways. Either the configuration and conformation of the sugar are such that its hydroxyl groups fit perfectly to the network of water, or they induce great disturbance in the nearest layers of water molecules. In the latter case, there is less order near the surface of the sugar than in bulk water, and a decrease of entropy should be observed when two such surfaces come into contact.

In the native GS-IV lectin, the combining site is strongly hydrated by seven water molecules. Further hydration of these molecules leads to the formation of zones of perturbation. Computer simulations indicated that the calculated water-to-water interaction energies in fact strengthen with increasing distance from the receptor surface. Thus water is perturbed in the vicinity of the receptor site, and this is attenuated over a distance of about three water molecules. To conclude, the negative $\Delta S^{\ominus}$ observed upon the binding of Le^{b}–OMe tetrasaccharide can be explained by the escape of water molecules from their unstable position in the perturbed layers near the lectin and the ligand to participate in the more rigid network of the hydrogen bonds of bulk water.

Other associations have been investigated by the same methods. Lectin UE-I from *Ulex europaeus* (Section 15.4.3) and PT-II from *Psophocarpus tetragonolobus* both bind the H-type trisaccharide glycoside α-L-Fuc-(1→2)-β-D-Gal-(1→4)-β- D-GlcNAc-(1-OMe). With the *Ulex* lectin, the thermodynamic parameters are $\Delta H^{\ominus}$ –29, $T\Delta S^{\ominus}$ –20.5 kcal mol^{-1}. Again a decrease in entropy is observed. However, with the *Psophocarpus* lectin, the figures are $\Delta H^{\ominus}$ – 5.4, TΔS° + 0.8 kcal mol^{-1}. In the latter case, the small increase in entropy seems to indicate an important hydrophobic effect.

References

Bourne, Y., Rougé, P., and Cambillau, C. (1990), *J. Biol. Chem.*, **265**, 18161–18165.

Bourne, Y., Rougé, P., and Cambillau, C. (1992), *J. Biol. Chem.*, **267**, 197–203.

Jeffrey, G. A. and Sundaralingam, M. (1981), *Adv. Carbohydr. Chem. Biochem.*, **38**, 417–529.

Lemieux, R. U. (1996), *Acc. Chem. Res.*, **29**, 373–380.

Quiocho, F. A. (1986), *Annu. Rev. Biochem.*, **55**, 287–315.

Quiocho, F. A. (1989), *Pure Appl. Chem.*, **61**, 1293–1306.

Sharff, A. J., Rodseth, L. E., Spurlino, J. C. and Quiocho, F. A. (1992), *Biochemistry*, **31**, 10657–10663.

Spurlino, J. C., Lu, G.-Y. and Quiocho, F. A. (1991), *J. Biol. Chem.*, **266**, 5202–5219.

15 Antigens and antibodies. Lectins

15.1 Foreword

Readers with a background in biochemistry or medicine will immediately recognize that Sections 15.2 and 15.3 are not on the same level as other parts of this book. This brief section on immunochemistry is above all designed for readers who have followed traditional teaching in organic chemistry. At the moment many sugar chemists work in more or less close collaboration with immunologists. In this book, expressions such as antigen, antibody, monoclonal antibody, determinant, and epitope are found in different sections; and certain analytical methods of quantitative immunology have also been mentioned. Reading Sections 15.2 and 15.3 of this chapter should give the organic chemist a more concrete view of these ideas and experiments. Our goal is to render the dialogue between the immunologist and the organic chemist somewhat easier.

15.2 Antigens and antibodies

The introduction of a foreign macromolecule into the blood stream of higher vertebrates such as mice, rabbits, guinea-pigs, sheep, or goats may provoke the appearance of a remarkable collection of proteins in the serum. In most cases, this is a heterogeneous mixture of molecules which share one property, which is the ability to bind non-covalently to the injected molecule which stimulated their appearance. Under these conditions, we would say that this macromolecule behaves as an *antigen* and the reactive proteins which have shown up in the serum are called *antibodies*. It is important to know the details of the experiment, such as the mode of injection and whether the molecule was a well-defined chemical species, a small molecule bound to a carrier, or part of a cell wall. Also the type of animal injected is important as the same molecule may or may not be antigenic, depending on the animal.

Knowledge of the structure of antibodies, obtained using a technique to be described later in this chapter, allows the preparation of homogeneous proteins. Antibodies are members of the globulin family of proteins, the so-called *immunoglobulins* (Ig). They are divided into classes and subclasses, which will be discussed later. All immunoglobulins are built according to a common pattern, outlined in Fig. 15.1 for an 'IgA'. There are two identical subunits linked by disulfide bridges. Each subunit comprises a heavy chain H (MW ≅ 50 000 D) and a light chain L (MW ≅ 23 000 D).

There are two domains in the light chain. There is a variable domain VL, the amino acid sequence of which differs according to the antigen's specificity, which extends over about half the chain from the amino end. The other domain on the carboxylic side is a constant domain, CL. In the middle of the variable domain there are 'hypervariable' regions, shown as black stripes. These hypervariable regions are in the loops which connect the β-sheets. The conformation of the polypeptide chain in the variable domain is such that these hypervariable regions are brought close to each other, and lie at one end of the fold.

In the same way, there is a variable domain VH in the heavy chain from the amino end, then a first constant domain CH1 up to the hinge region which contains two disufide bridges which link it to the other heavy chain, followed by two constant domains, CH2 and CH3. An oligosaccharide is bound to this part of the molecule. Limited proteolysis in the vicinity of the hinge separates two identical fragments called Fab, each built from an L chain, and domains VH and CH1 of the H chain. This Fab fragment retains the binding properties with antigens. The other half of the H chain, called Fc, is the part responsible for certain immunological properties of the immunoglobulin other than direct antigen binding.

The binding site is located at the hypervariable regions of the L and H chains, close to each other in the immunoglobulins. Thus there are two binding sites (two Fabs) in the immunoglobulin shown in Fig. 15.1. There are five classes of H chains, named μ, δ, γ, ε, and α, to which correspond the five classes of immunoglobulins, IgM, IgD, IgG, IgE, and IgA, and two types of light chains, κ and λ. Some of the immunoglobulins circulate while others play the role of receptors on the surface of certain blood cells, and some can do both.

The perfect fit of the antibody to the antigen is somewhat perplexing as the immunized animal reacts towards molecules which are completely foreign to it

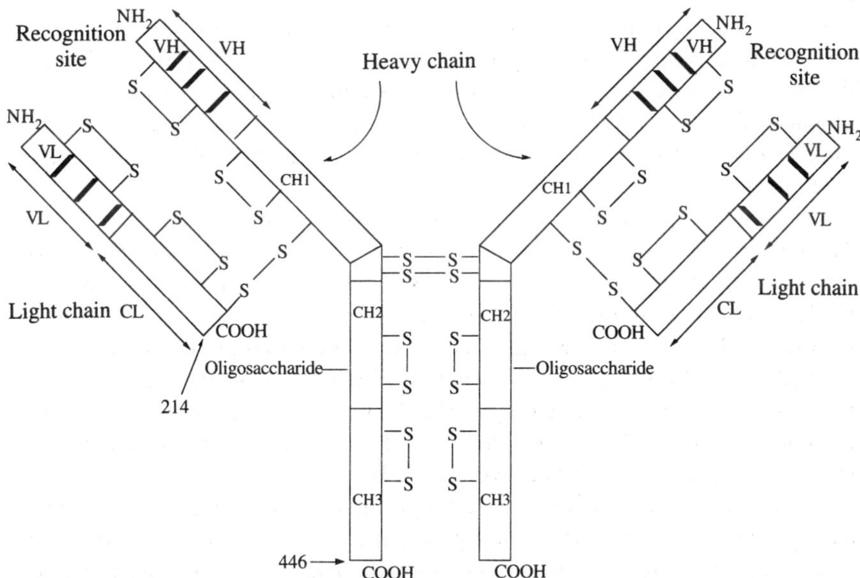

Fig. 15.1 Schematic representation of an immunoglobulin.

and sometimes even to chemicals very different from the constituents of living cells. The general impression is that of a suit made to measure by a very high-class tailor, but this is only an illusion because the suit was actually purchased ready-made although from a shop with an incredible collection of shapes and sizes. Furthermore, some alterations may have been made. We will briefly described the mechanism.

The cells involved in the synthesis of antibodies, especially carbohydrate-directed antibodies in which we are mainly interested, are the B lymphocytes. There are at least 10^8 clones of these lymphocytes, that is to say very small and restricted populations, each coming from the same mother cell. The cells, all identical in a particular clone, make and express a characteristic immunoglobulin on their surface. Cells from another clone express a different immunoglobulin. Thus there are as many different immunoglobulins as there are clones, that is around 10^8. The antigen binds itself to the immunoglobulin receptor with which it has the greatest affinity. This binding triggers the proliferation of the lymphocytes which carry these immunoglobulins and the expansion of the clone beyond its former dimensions. Activated B lymphocytes are further differentiated into memory cells and plasmocytes which secrete the complementary immunoglobulins of the antigens into the serum. Finally, an elevated mutation rate results in further refinement for the antigen–antibody fit. These are the so-called alterations whose mechanism will not be described here.

Under these conditions, it is easy to understand the heterogeneity of antibodies. Among the 10^8 immunoglobulin molecules, there will generally be more than one which can fit the antigen well enough to activate a proliferation of the carrying lymphocytes. The serum of an immunized animal contains a mixture of antibodies secreted by different clones of B lymphocytes, which is called a *polyclonal antibody*. Their properties change over time and with the injection schedule. Another source of heterogeneity is the presence of several chemically different antigenic sites on the antigen molecule. The heterogeneity of polyclonal antibodies may be demonstrated by several very precise techniques, but the separation of the monoclonal components into quantities large enough to study their properties is not feasible.

In fact the preparation of monoclonal antibodies rests on another principle. The reader will realize, as in all other more traditional branches of chemistry, that considerable progress is being made when pure compounds are obtained. As we have seen, after immunization with one antigen (Ag), there is a proliferation of B lymphocytes, each one specialized in the production of one of the anti-Ag antibodies, say X, X′, X″, etc. If it were possible to cultivate these cells *in vitro*, these clones could be separated by established methods. Unfortunately, it is not possible to cultivate B lymphocytes *in vitro*. The problem has been solved by using the properties of a line of myeloma cells.

Multiple myeloma is a disease resulting from the malignant transformation of a unique clone of plasmocytes which proliferate in the bone marrow. These cells synthesize and secrete a monoclonal immunoglobulin. They may be cultivated *in vitro* indefinitely. Fusion with antigen-activated B lymphocytes gives a *hybridoma* which can be cultivated *in vitro* and expresses one of the immunoglobulins X, X′, X″ ... The following steps are carried out for the preparation of monoclonal antibodies. Mice are immunized with the antigen. Lymphocytes are collected from the spleen of the animal and mixed with myeloma cells in 50% aqueous polyethyleneglycol at pH 8.0. After a few hours, a mixture of hybrid cells with the two types of original cells is obtained. The non-hybridized spleen cells die in 1 or 2 days. As for the myeloma cells, they were genetically programmed to allow their selective killing. Only the *hybridoma* expressing the immunoglobulins X, X′, X″, ... survive. At this stage it should be verified that these cells are actually producing antibodies. The last step is cloning, a technique which allows the preparation of homogeneous cultures, coming from only one cell, and each expressing a particular monoclonal antibody.

15.3 Immunochemical reactions *in vitro*

While perusing the preceding section, the reader will have understood that the immunization of an animal to obtain polyclonal antibodies and subsequent separation of them into homogeneous monoclonal components involves essentially biological methods. On the other hand, the *in vitro* study of the association between an antigen and an antibody is a typical branch of chemistry, the domain of which covers all types of bonding, covalent or not. Unless otherwise stated, only monoclonal antibodies will be considered henceforward.

15.3.1 Haptens

The above section describes the receptor site of immunoglobulins, built from an association of the hypervariable loops near the L and H chain ends. We shall now consider the complementary site on antigens. There may be several antigenic sites on the same antigen molecule, either different chemically or conformationally, and the corresponding antibodies are found in the antiserum. In certain cases, the chemical nature of these complementary sites is known. This part of the antigen is called the *antigenic determinant* or, sometimes in modern texts, the *epitope*. This epitope may be an oligosaccharide residue and the corresponding free oligosaccharide may come from either a natural source or by chemical synthesis. This relatively small molecule, when added to a solution of the corresponding monoclonal antibody raised against the larger antigen molecule, binds reversibly to this antibody. This has been demonstrated in several ways, one of which is the inhibition experiments described below. However, this small molecule is not antigenic when introduced into the blood stream. To be antigenic, a molecule must have a minimum molecular weight of 3000 D and preferably higher than 10 000 D. The problem may be solved by the covalent binding of the oligosaccharide to a macromolecular carrier as, for example, bovine serum albumin. Binding is achieved on the side chains of amino acid residues, almost always amino functions. There are obviously many ways this can be done. One is the preparation of glycoside **15.1** of the methyl ester of 9-hydroxynonanoic acid which is converted to acyl azide **15.2** by standard treatment with hydrazine and nitrous acid. Such derivatives react with primary amines to give amides and the oligosaccharide is bound in this way as in **15.3**. Of course, there are many side chains with amino functions in bovine serum albumin, hence many bound oligosaccharide molecules. Thus there are several epitopes on the same antigen. The antigen is said to be *multivalent*.

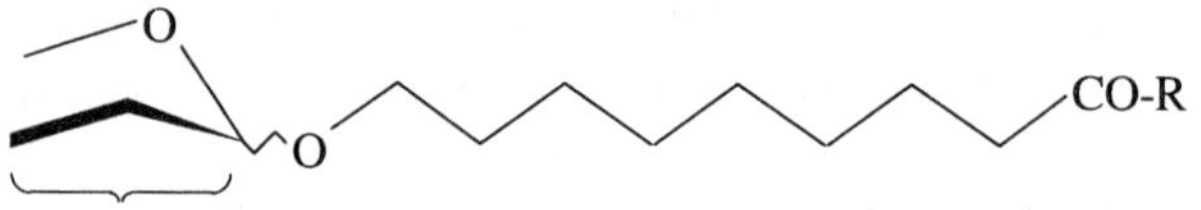

O

O CO-R

15.1 R=OMe

15.2 R=N_3

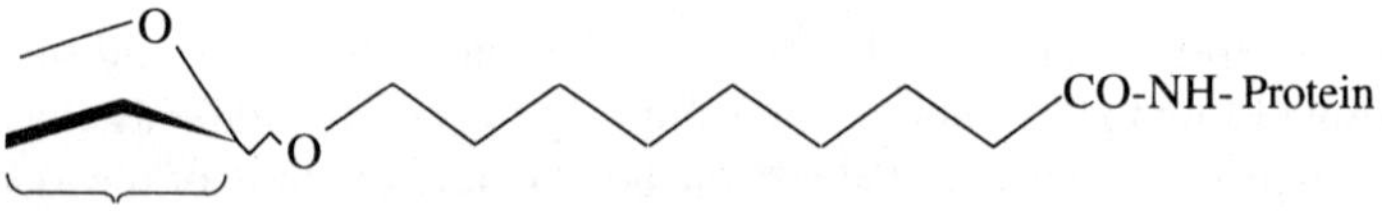

O

O CO-NH- Protein

15.3

This small molecule bound to a macromolecular carrier which elicits antibody formation is called a *hapten*. The method of covalent binding of a hapten to a macromolecular carrier allows the preparation of antibodies directed against a great number of chemical structures, some of them totally foreign to the living world. One disadvantage to this practice is that it also provokes the formation of antibodies directed against the carrier molecule and sometimes the linking arm.

15.3.2 Physical chemistry of the immune reaction

The binding forces are electrostatic attractions between sites with opposite charges such as $-NH_3^+$ carried by a lysine residue and a carboxylate, van der Waals forces, and hydrogen bonds. It is predicted that the first two types of bonds, which increase rapidly as the distances decrease, will be all the more efficient as the complementarity is better. The expression 'hydrophobic interactions' is used because an exact fit drives away water molecules present close to the 'hydrophobic' residues such as valine, leucine, isoleucine, etc. This concept, developed further in Chapter 11, implies that a decrease in surface contact with water is in itself a stabilizing factor. All these forces are reversible. When the hapten–antibody complex in solution is introduced into a dialysis bag, only the hapten can cross over the membrane, and its elimination from the interior bag causes complete dissociation of the complex through equilibrium displacement.

The reaction at one receptor site of an antibody Ab with a molecule of hapten H (or a monovalent antigen having the same specificity) is an equilibrium reaction, characterized by a constant k at a fixed temperature equation (15.1) and a variation of free enthapy given by equation (15.2).

$$\text{Ab} + \text{H} \underset{k_{12}}{\overset{k_{21}}{\rightleftharpoons}} \text{Ab–H} \qquad \frac{[\text{Ab–H}]}{[\text{Ab}][\text{H}]} = k \tag{15.1}$$

$$\Delta G^{\ominus} = -RT \log k \tag{15.2}$$

The rate of association is always very great, $k_{12} = 10^6–10^8$ mol^{-1} s^{-1}. As the equilibrium constants k span a wide range equal to k_{12}/k_{21}, there is a similar spreading of the rate of dissociation characterized by k_{21}.

The situation is actually far from being so simple. There are two antigen-combining sites on immunoglobulins IgG, IgD, and IgE, but for the other two, an association of several fundamental Ig units similar to that oulined in Fig. 15.1 is possible. Thus there are ten antigen combining sites on IgM and two or four on IgA. These antibodies are said to be deca-, di-, and tetravalent. Antigens and hapten conjugates may also be multivalent. For instance, on artificial antigens prepared by coupling a hapten to a protein, there may be a binding site on each lysine residue of the carrier molecule. Bacterial polysaccharides, the study of which is an important branch of immunochemistry, are built from repeating oligosaccharide units. There may be as many as 40 binding sites. The reaction between multivalent antigens and antibodies is highly complex and gives products with continually varying compositions, depending on the reagent proportions. The reaction is not stoichiometric. Equations (15.3) and (15.4) are not rigorous in a physicochemical sense, but rather describe the average of subreactions.

$$\text{Ag} + \text{Ab} \rightleftharpoons \text{Ag–Ab} \qquad \frac{[\text{Ag–Ab}]}{[\text{Ag}][\text{Ab}]} = k \tag{15.3}$$

$$\Delta G^{\ominus} = -RT \log k \tag{15.4}$$

From measurable concentrations of the free and bound antigen [Ag–Ab], it is possible to calculate k. These k values, from 10^3 to 10^{11} mol^{-1}, are a practical measure for the affinity of the antibody for the antigen. The word affinity is replaced by *avidity* in multi-binding cases, to stress the empirical nature of these figures. The multivalency for both an antibody and an antigen brings about a great increase in the association constant, a fact which may be understood in a qualitative manner by noting that the breaking of one bond at one site does not separate an antigen from an antibody still bound at other sites. It has been found that $\Delta G^{\ominus}$ varies from -6 to -11 kcal mol^{-1} and $\Delta H^{\ominus}$, from -4 to -13 kcal mol^{-1}. The entropy of association is nearly always positive.

Even a perfectly characterized epitope on an antigen gives rise to several monoclonal antibodies which differ by their combining sites and affinities. This is readily understandable in view of the proliferation mechanism explained in Section 15.2. It is important to note that the specificity of each is not perfect. While the affinity is generally high, the association of similar molecules may be observed, in which case the association constant is weaker. For instance, an oligosaccharide epitope may correspond to six monosaccharide residues (hexasaccharide). With the monoclonal antibody specific to this sequence, binding may still be observed, with rapidly decreasing affinities, to pentasaccharides, tetrasaccharides, etc., obtained by pruning one, two, or more monosaccharide residues from one end of the hexasaccharide. On the other hand, if deletion of a function such as a hydroxyl group at a given site on an oligosaccharide changes little or nothing of the binding to the specific antibody, the conclusion is that it does not participate in the binding. As the antibody only 'sees' the outer electronic cloud of the hapten, we may imagine a distribution of atoms, and hence a molecule radically different from the specific epitope, which will nevertheless give a cross reaction because its outer part is approximately the same. Examples are known but they have not been studied systematically. *Cross reactions* have been put to good use in the technique of catalytic antibodies. The basic idea is to stabilize the transition state of a reaction, thus decreasing the energy of activation by association with a specific antibody. Obviously, it is impossible to process a highly unstable structure such as a transition state as can be done for a hapten. It is replaced by a stable molecule whose periphery is as similar as possible to that of the transition state. Antibodies prepared in this way accelerate some reactions and work like enzymes. About 50 of these immunoglobulins, which are called 'abzymes', have been prepared to date.

Precipitation reaction

Polycondensations of bifunctional molecules with other at least trifunctional molecules have been done for a long time by organic chemists, for example the condensation of phthalic anhydride and glycerine. This leads to a tridimensional network. The situation is the same when a multivalent antigen and an immunoglobulin are brought together, resulting in a cluster of macromolecules which quickly become insoluble in water. However, there is a very important difference with industrial polycondensations, in that the binding of an antigen to an antibody is reversible under the reaction conditions. If the antigen is progressively added to an aqueous solution of the specific antibody, there will be an excess of antibody at first, and the associations will be of the Ab–Ag and Ab–Ag_2 types. The addition of more antigen brings about cross-linking and, finally, precipitation. The precipitate increases progressively while changing the composition, if it is at equilibrium with the solution, until the 'stoichiometric' composition is reached whereby all binding sites of both antigen and antibody are utilized. Addition of excess antigen beyond this stage leads to the progressive dissolution of the precipitate. This is a consequence of the reversibility of the binding. Reaction (15.5) shows how excess antigen causes the rupture of the association between two molecules of the antibody bound by at least a divalent antigen.

$$\text{(11.5)} \qquad \text{R–Ab–Ag–Ab–R}' + \text{Ag} \rightleftharpoons \text{R–Ab–Ag} + \text{Ag–Ab–R}'$$

This leads to complete depolymerization and dissolution of the precipitate. In agreement with this mechanism, precipitation is not observed with monovalent antigens, nor with certain immunoglobulins which have become monovalent because of the intermolecular complexing of one of their active sites by an oligosaccharide sequence covalently linked to the polypeptide.

Agglutination reaction

We come now to approaching biology but the agglutination reaction must not be omitted because of its extreme sensitivity. Antigens may sometimes be expressed naturally in very high density on erythrocytes (red blood cells). There are 10^6 blood group A determinants on the surface of erythrocytes A_1. Suspensions of these cells are moderately stable and they sediment slowly. In the presence of anti-A antibodies, cross-linking occurs at a number of points between the antigen attached to different cells causing agglomeration followed by rapid sedimentation. Decavalent IgMs are especially efficient in this respect.

Antibody and antigen estimations

Here we will only give the rudiments but the reader will find their applications in Chapters 16 and 17. The direct weighing of the centrifuged precipitate or the estimation of its weight from a nitrogen determination does not appear sensitive enough for modern research which involves microgram quantities of antigens and antibodies. However, a technique based on measuring the precipitates exists.

The estimation of an antibody in a liquid medium utilizes the specific antigen labelled with an isotope (generally ^{125}I) Ag* added in excess. The association Ag*–Ab, soluble under these conditions, is separated by precipitation with 50% ammonium sulfate or a general anti-Ig antibody. Radioactivity of the precipitate is measured.

Antigen estimation begins in the same way by making a soluble Ag*–Ab association with excess Ag*. The solution of Ag to be estimated is added, hence the equilibrium (15.6).

$$\text{(15.6)} \qquad \text{Ag}^*\text{–Ab} + \text{Ag} \rightleftharpoons \text{Ag–Ab} + \text{Ag}^*$$

What is measured is the ratio of free to bound radioactivity, and a calibration curve is drawn for a series of concentrations of Ag. This technique can also be used to compare the affinity of small molecules H for the antigen receptor on the antibody. The latter partially displace the antigen from association (15.7).

$$\text{(15.7)} \qquad \text{Ag}^*\text{–Ab} + \text{H} \rightleftharpoons \text{Ag}^* + \text{H–Ab}$$

For instance, the affinity of H for the antigen combining site on the Ab antibody will be characterized by the molarity necessary to observe 50% inhibition of the 'authentic' Ag–Ab association. In a more complete manner, results may be given as a curve giving the percentage of inhibition as a function of the molarity of H. It should be emphasized that for such assays, the chemist must provide the immunologist with the most pure possible samples since affinity can vary from one to a thousand in a series, i.e. one-thousand part of a very active impurity in a compound having little activity may lead to utterly false conclusions. Many assays are now carried out in solid phase; typically, the antigen is adsorbed on the surface of plastic wells and after addition of the antibody solution, the bound and free antibodies are separated by washing.

15.4 Lectins: definition and extraction

The name *lectin* was initially given to proteins or glycoproteins, almost exclusively extracted from plant seeds, able to bind reversibly to sugar molecules (Goldstein and Hayes 1978; Liener *et al.* 1986). From the beginning, certain preparations from fluids from animals such as eels, edible snails, or slugs were also classified as lectins. The term was then extended to a whole series of proteins having affinity for sugars from the most diversified sources. For example, there is a glycoprotein, ceruloplasmin, found in rabbit serum, the oligosaccharide chains of which end with a sialyl residue at the non-reducing end, carried by a penultimate galactose residue. Desialylation of this glycoprotein causes its rapid absorption by hepatic cells. This is due to the presence of a lectin in these cells, which retains the molecules having a galactose residue at the non-reducing end. These animal 'lectins' are rather widespread and probably possess important biological functions. There are also proteins capable of binding reversibly the sugars found in bacteria and mold. Thus, the term lectin covers a category more extensive than at the beginning; however, excluded from this category are enzymes from carbohydrate metabolism, anti-sugar antibodies, and sugar transport proteins. Only traditional lectins will be described in this chapter, whereas examples of animal lectins, the selectins, will be discussed in Section 17.6.

Nearly all current methods of isolation and purification of lectins rely on affinity chromatography. Naturally, the characteristic ligand must be determined in advance. The properties of lectins can be used to precipitate macromolecules and to agglutinate some types of cells, be they plant or animal. The driving force of this reaction is the association with certain bound residues, generally monosaccharides, from the macromolecule or the cellular periphery. When this type of reaction is observed, the problem is to find the sugar that can inhibit activity at the lowest possible molar concentration. As in the case of immunochemical precipitations, this inhibition is due to the occupation of the recognition site by the small soluble molecule.

Once the most efficient ligand is found, it is linked by a stable covalent bond to an insoluble and inert macromolecule and a chromatographic column is set up with this. In certain cases, columns can be directly filled with commercial absorbants which contain the ligand structure. Dextran-based 'Sephadex', a polysaccharide with α-glucopyranosyl residues, retains lectins specific for this configuration. Chitin, an insoluble polysaccharide with β-*N*-acetylglucosamine residues, or agarose can also be used directly. Among the supports used to produce conjugates, let us keep in mind agarose (sepharose), polyacrylamide (BioGel P), and starch.

To isolate a plant lectin, seeds are ground into very fine powder which is treated first with organic solvents (methanol, ether) in order to remove the lipids. The proteins are then extracted by salt solutions or buffers and the lectin is concentrated following usual precipitation techniques with ammonium sulfate, redisolving, etc. For final purification, the lectin solution is passed through an affinity column. The lectin is retained by its association with the combined bound ligand contained therein. After washing the column, the lectin is eluted by adding the same soluble ligand which displaces the equilibrium of the association. A certain number of lectins, extracted from inexpensive seeds by simple steps, are easily available. For example, three grams of the lectin 'concanavalin A', can be isolated from 100 grams of jack bean flour. This easy access to lectins, combined with their multiple properties, has led to numerous publications.

It does not seem possible to design a nomenclature of lectins from their structures. That proposed by Goldstein and Hayes (1978) links the botanic origins to recognized configurations classified by decreasing affinities. For example:

- lectin from *Canavalia ensiformis* (α-D-Man*p* > α-D-Glc*p* > α-D-GlcNAc*p*),
- lectin from *Triticum vulgaris* [β-D-GlcNAc*p*-(1→4)-β-D-GlcNAc*p*-(1→4)-D-GlcNAc > β-D-GlcNAc*p*-(1→4)-D-GlcNAc >> β-D-GlcNAc*p*).

The above lectins are from jack bean and wheat germ, respectively. Unfortunately, this nomenclature depends on experimental refinement. Just compare the name of the lectin from jack bean given here (Goldstein and Hayes 1978) with the one proposed in Section 15.5.3 (Liener *et al.* 1986).

15.4.1 Structure

We will use the example of concanavalin A, abundant in jack bean flour. Final purification is done by absorption on a Sephadex column from which the lectin is eluted by a glucose solution. Concanavalin A is formed from four subunits—MW = 26 500 daltons—associated in dimers below pH 5.6 and tetramers above. Each subunit contains Mn^{2+}, Ca^{2+}, and a sugar binding site. Figure 15.2 shows their mode of association, with the location of the ions and the binding site. A very important part of polypeptide chains are β-sheets. These are the regions where the associations between monomers take place.

Mineral ions can be removed by treatment with 0.1 N HCl, followed by dialysis against distilled water. This removal abolishes activity. The binding site was identified by X-ray analysis of the crystallized complex with methyl α-D-mannopyranoside. It is located at 7 and 11 Å from Ca^{2+} and Mn^{2+}, respec-

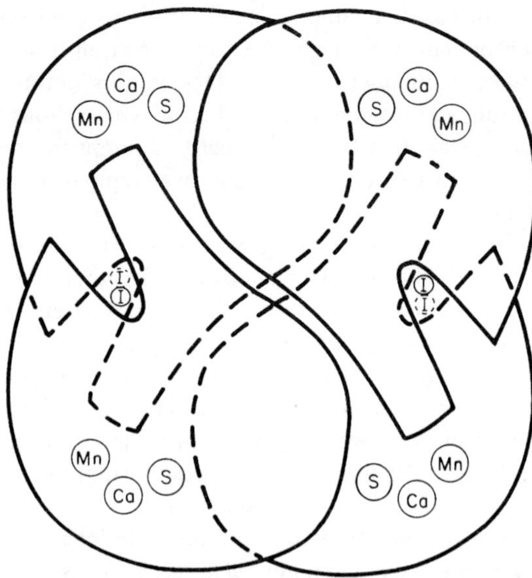

Fig. 15.2 Schematic representation of the tetramer of concanavalin A. The sites of Mn^{2+} and Ca^{2+} and the binding site of the sugar are indicated by Mn, Ca, and S, respectively (from Becler *et al.* 1976) (reproduced with kind permission from *Nature* and the authors;–1976 Macmillan Magazines Limited).

tively. The closest amino acid residues seem to be two tyrosines, two aspartic acids, one asparagine, one leucine, one serine, and one arginine. This suggests a clearly hydrophilic cavity. Indeed, the binding of the ligand hides two carboxyl groups by a subunit, as shown for example by acidimetric titration. The carboxyl groups in question could be those of aspartic acid residues close to the binding site.

Let us now look at the other lectins. The presence of isolectins is frequently observed. These molecules, isolated simultaneously from a given natural source, are slightly different but have the same association properties. Chromatographic separation of proteins with agglutinating activity from wheat germ (250–500 mg/kg) shows that this activity is divided into four fractions whose amino acid compositions are nearly identical. Neither wheat germ lectins nor concanavalin A are covalently bound to sugars, whereas many other lectins are glycoproteins. They sometimes show a very high content of sugars. For example, a lectin can be extracted from the potato (*Solanum tuberosum*) (38 mg for 4.5 kg of tubers) which is specific for oligosaccharides related to chitin:

$$\beta\text{-D-GlcNac}p\text{-}(1\rightarrow4)\text{-}[\beta\text{-D-GlcNAc}p\text{-}(1\rightarrow4)]_n\text{-}\beta\text{-D-GlcNAc} \qquad (n = 1 \text{ or } 2)$$

This lectin, a glycoprotein, contains around 50% of its weight as sugar.

Like concanavalin A, the majority of lectins are associations of subunits. Sometimes they are identical, sometimes different, often numbering four, but sometimes more or less (two). Each of the two isolectins of lentils (*lens culinaris*, synonym *esculenta*) is decomposed into four subunits in an acidic medium or in 8 M urea. There are 'heavy' chains (H) (MW = 17 570 D) and 'light' ones (L) (MW = 5710 D). Each of the two lectins are a non-covalently bound association of two H chains and two L chains, to give a molecular weight close to 46 000 D. Wheat germ lectins are dimers from units (*M* 21 600) dissociable by denaturing agents or under extreme pH conditions. On the other hand, the subunits of lectin from ricin (*Ricinus communis*) are covalently bound by disulfide bridges, and a reducing agent such as mercaptoethanol is needed to separate them.

Finally the presence of Mn^{2+} and Ca^{2+}, indispensible to the activity of concanavalin A, has been observed in a certain number of other lectins.

15.4.2 Specificity

The specificity of lectins is expressed by soluble associations with small sugar molecules or oligosaccharides, by precipitation reactions with oligosaccharides, and finally by agglutinating reactions of plant or animal cells. Their specificity varies as with blood cells for example. Some lectins agglutinate all of them, while some are specific to the animal species and others specific to its blood group. Associations which do not lead to precipitation with simple sugars and their derivatives, and oligosaccharides, can be analysed quantitatively using the equilibrium dialysis method. The ligand solution is poured into two compartments separated by a membrane and the lectin is added to only one of them. At equilibrium, the free ligand concentration is the same in both compartments and it is easily measured in the one containing no lectin. The association constants vary from 10^2 to 10^5 M^{-1}. It is interesting to compare the association constant of one monosaccharidic ligand or its alkyl glycoside having a suitable anomeric configuration to that of an oligosaccharide in which this ligand is located at the non-reducing end. Comparison seems to indicate that the recognition site of most lectins corresponds to a single glycosyl residue. However, with several lectins, di- and trisaccharides are better ligands than monosaccharides, which indicates an interaction that extends beyond a single residue. In this way, concanavalin A recognizes essentially the α-D-mannose; however, the associations are respectively four to twenty times stronger with di- and trisaccharides α-D-Man*p*-(1$\rightarrow$2)-D-Man and α-D-Man*p*-(1$\rightarrow$2)-α-D-Man*p*-(1$\rightarrow$2)-D-Man than with the methyl α-D-mannopyranoside. This type of sequence is found in glycoproteins. In nearly every case of binding between lectins and oligosaccharides or glycoproteins, the greatest association, from the point of view of binding energy, occurs with the sugar at the non-reducing end. More rarely, it occurs with monosaccharide units branched on a major chain. For example, concanavalin A not only recognizes α-D-mannopyranosyl and α-D-glucopyranosyl terminal units on an oligosaccharide chain, but also α-D-mannopyranosyl branches linked at position 2 of a sugar on the major chain. Certain lectins only bind efficiently with a given anomer, while others are more or less indifferent to the anomeric configuration of the residue. Lectin from soya (*Glycine max*) recognizes both the α and β anomers of *N*-acetylgalactosamine. Its definition by its ligands is written: α-D-GalNAc*p* $\geq$ β-D-GalNAc*p* >> α-D-Gal*p*. There is, however, a slight preference for the α-anomer. Likewise, a certain number of lectins tolerate variations at position 2. Above we saw that concanavalin A binds with both D-mannose and D-glucose. Some lectins recognize both D-galactose and *N*-acetylgalactosamine, although with a marked preference for one or the other of these two sugars. We have seen this above with the lectin of soya. On the other hand, lectins scarcely tolerate variations at positions 3 and 4. There are no 'cross reactions' between the D-*gluco* and D-*galacto* configurations, for example.

An aqueous mixture of a lectin with a polysaccharide comprising residues recognized by this lectin generally leads to precipitation, as does a mixture of an immunoglobulin with the corresponding antigen. Concanavalin A precipitates dextran, a polysaccharide built from α-D-glucopyranose units. This can be done quantitatively by adding increasing amounts of polysaccharide to aliquot portions of lectin, followed by centrifuging, washing of the precipitate, and final nitrogen estimation. A precipitation curve is observed exactly like that of an antigen–antibody reaction (see Section 15.3). There are three zones: at the beginning there is an excess of lectin and all the dextran is precipitated. In equal amounts, all dextran and lectin are precipitated. It is at this stage that the maximum precipitation is observed. Excess dextran solubilizes the precipate, which disappears.

15.4.3 Brief description of a few lectins (Goldstein 1986)

For each lectin, we will first give its description according to the recommended nomenclature. On the following line will be given its molecular weight and that of the subunits (α, β...) as well as the association mode of the latter, the number (n) of recognition sites, and the possible metals present. We will end with the association constant with a good substrate.

Lectin from jack bean (Canavalia ensiformis)

α-D-Man*p*-(1$\rightarrow$2)-α-D-Man*p*-(1$\rightarrow$2)-D-Man > α-D-Man*p*-(1$\rightarrow$2)-D-Man > α-D-Man > α-D-Glc > α-D-GlcNAc

MW 106 000; α: 26 500; pH 7: α_4; pH 5: α_2; $n = 4$; Ca^{2+}, Mn^{2+}

K_a 2.06 × 10^4 M^{-1} (2°C) (methyl α-D-mannopyranoside; 3,6-di-*O*-α-D-mannosyl-D-mannose)

Lectins from Griffonia simplicifolia

Isolectin B_4: α-D-Gal >> α-GalNAc; Isolectin A_4: α-D-GalNAc >> α-D-Gal

MW 114 000; α: 32 000; β: 33 000; five isolectins: A_4 (α_4), A_3B ($\alpha_3\beta$), A_2B_2 ($\alpha_2\beta_2$), AB_3 ($\alpha\beta_3$), and B_4 (β_4); $n = 4$; Ca^{2+}

Isolectin B_4: K_a 2.06 × 10^4 M^{-1} (methyl α-D-galactopyranoside); isolectin A_4: K_a 1.87 × 10^5 M^{-1} (α-D-GalNAc residue at the non-reducing end, reacts strictly with erythrocytes of blood group A)

Lectins from wheat germ (Triticum vulgare)

β-D-GlcNAc*p*-(1$\rightarrow$4)-β-D-GlcNAc*p*-(1$\rightarrow$4)-D-GlcNAc > β-D-GlcNAc*p*-(1$\rightarrow$4)-D-GlcNAc >> D-GlcNAc

MW 43 200; α: 21 600; at least three isolectins; pH > 4: α_2; acidic solution: α; $n = 2$; Ca^{2+}
K_a 5.3 × 10^4 M^{-1} (20°C) {(H→4)-[β-D-GlcNAc*p*-(1→4)]$_4$-(1→OH)}; K_a 1.3 × 10^3 M^{-1} (4°C) (D-GlcNAc)

Lectins from soya (Glycine max)

α-D-GalNAc*p* = β-D-GalNAc*p* > α-D-Gal*p*
MW 122 000; α 30 000; α_4; $n = 4$; Ca^{2+}, Mn^{2+}
K_a 3.0 × 10^4 M^{-1} (4°C)

Lectins from snails (Helix pomatia)

α-D-GalNAc-(1→3)-D-GalNAc > α-D-GlcNAc*p* >> α-D-Gal
MW 79 000; α 13 000; at least twelve isolectins; α_6; $n = 6$; K_a 5 × 10^3 M^{-1} (pentasaccharide from blood group A: α-D-GalNAc*p*-(1→3)-[α-L-Fuc*p*-(1→2)]-β-D-Gal*p*-(1→4)-β-D-GlcNAc*p*-(1→6)-R)

Lectins I from broom (Ulex europaeus)

α-L-Fuc
MW 60 000–68 000; α 29 000, β 31 000; $\alpha\beta$; Ca^{2+}, Mn^{2+}, Zn^{2+}
K_a 3.1 × 10^{-3} M^{-1} (L-fucose)

Lectins from the slug (Limax flavus) (Knibbs *et al.* 1993)

Neu5Ac
MW 44 000 (2 × 22 000); $n = 2$
K_a 3.8 × 10^4 M^{-1} (Neu5Ac)

15.4.4 Biological properties of lectins (Lis and Sharon 1986)

Here we will not reconsider the properties of cell agglutination, which we have already discussed in Section 15.4.2. One of the lectins most impressive characteristics is their mitogenic power, which is to say that in a lymphocyte colony in a resting and non-dividing state, they activate growth and proliferation. This mitogenic effect is nearly always inhibited by simple sugars in a reversible fashion as is in the case of concanavalin A. While on the subject, it is important to note that a certain number of agents, other than lectins but just as capable of reacting on peripheral sugars of lymphocytes, also show mitogenic activity. The lectins most widely used as mitogens are concanavalin A and the lectin of *Phaseolus vulgaris*. Contrary to antigens which activate specific clones, lectins behave indiscriminately on a suitable population and the proportion of stimulated cells may reach 80%. Not only is stimulation observed simultaneously in all metabolic activities, but also in the secretion of a family of biologically active polypeptides, the lymphokines. It is probable that the initial phase of stimulation is the association of the lectin with cell-surface sugars, but association does not appear to be sufficient in every case.

There are a certain number of very characteristic properties, but explaining them would suppose a certain background in hematology which is outside the scope of this work.

15.4.5 Comparison of anti-sugar antibodies and lectins

Comparing anti-sugar antibodies and lectins is obvisouly essential to this chapter. Their similarities can be explained in a few words: lectins and antibodies are proteins (or glycoproteins) bearing several reversible recognition sites that make them reversible cross-linking reagents. Thus with these two families one can observe the association with mono- and oligosaccharides, precipitation with polysaccharide and glycoprotein macromolecules with dissolution of the precipitates in the presence of excess polysaccharides, and finally the agglutination of cells. Precipitation and agglutination are the result of the multivalency of the two complementary reagents and are inhibited in the presence of a specific mono- or oligosaccharidic ligand. Another consequence of this multivalency, observed with lectins and antibodies, is the increase in the apparent affinity when the ligands permit cross-linking. We have to realize that immunoglobulins are built from a uniform model or by association of molecules constructed after the same model, whereas lectins seem to display a great variety of structures. In the basic model of immunoglobulins, two recognition sites belong to two identical half-molecules linked to each other by disulfide bridges. A lectin is an association of subunits which may or may not carry a recognition site and are non-covalently linked, probably through contacts between extended β-sheets.

References

Becler, J. W., Reeke, Jr., G. N., Cunningham, B. A., and Edelman, G. M. (1976), *Nature*, **259**, 406–409.

Goldstein, I. J. and Hayes, C. E. (1978), *Adv. Carbohydr. Chem. Biochem.*, **35**, 127–340.

Goldstein, I. J. (1986), *The lectins, properties, functions and applications in biology and medicine*, p. 35, Academic Press, New York.

Knibbs, R. N., Osborne, S. E., Glick, G. D., and Goldstein, I. J. (1993), *J. Biol. Chem.*, **268**, 18524–18531.

Leiner, I. E., Sharon, N., and Goldstein, I. J. (1986), *The lectins, properties, functions and applications in biology and medicine*, Academic Press, New York.

Lis, H. and Sharon, N. (1986), *The lectins, properties, functions and applications in biology and medicine*, p. 266, Academic Press, New York.

16 ABH and related blood group antigens

16.1 ABH antigens (Clausen and Hakomori 1989; Hakomori 1991; 1993)

16.1.1 General background and polymorphism

Antigenic determinants are trisaccharide A, **16.1**, trisaccharide B, **16.2**, and disaccharide H, **16.3**. The reader will recognize that trisaccharides A and B are glycosidation products of disaccharide H at position 3 of galactose by an *N*-acetylgalactosamine unit and a galactose unit, respectively, and in the two cases by an α-anomeric linkage. Thus the A and B determinants differ only by their substitution at position 2 on their D-*galacto* non-reducing terminal end, *N*-acetyl in the A substance, and hydroxyl in the B substance. Note as well the participation of the deoxygenated sugar fucose and finally the α-1,2-*cis* bonds, not as common as the β-1,2-*trans* bonds in glycoconjugates. These disaccharides and trisaccharides are located at the non-reducing terminal ends of the oligosaccharide chains of glycoproteins and glycolipids and possibly their branches. Blood group A individuals have the A determinant and a certain quantity of H, but not B, whereas those of blood group B have B and H but not A, in a symmetrical fashion. Blood group O individuals only have the H determinant. This immediately implies an incomplete biosynthesis due to the absence or the non-expression of genes which code for the A or B glycosyltransferases. Carrier molecules are found in the membrane of erythrocytes and determinants are exposed towards the outside. In blood group A individuals, the B molecule is recognized as a foreign substance and gives rise to the appearance of anti-B antibodies. For the same reason, anti-A antibodies are found in people with blood group B and and anti-A and anti-B antibodies in those with blood group O. Problems observed during blood transfusions were due to the presence of these antibodies. Thus the donor of blood group A who has anti-B antibodies causes the agglutination of erythrocytes in a receiver of blood group B. ABH antigens are also present on cell surfaces in the majority of organs and in secretions. Their presence is one of the major causes of the failure of organ transplants between a donor and receiver of different blood groups. For this reason, they should also be named histo-blood group antigens.

ABH antigens display to a high degree the phenomenon called *polymorphism*. This means that the ABH determinants can be carried by a multitude of different chemical molecules. Structural differences far from the determinants have no effect on the immunochemical A–anti-A reaction, for example, but differences in the vicinity can show up using the fine techniques of monoclonal antibodies.

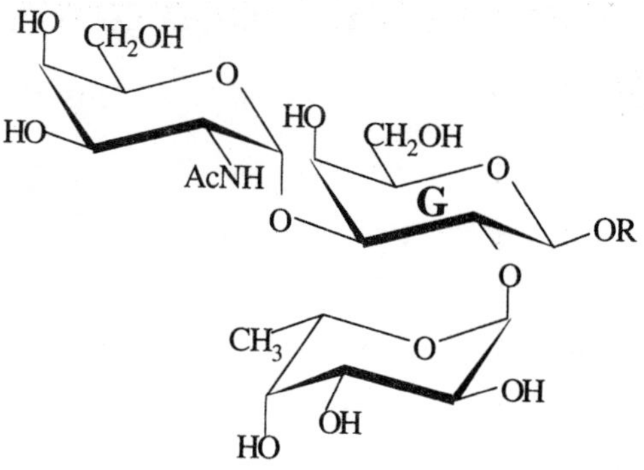

16.1

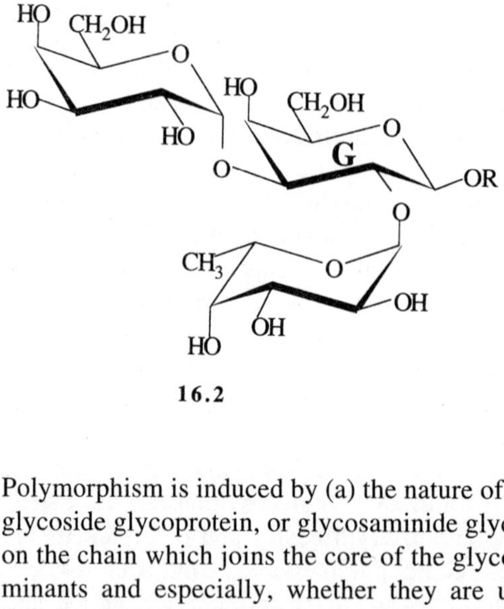

16.2

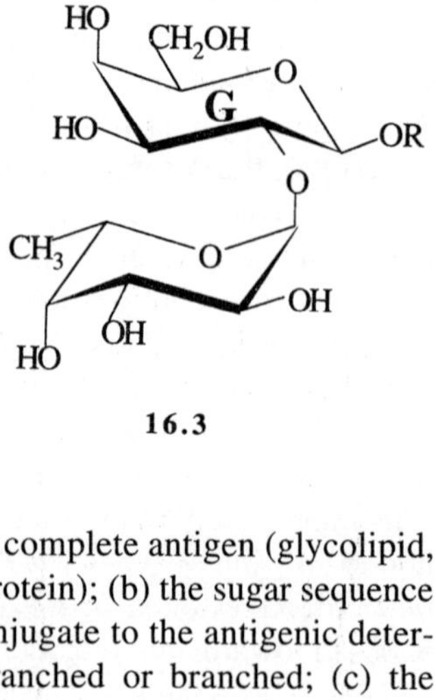

16.3

Polymorphism is induced by (a) the nature of the complete antigen (glycolipid, glycoside glycoprotein, or glycosaminide glycoprotein); (b) the sugar sequence on the chain which joins the core of the glycoconjugate to the antigenic determinants and especially, whether they are unbranched or branched; (c) the linkage of the determinant to the oligosaccharide.

The reader who feels confused by these formulas or those presented further on can refer to the representation in Fig. 16.6 at the end of this chapter.

16.1.2 Linkage types. The Lewis system

The G galactose of formulas **16.1** to **16.3** may be linked in four different ways to the carrier chain of the glycoconjugate, in the so-called 'types' 1 to 4, shown in Table 16.1. We would like to point out that the galactose on the left in Table 16.1 is the G galactose of the ABH antigens. The distribution of these types of disaccharide linkages depends on the family of the glyconjugate carrier.

Type 1 and 2 chains are found on glycoproteins, glycosides, or glycosaminides as well as glycolipids, type 3 chain especially on glycoside proteins, and type 4 chain only on glycolipids.

Table 16.1 The four types of linkage for the ABH determinants.

Type 1: Gal-β-(1→3)-GlcNAc-β-1(1→R)
Type 2: Gal-β-(1→4)-GlcNAc-β-1(1→R)
Type 3: Gal-β-(1→3)-GalNAc-α-1(1→R)
Type 4: Gal-β-(1→3)-GalNAc-β-1(1→R)

Enzymic fucosylation of type 1 chain gives antigens from the Lewis group. For example, in antigen Le^a, there is a fucosyl residue at position 4 of GlcNAc, **16.4**. A second fucosylation gives the Le^b antigen, which is therefore a difucosylated product, **16.5**. Type 1 antigens are the major carriers of the ABH determinants in body fluids and secretions. They are not synthetized by erythrocytes and lymphocytes which, nevertheless, absorb them from plasma and, for this reason, express them.

```
Gal-β-(1→3)-GlcNAc-β-(1→ R)
                ↑
       Fuc-α-(1→4)
```

16.4

```
Fuc-α-(1→2)-Gal-β-(1→3)-GlcNAc-β-(1→ R)
                             ↑
                    Fuc-α-(1→4)
```

16.5

Type 2 antigens are seen on the skin and erythrocytes. Fucosylation transforms the type 2 disaccharide into the H determinant, but other fucosylation products are observed as well (**16.6** and **16.7**) which are respectively isomers of Le^a and Le^b antigens and have been given the name of Le^x and Le^y antigens. The reader will recognize *N*-acetyllactosamine in the type 2 chain.

```
Gal-β-(1→4)-GlcNAc-β-(1→ R)
                ↑
       Fuc-α-(1→3)
```

16.6

```
Fuc-α-(1→2)-Gal-β-(1→4)-GlcNAc-β-(1→ R)
                             ↑
                    Fuc-α-(1→3)
```

16.7

Type 3 chain is the core disaccharide of glycoside glycoproteins, **16.8**. The disaccharide serine (threonine) is the T antigen and its monosaccharide precursor the T_n antigen. ABH determinants are linked directly to the core disaccharide. They have been found, for example, in the gastric mucous membrane and in the glycoprotein of ovarian cysts.

```
                T
 ______________________________
                  Tn
           ____________________
Gal-β-(1→3)-GalNAcα-(1→O)-Ser/Thr
```

16.8

This is the category in which the 'repetitive A' antigen should be classified corresponding to the sequence **16.9**. The A determinant is constructed on the disaccharide of type 3. In fact, this disaccharide is the galactosylation product of the α-GalNAc residue of an initial A determinant. A certain number of repetitive A glycolipid antigens have been isolated and it is possible that they constitute more than half of the A-active substances of erythrocytes. Naturally, this antigen is only found in blood group A individuals.

GalNAc-α-(1→3)-Gal-β-(1→3)-GalNAc-α-(1→3)-Gal-β-(1→4)-GlcNAc-β-(1→R)

Fuc-α-(1→2) (on the first Gal)   Fuc-α-(1→2) (on the second Gal)

16.9

Concerning type 4 chain, it appears by galactosylation at the non-reducing terminal end of certain glycolipids such as globoside **16.10** which gives the precursor **16.11**. The structures below brackets P, p^k, and p in formula **16.10** are three of the four antigenic determinants of the P blood group system. The ABH determinants can be constructed on the non-reducing galactose end of the elongated globoside **16.11**. The majority of the ABH antigens from kidneys belong to this family.

P (bracket over GalNAc-β-(1→3)-Gal-α-(1→4)-Gal-β-(1→4)-Glc)

p^k (bracket over Gal-α-(1→4)-Gal-β-(1→4)-Glc)

p (bracket over Gal-β-(1→4)-Glc)

GalNAc-β-(1→3)-Gal-α-(1→4)-Gal-β-(1→4)-Glc-β-(1→Cer)

16.10

Gal-β-(1→3)-GalNAc-β-(1→3)-Gal-α-(1→4)-Gal-β-(1→4)-Glc-β-(1→Cer)

16.11

According to what has just been said, the reader might believe that the ABH antigenic determinants are reduced to di- and trisaccharides. This might be possible, but we have reasons to believe that, in general, antigenic determinants can be extended to longer sequences. We should therefore distinguish between the variants due to polymorphism. The latter was possible through immunization with carefully purified antigens, by preparing thereafter monoclonal antibodies. For example, among the anti-A monoclonal antibodies, one recognizes specifically **16.12** (type 1 chain), another recognizes **16.13** (type 2 chain), but gives a cross reaction with **16.14** (type 2, difucosylated, called A/Le^y).

GalNAc-α-(1→3)-Gal-β-(1→3)-GlcNAc-β-(1→3)-Gal
↑
Fuc-α-(1→2)

16.12

GalNAc-α-(1→3)-Gal-β-(1→4)-GlcNAc-β-(1→3)-Gal
↑
Fuc-α-(1→2)

16.13

GalNAc-α-(1→3)-Gal-β-(1→4)-GlcNAc-β-(1→3)-Gal
↑ ↑
Fuc-α-(1→2) Fuc-α-(1→3)

16.14

Serological and genetic studies have shown that group A is subdivided into two major groups, A_1 and A_2. There has not yet been a definitive explanation given in terms of structure, but strong indications have been given concerning the subgroup A_1. Above we pointed out the discovery of a repetitive A antigen, related to type 3, only expressed on glycolipids—on globoside for type 4. These structures are only found on A_1 erythrocytes.

16.2 The Ii system and effect of branching

The determinant of the i antigen is the linear hexasaccharide **16.15**, a trimer of *N*-acetyllactosamine. The I determinant, **16.16**, is an *N*-acetyllactosamine residue linked to the primary alcohol position of a hexopyranose (Gooi *et al.* 1984). It covers a *branching site*. The i determinant appears as a fragment of long polycondensed chains of poly-*N*-acetyllactosamine linked to core mannoses of glycosaminide glycoproteins and to GalNAc (or Gal) core residues of glycoside glycoproteins. Very long chains of this type, up to 40 monosaccharides, can be found in the glycolipids of erythrocytes in rabbits with seven branches. A glycolipid from human placenta has been isolated with 20 residues and three branches. In general, these are the branches which make up the combining site of the I determinant.

Gal-β-(1→4)-GlcNAc-β-(1→3)-Gal-β-(1→4)-GlcNAc-β-(1→3)-Gal-β-(1→4)-GlcNAc-β-(1→R)

16.15

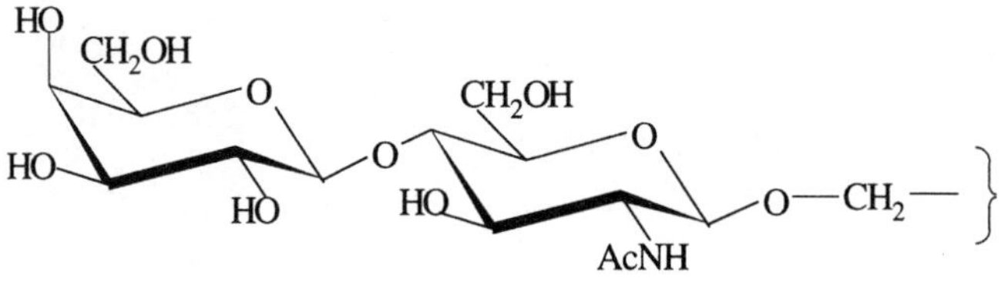

16.16

These antigens, present in every human, should not be recognized as nonself and so should not induce the production of antibodies. However, there are certain pathological conditions, called autoimmune diseases, whereby antigens and antibodies coexist in the same person. The antibodies which we will discuss are called cold agglutinins. The agglutination of erythrocytes happens when the blood is chilled to 4°C, which then disappears when it returns to 37°C. Sera from such patients are a source of anti-i and anti-I antibodies. These antibodies are naturally monoclonal and serve as reagents in the study of the distribution of I and i antigens in healthy individuals. Thus the blood of a human fetus expresses the i antigen, seen on the erythrocytes of the umbilical cord. During the first 18 months of an infant's life, the i antigen disappears to make room for the I antigen. In terms of structure, this can be interpreted simply by the branching of the linear chain i, by the coupling of the β-anomer of the *N*-acetyllactosamine disaccharide to the primary alcohol functions of the galactose residues. In Chapter 17, we will return to the probable role of the I and i antigens in the embryogenesis of mammals. These questions on branching will serve as a brief review on the ABH antibodies. The anti-sugar antibodies generally show weak or moderate affinity. The affinity is increased when there are several determinants on a single carrier molecule, which is made possible if it is branched. The combining with an antibody with two receptor sites multiplies affinity in the order of 10^3 to 10^4.

16.3 Synthesis of oligosaccharide determinants

16.3.1 ABH determinants

These syntheses pose the difficult problem of the 1,2-*cis* glycosidation leading to α-glycosides. It was specifically for the purpose of preparing some of these oligosaccharides that the common ion method was elaborated (see Section 10.3.3). Figure 16.1 shows the preparation of the Le^a trisaccharide by fucosylation of a protected derivative of *N*-acetyllactosamine (Lemieux and Driguez 1975*a*). Figure 16.2 illustrates the use of the same method in the preparation of an α-D-galactopyranoside such as trisaccharide B (Lemieux and Driguez 1975*b*). In recent syntheses of the H antigen, the Le^x and Le^y antigens and their dimers, α-fucosylation was achieved in a simple and nearly quantitative manner with the α-trichloroacetimidate of 3,4-di-*O*-acetyl-2-*O*-benzyl-α-L-fucopyranose, in

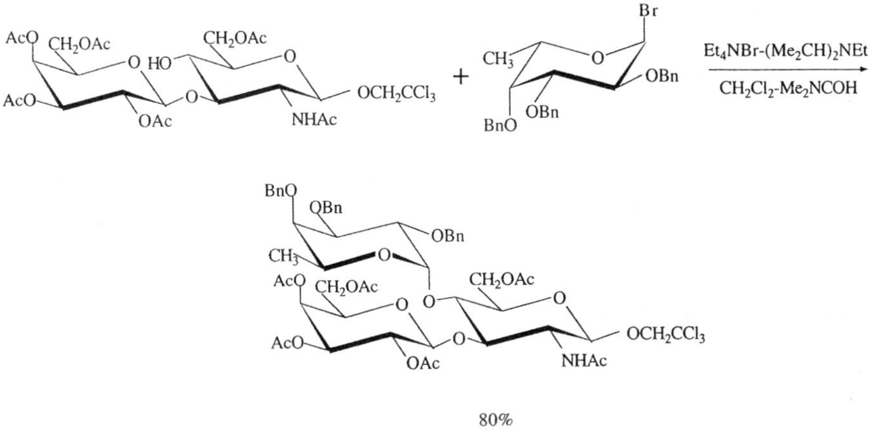

Et_4NBr-$(Me_2CH)_2NEt$

CH_2Cl_2-Me_2NCOH

80%

Fig. 16.1 Synthesis of the Le^a trisaccharide.

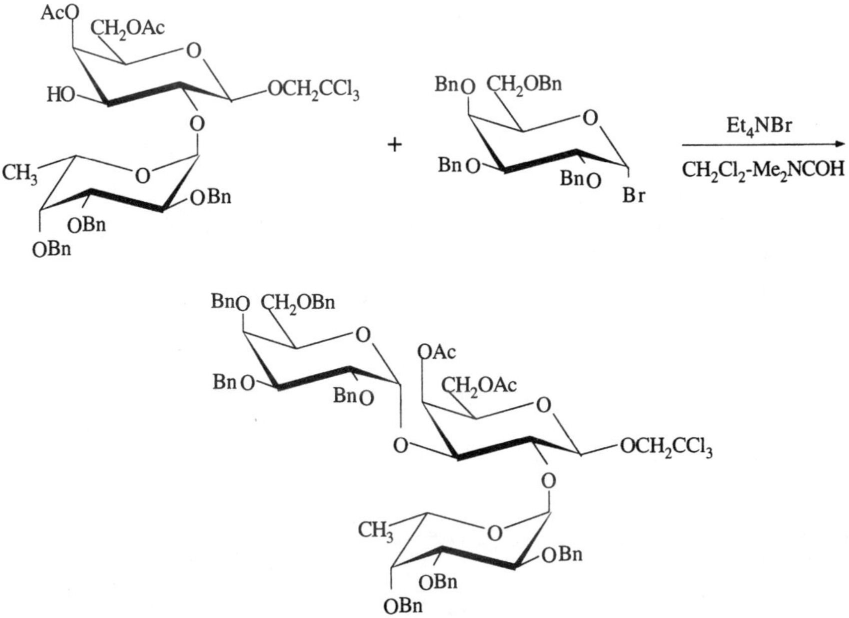

Et_4NBr

CH_2Cl_2-Me_2NCOH

Fig. 16.2 Synthesis of the B trisaccharide.

dichloromethane solution at room temperature in the presence of zinc dichloride etherate (Windmüller and Schmidt 1994).

We have reported in Section 12.4 the enzymic synthesis of the 'sialyl Lewis^x' tetrasaccharide **12.24** in water solution with the use of three appropriate glycosyltransferases and regeneration of cofactors, in sucession, a technique which allowed kilogram-scale synthesis.

In order to introduce an α-1,2-*cis* *N*-acetylgalactosamine unit, the starting reagent must not have a participating group. 2-Azido-2-deoxy-α-D-galactopyranosyl halides **16.17** and **16.18** are used, whose most common preparation utilizes the 'azidonitration' of the peracetylated galactal **16.19** (Fig. 16.3) (see Section 10.3.7). Condensation of bromide **16.18** with trisaccharide **16.20** is the key step in the synthesis of the A determinant carried by a type 2 chain. Deprotection is then carried out by reducing the azide followed by *N*-acetylation (Paulsen 1982) (Fig. 16.4).

16.3.2 The Ii determinants

The synthesis of the I determinant and related compounds was first carried out by coupling oxazoline **16.21** derived from peracetylated lactosamine with a galactose having the primary alcohol function free (Fig. 16.5). Ever since,

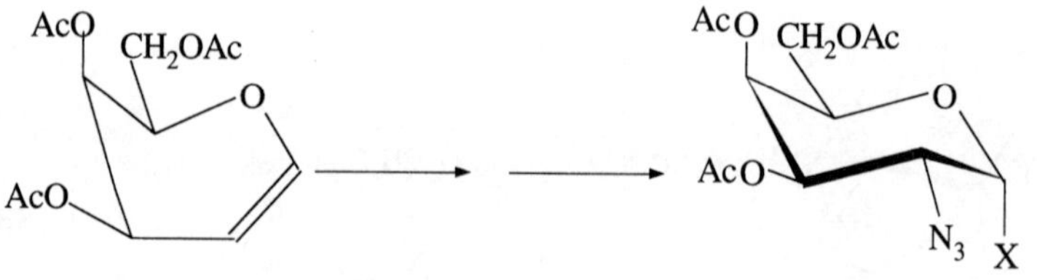

AcO, CH_2OAc, O, AcO

16.19

AcO, CH_2OAc, O, AcO, N_3, X

16.17 X = Cl
16.18 X = Br

Fig. 16.3 Azidonitration reaction.

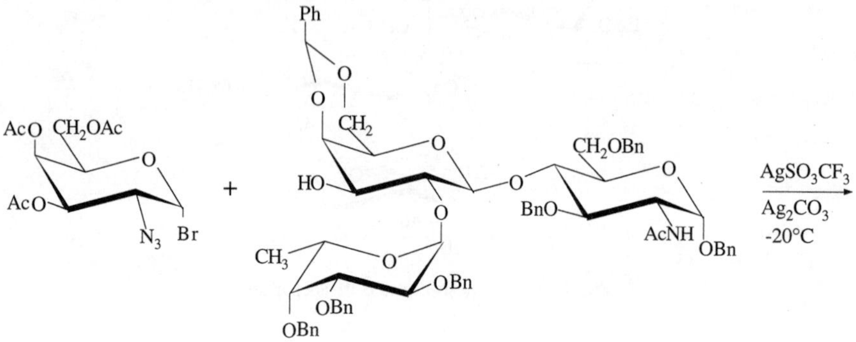

AcO CH_2OAc, O, AcO, N_3, Br

+

Ph, O, O, CH_2, O, HO, O, CH_3, O, OBn, OBn, OBn, O, CH_2OBn, O, BnO, AcNH, OBn

16.20

$AgSO_3CF_3$
Ag_2CO_3
-20°C

GalNAc-α-(1→3)-Gal-β-(1→4)-GlcNAc-β-(1→R)
↑
Fuc-α-(1→2)

Fig. 16.4 Synthesis of an A tetrasaccharide.

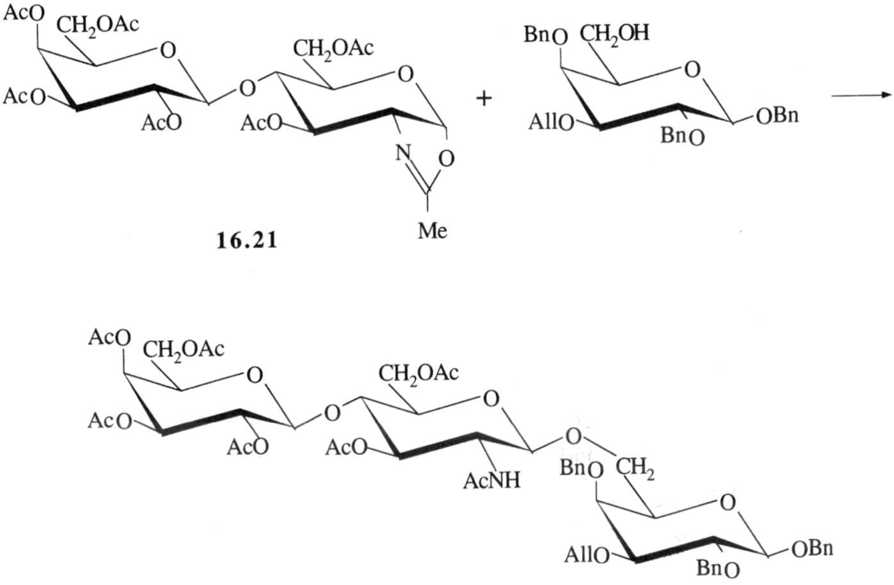

Fig. 16.5 Synthesis of the I trisaccharide.

oxazolines have been largely replaced by phthalimido chlorides for the β-1,2-*trans* coupling of amino sugars. Figure 16.5 shows that the alcohol function at position 3 of the galactose acceptor is temporarily protected by allylation. Selective deprotection of this position frees a hydroxyl group which can be used to extend the chain in another direction (Augé *et al.* 1979).

The synthesis of oligosaccharides of the i determinant family illustrates a special strategy of doubling (Alais and Veyrières 1990). Starting from a portion of the conveniently derivatized phthalimidolactosamine **16.22**, the anomeric benzyl group is hydrogenolysed and the anomeric hydroxyl group is converted to the trichloroacetimidate. In this way the glycosylating donor is obtained. Moderate acid hydrolysis is carried out on another portion of **16.22**, which exposes the hydroxyl groups at positions 3 and 4 of the galactose unit. Only the hydroxyl group at position 3 reacts under these coupling conditions. Coupling leads to tetrasaccharide **16.23**. A portion of **16.23** is activated on the anomeric carbon of the reducing unit exactly as above and the other portion, the alcohol function at position 3 of the non-reducing galactose end, is exposed by hydrolysis. Coupling gives the protected octasaccharide **16.24**. Immunochemical studies done using this group of products after deprotection seem to indicate that the i determinant is a hexasaccharide chain. Figure 16.6 represents the major antigenic determinants mentioned in this chapter.

16.3.3 One-pot synthesis of sialyl T-antigen (Kren 1995)

A sialylated epitope **16.26** of the T-antigen (see Section 16.1.2) is found in various tissues and was needed in higher quantities for immunological studies.

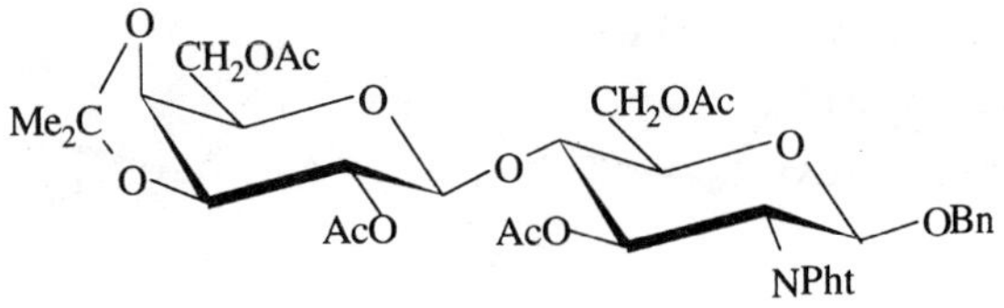

16.22

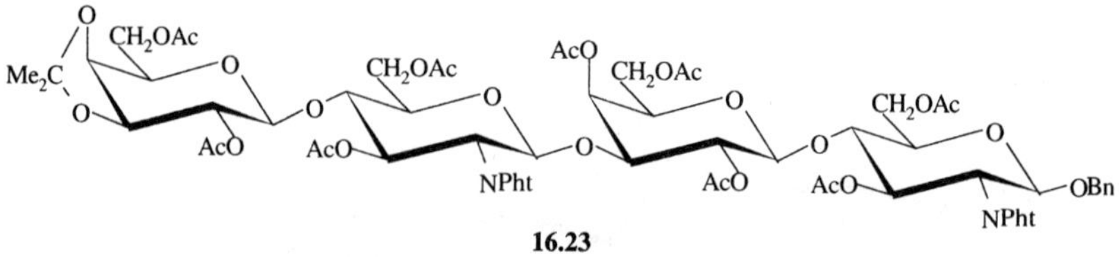

16.23

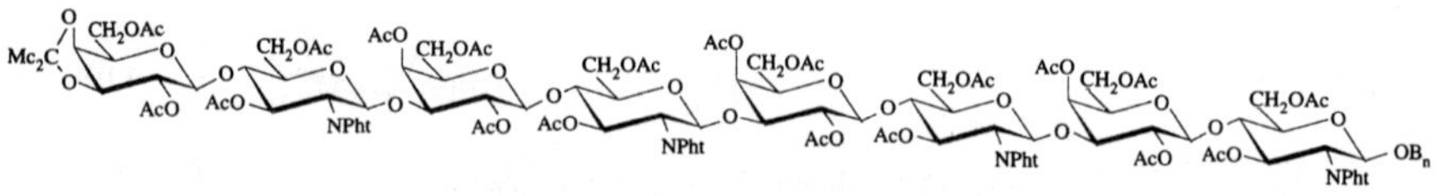

16.24

Disaccharide **16.25** is first made by *trans*-galactosylation in the presence of a β-galactosidase from bovine testes, according to reaction (16.1). This is followed by sialylation in the presence of ST3 sialyltransferase from porcine liver, according to reaction (16.2). CMPNeu5Ac was regenerated by the multi-enzyme system described in Section 12.4.

(16.1) Gal-NAc + paranitrophenyl-β-D-galactopyranoside ⇄

Gal-β-(1→3)-GalNAc + paranitrophenol

16.25

(16.2) CMPNeu5Ac + Gal-β-(1→3)-GalNAc ⟶

Neu5Ac-α-(2→3)-Gal-β-(1→3)-GalNAc + CP

16.26

All six enzymes, substrates and cofactors are mixed together to achieve a one-pot reaction. The pH adopted, 7.5, is suitable for all enzymes involved in sialylation, but far from the optimal pH for the galactosidase which is 4.3. However, the reaction could be conducted at pH 7.5 because galactosyl transfer to GalNAc is an efficient process and transglycosidation in synthesis is quite often more

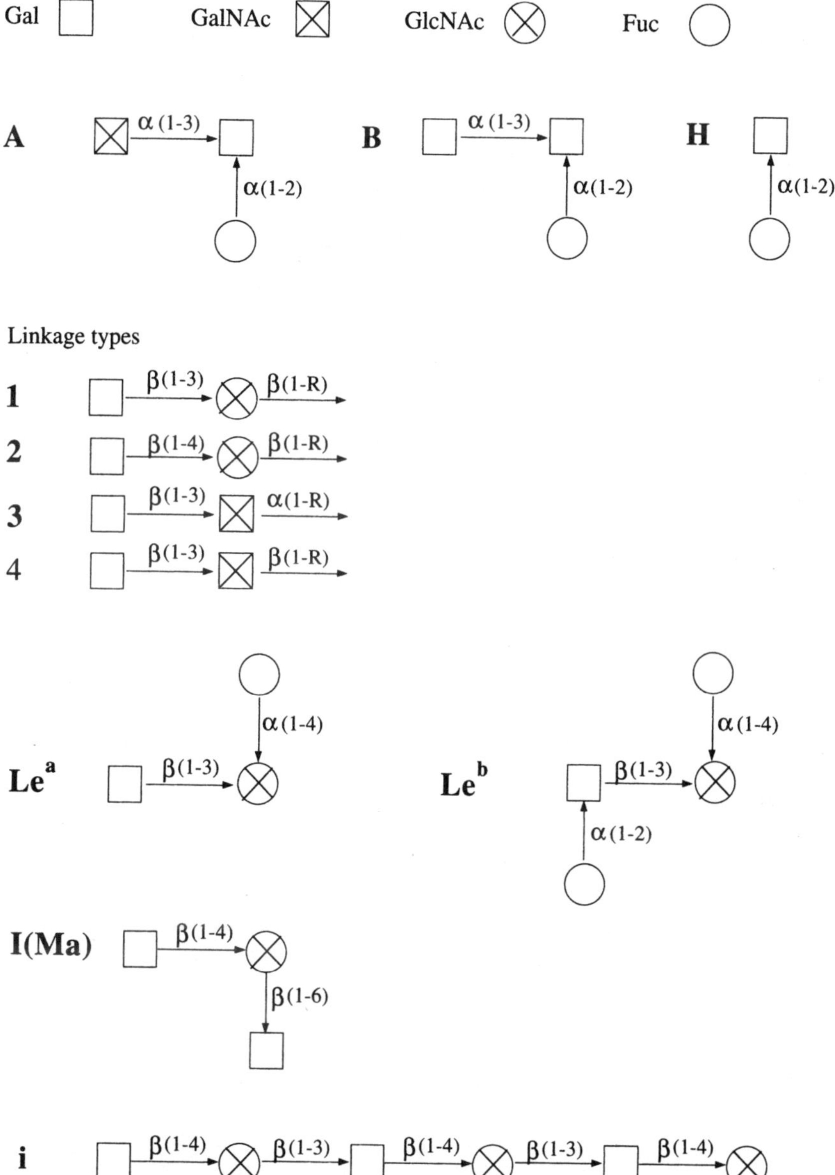

Fig. 16.6 Representation of certain sequences considered in Chapter 16.

pronounced under more basic conditions than at the pH optimal for hydrolysis. The isolated yield of trisaccharide **16.26** was 36%.

A drawback in the synthesis of disaccharides by transglycosidation is the enzymic hydrolysis of the product by the glycosidase. In Section 10.4.3, we saw a mechanical device, the membrane reactor, to deal with this problem. In the present synthesis, the final product **16.26** is no longer a substrate for the galactosidase, a fact which drives equilibrium (16.1) to the right.

References

Alais, J. and Veyrières, A. (1990), *Carbohydr. Res.* **207**, 11–31.

Augé, C., David, S. and Veyrières, A. (1979), *Nouv. J. Chim*, **3**, 491–497.

Clausen, H. and Hakomori, S.-i. (1989), *Vox Sanguinis* **56**, 1–20.

Gooi, H. C., Veyrières, A., Alais, J., Scudder, P., Hounsell, E. F., and Feizi, T. (1984), *Mol. Immunol.*, **21**, 1099–1104.

Hakomori, S.-i. (1991), *Baillère's Clinical Haematology* **4**, 957–974.

Hakomori, S.-i. (1993), *La Recherche*, 548–554.

Kren, V. and Thiem, J. (1995), *Angew. Chem., Int. Ed. Engl.*, **34**, 893–895.

Lemieux, R. U. and Driguez, H. (1975*a*), *J. Am. Chem. Soc.*, **97**, 4063–4069.

Lemieux, R. U. and Driguez, H. (1975*b*), *J. Am. Chem. Soc.*, **97**, 4069–4075.

Paulsen, H. (1982), *Angew. Chem., Int. Ed. Engl.*, **21**, 155–173.

Windmüller, R. and Schmidt, R. R. (1994), *Tetrahedron Lett.*, **35**, 7927–7930.

17 Important recognition events involving oligosaccharides in the living world

17.1 Introduction

The recognition events of oligosaccharides in this section were not only chosen for their importance, but also because they illustrate good examples. The Ciba Foundation Symposium (Bock and Harnett 1989) published a volume which gives an account of a number of similar observations in the most varied domains. The reader interested in the questions not discussed here, such as the intervention of oligosaccharides in the fertilization of mammals and the embryonic development of the nervous system may refer to this book. Within the more limited framework of this chapter, all examples, apart from the first, have been taken from the animal kingdom. Although the subject may appear to be somewhat disparate, the author is precisely trying to show the reader the general character and great variety of supramolecular protein–sugar associations in the living world.

17.2 *Rhizobium* nodulation signals

Tetrasaccharides **17.1** and **17.2** play a fundamental role in the *Rhizobium*–legume symbiosis. They induce the formation of nitrogen-fixing root nodules and root hair deformation on alfalfa (Roche *et al.* 1991). They are inactive on vetch which, on the other hand, responds to non-sulfated analogues. The tetrasaccharide is a fragment of the chitin polymer, **17.3**, a structural polysaccharide in the shells of crustacea. Acetolysis of chitin (Ac_2O–H_2SO_4) decomposes it to peracetylated *N*-acetylglucosamine and oligosaccharides (n = 2, 3, 4 ...). Compounds **17.1** and **17.2** can be distinguished from 'chitotetraose' **17.3** ($n = 4$) by the following substitutions: sulfation of the primary alcohol function of the reducing unit and, at the non-reducing end, replacement of the acetyl group by one derived from an unsaturated fatty acid, and possible acetylation of the primary hydroxyl group. The synthesis of tetrasaccharides **17.1** and **17.2** was done by successive addition of suitably protected glucosamine units (Nicolaou *et al.* 1992). The precursor **17.4** of the reducing unit contains four types of protecting groups (see Section 5.1.1). A fifth type, *t*-$BuMe_2Si$, protects the primary alcohol function of the non-reducing end precursor. Coupling uses activated glycosyl donors in the form of anomeric fluorides. A tetrasaccharide glycoside is finally obtained in which all the alcohol functions are protected by the MP or PMB groups, except for two, protected by silylation. The *t*-butyldimethylsilyl ether is hydrolysed selectively on the primary alcohol function of the non-reducing terminal unit in a moderately acidic medium (pyridinium tosylate) and then acetylated. The *t*-butyldiphenylsilyl ether is hydrolysed with tetrabutylammonium fluoride and the alcohol is converted to a sulfate with $SO_3 \cdot NMe_3$ in pyridine. Finally, **17.2** is obtained by eliminating the remaining ethers by oxidation with ceric ammonium nitrate.

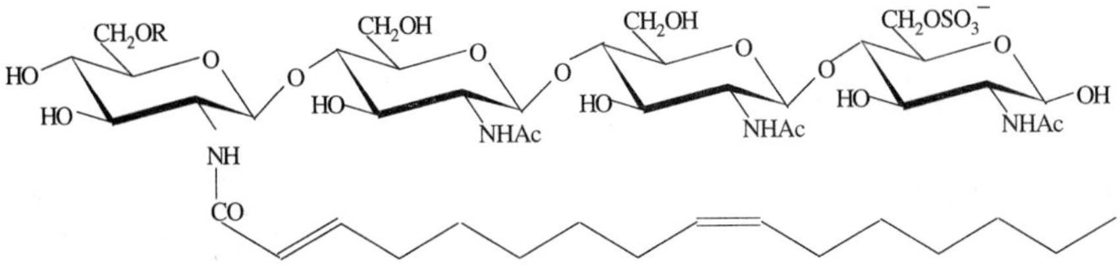

17.1 R = H

17.2 R = Ac

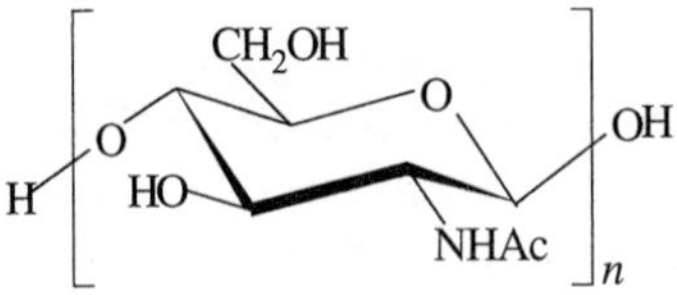

17.3

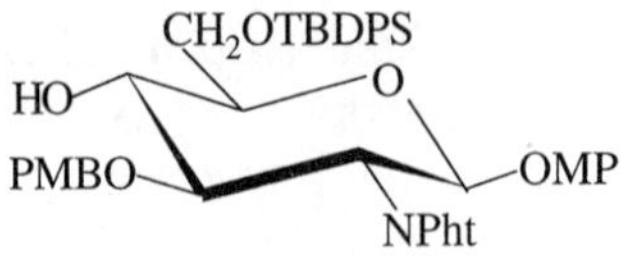

17.4

MP 4-methoxyphenyl

PMB 4-methoxybenzyl

TBDPS t-$BuPh_2Si$

Pht phthalimido

17.3 Active pentasaccharide of heparin

17.3.1 Isolation of heparin (Casu 1985; Casu 1989)

The most common commercial source of heparin comes from pig intestinal mucosa from which more than 100 mg per kilogram of tissue can be obtained. This is a polyacid which is extracted from autolysed tissues with an alkaline solution. The proteins are then coagulated upon heating. Cooling precipitates a heparin–protein association from which the contaminating lipids are removed by extraction with ethanol or acetone. After redissolution, the proteins are removed by tryptic digestion. Contaminants from the same family are separated by conversion to a crystalline barium salt or insoluble quaternary ammonium salts. Heparin can be purified on an anion-exchange column. This is the most highly charged molecule of the series and therefore requires the highest concentration of salts (≥ M) for its elution.

17.3.2 Biosynthesis of heparin

The biosynthesis of heparin is particularly interesting as it shows how heterogeneity is introduced into this originally regular molecule. At the start a polypeptide backbone is formed composed of about 15 glycine and L-serine residues in alternating sequence, outline **17.5** showing the dipeptide. To most of the serine residues are added successively, by appropriate transferases, one β-D-xylopyranosyl, two β-D-galactopyranosyl, and one β-D-glucuronosyl units to give glycopeptide **17.6**. The glycosaminoglycan chain is now constructed from the β-D-GlcUA unit by alternate addition of the α-D-GlcNAc and β-D-GlcUA units. The repeating motif is thus **17.7**. This regular chain undergoes the following modifications: (a) *N*-deacetylation and nitrogen sulfation which replaces –NHAc by –$NHSO_3H$; (b) C-5 epimerization which transforms the D-glucuronic acid residue into the L-iduronic residue; (c) sulfation at O-2 of the iduronic residue; (d) sulfation of the primary alcohol function. The new chain then consists mainly of the repeating disaccharide, **17.8**, but not all the original disaccharide motifs of the chain are modified by the above biosynthetic sequence so that the chains become heterogeneous. The molecular weight is between 60 000 and 100 000, but hydrolysis of certain peptide and glycoside bonds gives a final mixture of molecular weights from 5 000 to 25 000.

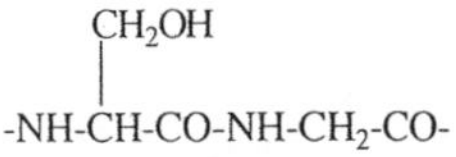

-NH-CH(CH_2OH)-CO-NH-CH_2-CO-

17.5

GlcUA-β-(1→3)-Gal-β-(1→3)-Gal-β-(1→4)-Xyl-β-(1→OCH_2-CH(NH–)(CO–)

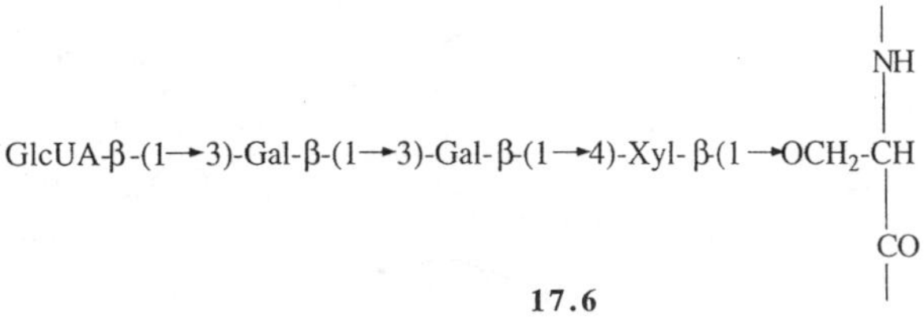

17.6

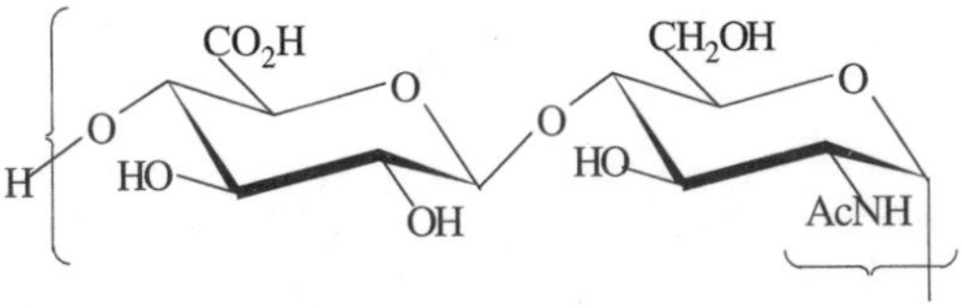

17.7

17.3.3 Degradation of heparin

The structural motif **17.8** remains dominant. Acid hydrolysis of the sulfamic and glycosidic bonds gives glucosamine hydrochloride as the only amino sugar. Treatment with nitrous acid gives mainly disaccharide **17.9**. Cleavage of the glycosidic bond and ring contraction is also observed on methyl α-D-glucosaminide, the schematic representation of which is shown in Fig. 17.1. Because of the

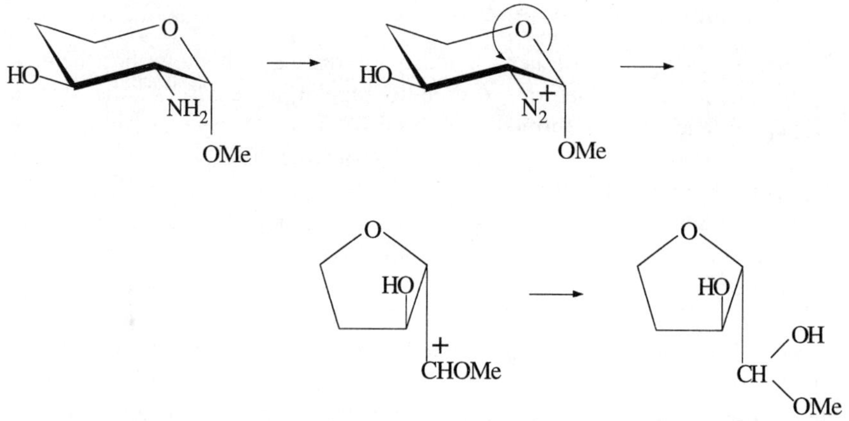

Fig. 17.1 Rearrangement of a glycoside of an amino sugar by nitrous desamination.

instability of a carbenium ion at C-2, the leaving of nitrogen is probably concerted with attack by the C-1–O-5 bond. Ring contraction gives rise to an oxo-carbenium ion to furnish, with water, an unstable hemi-acetal under these conditions. The latter is quickly hydrolysed to an aldehyde and methanol.

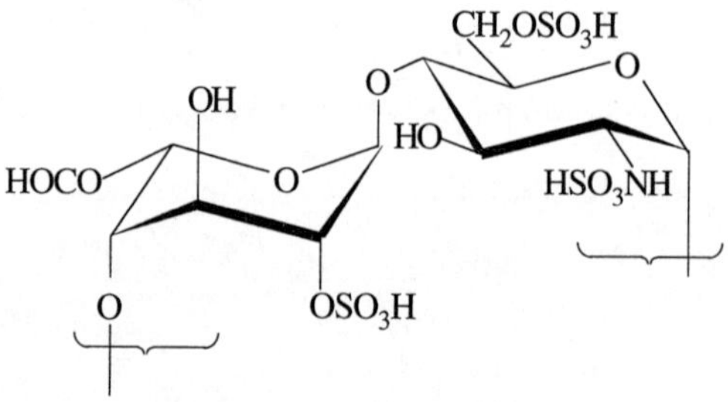

17.8

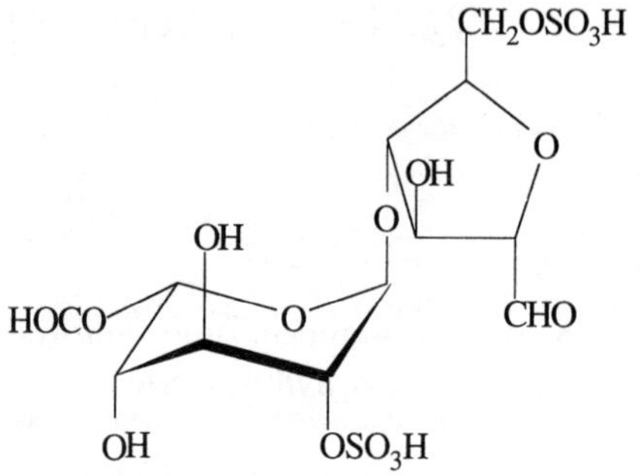

17.9

It is possible to induce enzymes capable of decomposing heparin, heparinase and heparanase in *Flavobacterium heparinum*, by cultivating this organism on polysaccharides of the same family. The cleavage product is disaccharide **17.10**, which, in the most favourable case, is obtained in 80% yield. The reader will note a β-elimination, very comprehensible at first sight, of an oxygen atom in the β-position to a carboxyl group. It is even facilitated by the *trans* diaxial disposition of the two substituents which eliminate as shown, **17.11**. However, on reading the protocols for extraction, we can conclude that heparin is stable in warm concentrated alkaline solutions. It is possible that the overabundance of negative charges close to H-5 under these conditions prevents the approach from a base and deprotonation.

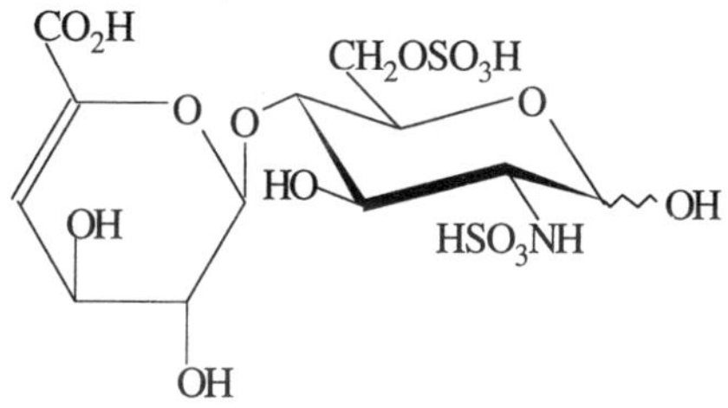

CO_2H CH_2OSO_3H O O O HO OH HSO_3NH OH OH

17.10

HOH HOCO O O OSO_3H O $NHSO_3H$ HSO_3-OCH_2 OH O

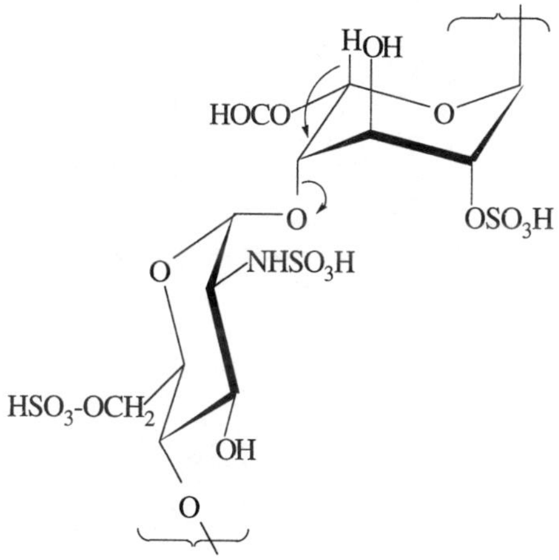

17.11

17.3.4 The active pentasaccharide

Heparin prevents blood coagulation by combining with a plasma protein, antithrombin III (AT III). This association causes a certain number of enzymes involved in the coagulation process to become inactive. This association has been shown *in vitro* by affinity chromatography whereby the AT III protein is covalently bound to activated agarose with cyanogen bromide (see Section 10.4) (Höök *et al.* 1976). Coupling is done in the presence of excess heparin in order to protect the AT III sites involved in the recognition of heparin. When a solution of heparin in 0.2 M NaCl at pH 7.5 (in the form of sodium salt) is put through a column of this agarose–AT III adsorbant, heparin is retained. This complex can be dissociated by forcing the saline concentration of the eluent (0.2 M $\rightarrow$ 3 M). Similar experiments with heparin fragments produced by desamination or enzymic elimination have shown that the recognition site on heparin is reduced to a sequence of five monosaccharide residues, arranged as in the (synthetic) pentasaccharide **17.12**. The sequence of **17.12** can be distinguished from the prevailing sequence by a glucuronic residue, diagnostic of an incompleted biosynthesis, and a trisulfated glucosamine. This third sulfate is indispensible for biological activity. The pentasaccharide **17.12**-AT III combination can be seen by chromatography on a Sephadex column (Choay *et al.* 1983). This type of adsorbant has cavities of standard dimensions that retain molecules that can enter them. Proteins are excluded and follow virtually the front of the eluent. Figure 17.2 shows the elution profiles with 0.15 M NaCl (pH 7.5) as eluent. Protein AT III is quickly eluted (peak A). Pentasaccharide **17.12**, is delayed *when alone*, and leaves the column later (peak B). Chromatography of an equimolar mixture of AT III and **17.12** shows that only one elution peak is observed (peak A), the **17.12**–AT III complex, which is excluded as AT III.

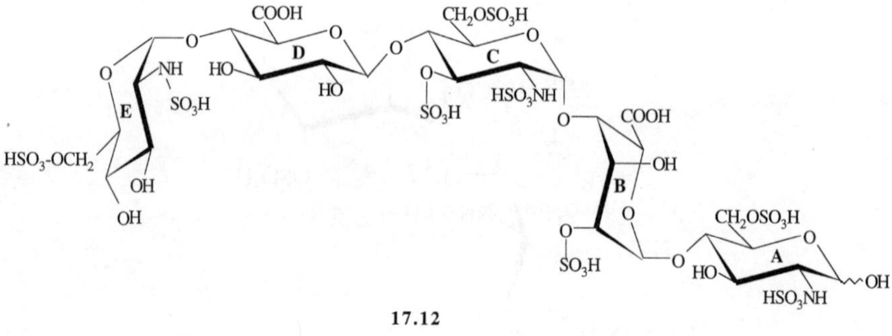

17.12

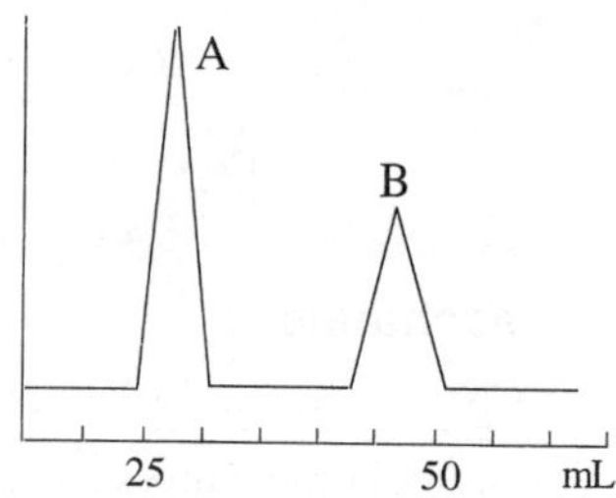

Fig. 17.2 Chromatographic profiles from a Sephadex G-60 column of AT III and/or its ligand **17.12** with 0.15 M NaCl (pH 7.5) as eluent.

Here is an outline of the major steps in the synthesis of **17.12**. The L-*ido* configuration is achieved by inversion of configuration of C-5 on furanose **17.13**. Acidic deacetalation of **17.14** gives the pyranose which, after a certain number of steps, is activated by conversion to orthoester **17.15**. The precursor of the **AB** fragment of **17.12** is obtained by condensation of this orthoester with the protected glucosamine **17.16**. Condensing bromide **17.17** on 1,6-anhydro sugar **17.18** provides a precursor of the **CD** fragment. Condensation of the two protected **AB** and **CD** disaccharides gives the tetrasaccharide precursor of the **ABCD** fragment to which the monosaccharide unit **E** is added at the end of the synthesis. This synthesis leads to a pentasaccharide in which all the hydroxyl groups to be sulfated are acetylated, whereas the precursors of the amino groups, the azide (introduced with **17.18**) and the benzyloxycarbonyl functions (introduced with **17.16**) remain intact. Sulfation is achieved in two steps. After deacetylation, the five free hydroxyl groups are sulfated using the $SO_3{\cdot}NMe_3$ system in dimethylformamide. The amino functions are then unmasked by catalytic hydrogenation, followed by *N*-sulfation in water (pH 9.5) with the same $SO_3{\cdot}NMe_3$ complex (Jacquinet *et al.* 1984; Petitou *et al.* 1986).

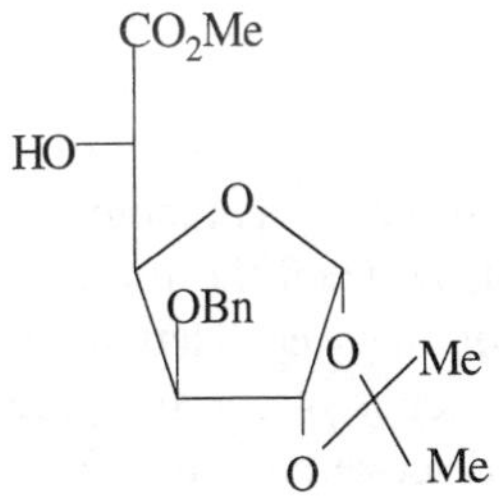

CO_2Me HO O OBn O Me O Me

17.13

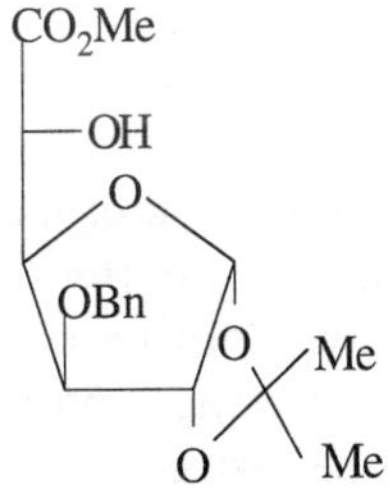

CO_2Me OH O OBn O Me O Me

17.14

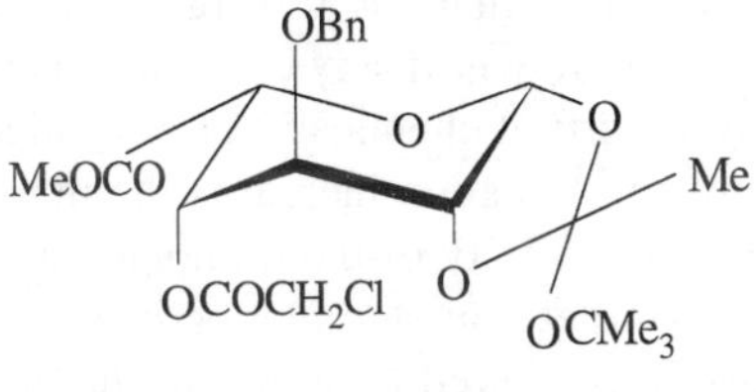

OBn O O MeOCO Me $OCOCH_2Cl$ O $OCMe_3$

17.15

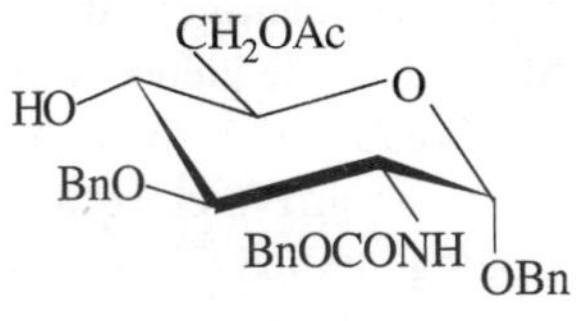

CH_2OAc O HO BnO BnOCONH OBn

17.16

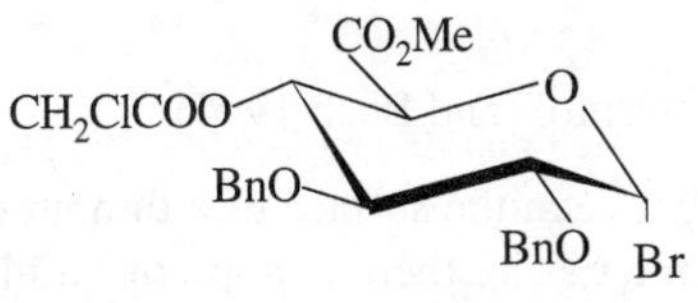

CO_2Me O CH_2ClCOO BnO BnO Br

17.17

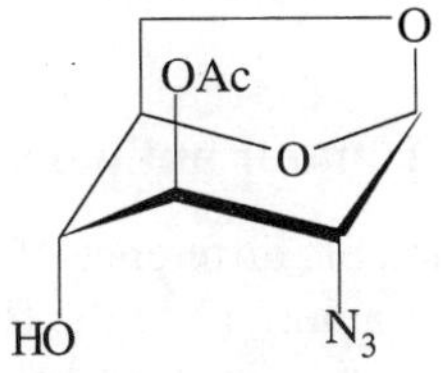

O OAc O HO N_3

17.18

17.4 Tumour markers

17.4.1 Research method

The research described in this section and in the following (Feizi 1985*a*) boil down to analyses of the various molecules found on cell surfaces. If this appears to be on the borderline of organic chemistry, it is because the problem's complexity and the experimental difficulties require resorting to fine immunochemistry methods. Also the major goal is to resolve biological and medical problems. The most frequently used approach is to stimulate the formation of antibodies in animals through immunization with whole cells or their membrane. With a given cell, a mixture of antibodies can be obtained which are then separated and reproduced on a large scale using the *hybridoma* technique (see Section 15.2). A great number of these monoclonal antibodies have been prepared in this fashion. Their specificity needs to be determined, which is not a simple problem. It requires having at ones disposal a collection of antigens whose structure is known and trying to find, one after another, those which combine with a given monoclonal antibody. Chemical analogy and biological or medical intuition can help in this search. When an antigenic molecule is found, it is necessary to recognize the determinant, its active part in the immunochemical reaction. On cell surfaces, the determinant can be expressed on molecules different from those of the antigen which has served to characterize the antibody. Finally, an optimistic hypothesis would be when the immunochemists have at their disposal a series of monoclonal antibodies, each having a particular specificity, making it possible to prove the existence of certain structural elements on the surface of a cell. All specificities discovered in this fashion correspond to oligosaccharide antigens, close to the ABH blood group antigens. The determinants are carried by either glycolipids or by glycoproteins.

The reader, doubting that this method is sufficiently rigorous, may think that it is not possible to discover a radically new determinant by testing monoclonal antibodies with antigens that are already well known. In any case, after having found an active sequence, it is possible to modify it chemically or enzymically so as to create an original structure. If this modification increases the immune response, we arrive closer to the authentic specificity of the monoclonal antibody being studied. This is how an unknown determinant may be discovered. Whatever may be, monoclonal antibodies have served to point out the differences between embryonic and adult cells, between normal and cancerous cells. In this chapter we present a few characteristic examples and a brief survey of certain experimental techniques.

17.4.2 Tumour antigens (Feizi 1985*b*; Ørntoft and Bech 1995)

When a notable difference of composition between the surface of a tumour cell and that of a normal cell is found in the same tissue, there is hope of building a diagnostic test on this difference, or possibly a therapeutic strategy. Using immunizations with various types of tumour cells, a great number of monoclonal antibodies have been prepared, directed against surface molecules. The determinants discovered so far are all oligosaccharides similar to blood group substances and derived from either the core or the periphery of these substances.

Thus an accumulation of I and i antigens, **17.19** and **17.20**, has been observed on the metastasis of colon tumours (but not on tumours). They are linked to glycoproteins of the mucin type, extracted from the corresponding tissues. Antigen 19.9 was identified by means of a monoclonal antibody obtained after the immunization of a mouse with adenocarcinoma cells from a human colon. Attachment of this antibody to these cells is inhibited by preliminary treatment with neuraminidase which indicates the presence of a sialyl terminal residue in the antigenic determinant and suggests looking for the antigen among the gangliosides. Glycolipids are extracted from cells and separated by thin-layer chromatography. The developer is the monoclonal antibody which is only retained on the antigen spot. The presence of an antibody at a particular spot is indicated by a specific combination with a molecule marked by ^{125}I. The antigen present on the tumour cell used in the immunization is a ganglioside, **17.21**, whereby the antigenic determinant of Lewis^a blood groups is recognized, sialylated at the non-reducing end. In the serum of patients, the antigenic determinant is linked to a glycoprotein. Antigen 19.9 is absent in normal colon and is used in the diagnosis.

Galβ (1→4) GlcNAc (1→6) Gal...

17.19

H – [–3) Galβ (1→ 4) GlcNAcβ (1 –]$_3$ – OH

17.20

NeuAc-α-(2→3)-Gal-β-(1→3)-GlcNAc-β-(1→3)-Gal-β-(1→ 4)-Glc-β-(1→ Cer)
↑
Fuc-α-(1→4)

17.21

An antibody was obtained by immunization with a line of rat cells which reacts specifically with the embryonic brain of a rat. The corresponding antigen belongs to the group of differentiation antigens dealt with in Section 17.5. Because of the abundance of polysialylated structures in these tissues, it is not surprising that the antigenic determinant is a sialylated oligosaccharide. This antigen is also present in the gangliosides of melanoma cells, where it is very abundant—or very exposed. It is remarkable because of the presence of sialic acid acetylated at position 9, **17.22**, as a non-reducing terminal unit.

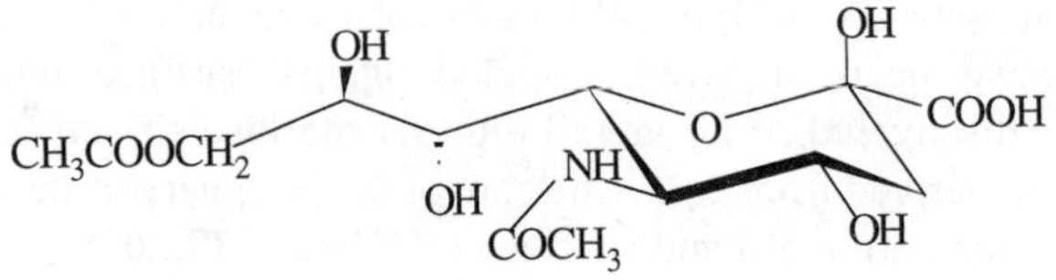

17.22

A good number of other antigens of tumour cells have been described. The reader can refer to the tables given in the references (Feizi 1985*a*; 1985*b*). Nevertheless, the corresponding antibodies should only be used in diagnoses with proper judgement. As the last example shows well, the presence of a determined antigen can be indicative of the presence of a tumour in one type of organ and be normal in another. Perfecting a diagnostic protocol should eliminate this cause for error.

17.5 Differentiation antigens (Feizi and Childs 1985; 1987; Feizi 1991)

Differentiation antigens are those whose expression on cell surfaces varies during successive steps of embryonic development. Immunization of an animal against whole embryonic cells has made it possible to obtain a great number of monoclonal antibodies directed against surface antigens. Thus the Forssman antigen has been found in the embryo of the mouse in its very first stage of development. This antigen corresponds to the oligosaccharide sequence of glycolipid **17.23**, but the carrying structure could well be a glycoprotein. This antigen is only expressed by a few types of cells in adult mice.

GalNAc-α-(1→3)-GalNAc-β-(1→3)-Gal-α-(1→4)-Gal-β-(1→4)-Glc-β-(1→Cer)

17.23

Other steps have been made clear with the help of natural monoclonal antibodies anti-I and anti-i. The role of Ii antigens as differentiation antigens is strongly suggested by their order of appearance in man. The i-antigen, whose epitope is **17.20**, is expressed on the erythrocytes of newborns, as for example in the blood of the umbilical cord. It disappears during the first year of life to the benefit of the I antigen, epitope **17.19**. These structures are without a doubt carried at the same time by glycolipids and glycoproteins. The order of appearance seems rational, if we assume that branching results from the action of transferases. However, the inverse order is observed during the embryonic development of the mouse, as described below.

Differentiation antigens can be found on the very site where they appear on the embryo, using microscopic slices. A specific antibody will bind to the sites where the antigen is found. Excess antibodies are removed by washing and the combined one can be visualized on the antigen sites with a fluorescent molecule. The I antigen appears very early, from the zygote stage onwards, and the i antigen is only detectable after 5 days, in the first differentiated cells (Kapadia *et al.* 1981). This early appearance of the branched oligosaccharide is surprising and it has been suggested, among the possible explanations, that there is a persistence of a glycosyltransferase of maternal origin in the oocyte. A new antigen, *stage-specific embryonic antigen* (SSEA-1), whose epitope is **17.24**, is expressed at the eight-cell stage in the mouse embryo (Gooi *et al.* 1981). Its expression is transitory as it is only expressed at a restricted number of sites in the adult mouse. Its structure, **17.24**, implies that its expression is the result of the fucosylation of the non-reducing terminal *N*-acetyllactosamine units of i or I antigens. A second fucosylation, giving **17.25**, eliminates the reactivity with respect to the anti-SSEA-I antibody. These observations suggest a mechanism for the appearance, disappearance, and reappearance of antigens during the different stages of embryogenesis, namely the addition (by transferases) or elimination (by hydrolyses) of monosaccharides. Likewise, in humans, ABH blood group antigens appear on various tissues at different stages of development, and, sometimes, are finally no longer detectable.

$$\begin{array}{c}\text{Gal-}\beta\text{-(1}\rightarrow\text{4)-GlcNAc}\\ \qquad\qquad | \\ \text{Fuc-}\alpha\text{-(1}\rightarrow\text{3)}\end{array}$$

17.24

$$\begin{array}{cc}\text{Gal-}\beta\text{-(1}\rightarrow\text{4)-} & \text{GlcNAc}\\ | & | \\ \text{Fuc-}\alpha\text{-(1}\rightarrow\text{2)} & \text{Fuc-}\alpha\text{-(1}\rightarrow\text{3)}\end{array}$$

17.25

One opinion is that these differentiation antigens act as a postal code. By an adhesion system, they direct the route that embryonic cells should follow in order to arrive at their destination, notably to assemble in large organs. But the traffic must be organized by 'postmen' or 'traffic controllers'. Therefore, other molecules are needed to recognize these antigens, a role which neighbouring lectins could carry out.

17.6 Selectins

17.6.1 Inflammatory response and selectins

In damaged tissues, mobilization of leucocytes takes place. Leucocytes present in the bloodstream slow down, then stop along the inside endothelial wall of the venule, and finally cross it by embedding themselves between the cells next to the injury: this is the *inflammatory response*. This mobilization involves a multiplicity of *adhesion molecules*, divided into three groups: integrins, super-immunoglobulins, and selectins (Paulson and Lasky 1992). We will give particular attention to the latter group, the selectin family, of which there are three types, E (endothelial), P (platelet), and L (leucocyte). The E-selectin appears on endothelial cells whereas the P- and L-selectins are in the majority of leucocytes.

Selectins are proteins whose primary structure was elucidated by cloning. They are constructed in a similar fashion (Fig. 17.3). These are transmembrane proteins whose acid C-terminal domain bathes in the cytoplasma while the N-terminal domain is directed towards the exterior. Going from the latter, there is first a chain of around 120 residues, the 'lectin domain', which is similar to animal lectins (whose effect depends on Ca^{2+}). This immediately gave rise to the suspicion that oligosaccharide ligands existed. Then comes a sequence of 30 amino acid residues, called *EGF*, then a structure built from 62 residues, which is repeated 6 times for E, 9 times for P, and twice for L. This constitutes the portion of selectins outside the cell. The rest of the sequence corresponds to the portion that crosses the membrane and the cytoplasmic part, which is much shorter. The name selectin reminds us of the analogy between these proteins and animal lectins. Among the abundant literature published on selectins, we have chosen that which concerns the oligosaccharide ligands of E- and L-selectins.

17.6.2 E-Selectins

E-Selectin is not expressed on resting endothelial vascular cells. It appears in a transitory fashion when the aggression of tissues in their neighbourhood causes the secretion of a cytokine, an active polypeptide, which stimulates its biosynthesis. Bacterial endotoxins produce the same effect. On the other hand, E-selectin would be expressed in a permanent way in rheumatoid arthritis. The activated cells adheres to a leucocyte by interaction of its E-selectin molecules

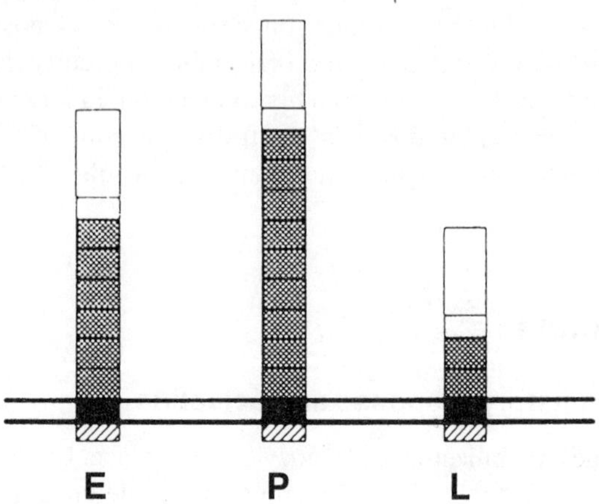

Fig. 17.3 Domain structures of the E-, P-, and L-selectins (Paulson and Lasky 1992; copyright © 1977 by W. H. Freeman and Company; reproduced with kind permission from the editors).

with a complementary oligosaccharide present at the cell surface of the leucocyte. Endothelial cells can also be activated in culture with cytokines. This observation has allowed the cloning of E-selectin and, finally, its stable expression in another type of cells in culture, the *Chinese hamster ovary* (CHO) cells. This lineage, CHO-E, was used as the research tool in discovering the complementary oligosaccharide ligand.

A logical method would consist in separating and isolating all the oligosaccharides from the leucocyte membrane and in testing them separately. This is absolutely impractical. It is therefore a matter of intelligently guessing the best candidates among the available natural sources. One method consists in adsorbing a known quantity of the product to be tested on the walls of a plastic well, covering with a CHO-E cell suspension, possibly labelled with tritium, and estimating the radioactivity remaining after washing. This method can be used in semi-quantitative determinations. Whatever may be, it requires multiplying the trials, thus the tests should be fast and easy. The 'neoglycolipid' method is good for this kind of exploration. It is based on the observation of remarkable analytical properties of glycolipids which can be separated very well one from another by thin-layer chromatography on silica gel. It is also easy to analyse their structure by mass spectrometry (see Section 10.5.1). If the ligands to be tested are not already glycolipids but rather oligosaccharides obtained by synthesis or degradation of glycoproteins, it is necessary to convert them first into glycolipid analogues, the neoglycolipids. For this, they are combined with dipalmitoyl phosphatidylethanolamine **17.26** by reductive amination.

$$
\begin{array}{l}
CH_2OCOC_{15}H_{31} \\
| \\
CHOCOC_{15}H_{31} \\
| \\
CH_2OPO\ (OH)\ CH_2CH_2NH_2
\end{array}
$$

17.26

This supposes that the oligosaccharide has an intact potential aldehyde function at the reducing end. If this is not the case, for example when the oligosaccharide is obtained by reductive alkaline β-elimination, it is necessary to generate an aldehyde function by periodic oxidation (to the detriment of the oligosaccharide's integrity). These neoglycolipids are separated by a technique familiar to organic chemists, thin-layer chromatography on silica gel, with a rather polar eluent, $CHCl_3$–MeOH–H_2O, 60:35:8, v/v/v. After migration, the plate is 'revealed' by covering it with a CHO-E cell suspension which expresses E-selectin. If there is an oligosaccharide ligand of E-selectin in the mixture, the cells adhere to the corresponding spot. The excess cells are eliminated by washing and the adhesion sites are seen by locating the fixed cells. Various methods of visualization are possible. One of them uses tritium-labelled cells and detection of the ligand by autoradiography. The spots can be analysed directly by mass spectrometry. Timewise, this method is very economical as only the active structures are analysed.

The first observation is that the CHO-E cells are greatly retained on a glycoprotein rich in carbohydrates of ovarian origin. There are oligosaccharide sequences on this glycoprotein which bind with E-selectin. Oligosaccharides are isolated by acid hydrolysis and converted into neoglycolipids. The technique described above makes it possible to show a particularly active fraction in the mixture. Mass spectrometry analysis and methylation indicates that this is an equimolecular mixture of neoglycolipids presenting the sulfated trisaccharides **17.27** and **17.28** at the non-reducing end. An earlier study had shown that the sequences of the sialylated oligosaccharides **17.29** '3′-sialyl-Le^a' and **17.30** '3′-sialyl-Le^x' were also recognized by E-selectin, but the affinity of **17.27** and **17.28** was higher.

HSO_3→ 3)-Gal-β-(1→3)-GlcNAc-β-(1→3)-Gal
↑
Fuc-α-(1→4)

17.27

HSO_3→ 3)-Gal-β-(1→4)-GlcNAc-β-(1→3)-Gal
↑
Fuc-α-(1→3)

17.28

In any case, this underlines the importance of the anionic charge in this particular recognition since the detachment of the sialic acid residue in **17.29** and **17.30** by neuraminidase abolishes this reaction. Modelling of the structure of tetrasaccharides **17.27** and **17.29** indicates that their anionic groups ($-SO_3^-$ and $-CO_2^-$) are found more or less in the same region of the molecule (Yuen *et al.* 1992). Finally, modifications of the sialic acid residue which leave the carboxyl group intact, such as the replacement of the acetylamido group by a glycolylamido group or the destruction of the side chain by periodic oxidation, do not affect the reactivity.

NeuAc-α-(2→3)-Gal-β-(1→3)-GlcNAc-β-(1→3)-Gal
↑
Fuc-α-(1→4)

17.29

NeuAc-α-(2→3)-Gal-β-(1→4)-GlcNAc-β-(1→3)-Gal

↑

Fuc-α-(1→3)

17.30

17.6.3 L-selectin

This adhesion molecule is involved in the migration of lymphocytes in the peripheral lymph nodes. Using methods similar to those in the preceding section, greater affinities were observed in a mixture whose compounds have sulfate structures **17.27** and **17.28**. There are, however, differences compared with E-selectin (Green *et al.* 1992).

17.6.4 Synthesis of oligosaccharide ligands of selectins

The synthesis of a collection of sulfated or sialylated oligosaccharides has confirmed that sulfates are the best inhibitors. The most active is the sulfated pentasaccharide **17.31** which is, until the present, the best ligand known of E-selectin, giving 50% inhibition at the very low concentration of 5×10^{-8} M (Lubineau *et al.* 1994).

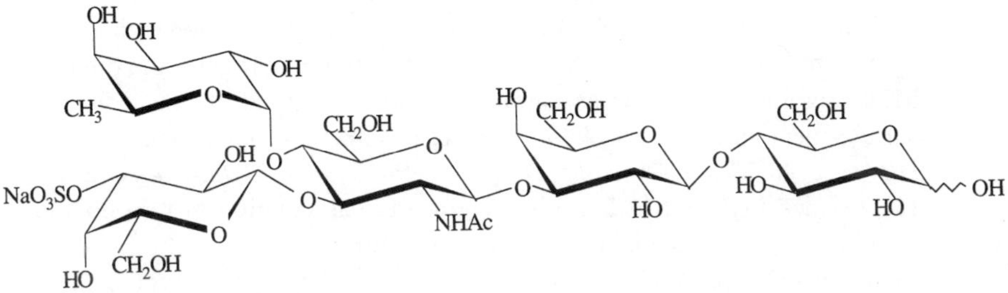

OH OH OH CH_3 O NaO_3SO OH O O HO CH_2OH O CH_2OH O O NHAc HO CH_2OH O O HO O HO CH_2OH O HO OH

17.31

17.7 Overall perspective on the biological role of oligosaccharides (Varki 1993)

Here and there in this book, and more specifically in this chapter, interactions of oligosaccharides with living cells have been mentioned. The author has focused on the chemical structure and syntheses of the ligands, considering that their association with macromolecules is also part of organic chemistry, so-called supramolecular chemistry. However, this knowledge could also be organized from the standpoint of biology. An attempt will be made in this section, with the help of a system of classification proposed in a recent review (Varki 1993). Following each heading, the reader will find the number of examples tabulated by the reviewer, and thus appreciate the great generality of these phenomena. Examples of interactions already given in this book are recollected under the relevant heading.

1. *Protective and stabilizing role.* There are more than 100 examples for *N*-linked, and nine examples for *O*-linked oligosaccharides. This difference may only reflect the fact that the former are easier to manipulate. Glycosylation helps to maintain correct polypeptide folding and solubility properties.
2. *Structural organization and barrier function* (10 examples). It is well known that polysaccharides such as cellulose or chitin **17.3** help to maintain tissue structure. On the other hand, some proteoglycans (Section 13.3) contribute to the organization of the extracellular matrix by association with oligosaccharide-binding domains of glycoproteins in their neighbourhood.
3. *Oligosaccharides as specific receptors for noxious agents* (14 examples). The trisaccharide sequence Neu5,9Ac_2-α-(2$\rightarrow$6)-Gal-β-(1$\rightarrow$4)-GlcNAc is part of the receptor of influenza C viral agglutinin. The synthesis of the corresponding trisaccharide, **12.22**, has been described in Section 12.4. The antigens of the Ii system of blood groups in man described in Section 16.2 are the antigens responsible for the spontaneous appearance of cold agglutinins and other autoimmune antibodies with, as a consequence, autoimmune destruction of cells and autoimmune haemolytic anemia.
4. *Protection from microorganisms and antibodies* (9 examples). A terminal sialic acid residue on red cell glycoproteins acts as receptor to the highly pathogenic influenza A viruses. Acetylation at position 9 abrogates the binding. Milk oligosaccharides act as inhibitors of the recognition of gut epithelial surface membrane oligosaccharides by various microbial pathogens.
5. *Oligosaccharides as specific receptors for symbiotic functions.* Tetrasaccharides **17.1** and **17.2** are involved in the association of root nodule-forming bacteria with plants.
6. *Effects of oligosaccharides on the biological functions of proteins* (15 examples). The capacity for homophilic binding of the neural cell adhesion molecule (N-CAM, Section 12.5.3) is modulated by the extension of its polysialic chains. The binding of a highly specific heparin pentasaccharide sequence to the protein antithrombin III converts it to a potent anticoagulant (Section 17.3).
7. *Cell–cell and cell–matrix recognition* (*25 examples*). The intervention of sulfated oligosaccharides in the adhesion of leucocytes to endothelial cells, mentioned in Section 17.6, comes under this heading, as do the properties of differentiation antigens (Section 17.5).
8. *Hormonal action of oligosaccharides* (10 examples). Some free oligosaccharides occur in some plants as a response to contamination, and elicit the production of phytoalexines, plant antibiotics against fungi.

Other functions of oligosaccharides which have been tabulated are the '*sink and depot*' function (12 examples). Thus calcium and water are bound and locally concentrated by sialyl oligosaccharides. Finally, oligosaccharides play an important role in the intra- and intercellular trafficking of proteins: the '*targeting*' and '*clearance*' functions.

Most probably all the functions summarily described above depend on non-covalent binding of oligosaccharides to receptor molecules, generally macromolecules, and in this case involving restricted binding areas with specific complementary structures. Even increasing the water solubility of a glycoprotein may be considered in this way, water being the simplest and most universal receptor for sugars (Section 11.6). Interactions may occur within a cell, or at the periphery with foreign molecules, these being either free or bound to another cell, or part of the extracellular matrix. All the theories pertaining to the role of oligosaccharides are correct, but there are exceptions to all of them. Thus the reader will find in the extensive tables compiled by Varki (1993) many instances when lack or alterations of the oligosaccharides of a glycoprotein has no obvious biological consequence. There are other puzzling observations.

The complexity of some sequences, a formidable synthetic challenge, sometimes appears pointless, as they seem to act only by a few residues at the non-reducing end. One explanation is that they have evolved as a response to aggression. When the non-reducing sugar end acted as a receptor to a new invader, the host could only protect itself by masking it with a different sugar. Thus consecutive layers would only be witness to a succession of defensive wars.

The same oligosaccharides may play different roles, depending on their localization in an organism. Conversely, important variations from species to species may be observed in sequences with similar functions. At first sight the detailed structure of these chains should not be very important, but we really do not know.

There is also diversity within one particular species, such as is observed in blood group phenomena in man. While the differences in blood group substances are most important in the practice of blood transfusions, they appear to be of little consequence in non-artificial human biology. Their *raison d'être* is not obvious. However, blood group substances are expressed in many places in the human body other than the periphery of red cells (Section 16.1.1), so that they may have some unsuspected utility.

Finally, we recall that oligosaccharide ligands are not systematically non-reducing terminal sequences. We have mentioned the 'conformational epitope' of an α-(2→8) sialic acid in Section 12.5.2. Another very well known example is the heparin pentasaccharide (Section 17.3.4). In this case we note that a succession of metabolic transformations has greatly altered a five-sugar sequence in a monotonous polysaccharide chain.

17.8 Towards a combinatorial chemistry of oligosaccharides

Let us go back to the experiments which allowed the characterization of the E-selectin ligand (Section 17.6.2). Expressed in general terms, the applied method consists of the following steps: preparation of a mixture of analogous molecules from a natural source, physical separation of the constituents, locating the active molecules, and determination of their structure from very small quantities. The same steps can be found in combinatorial chemistry with, however, one essential difference, instead of using a natural source, the organic synthesis is used to prepare the starting mixture, called the 'library'. The chemist, thinking about his or her setbacks, may think that it is only too easy to prepare a mixture instead of a pure compound. In truth, the addition of a mixture of reagents to a mixture of substrates does not necessarily give all possible combinations in the same amounts because of kinetic differences. Various protocols have been devised to ensure chemical diversity. In the so-called split and mix technology, a solution of substrate S is divided into n equal portions, which are treated separately with reagents R_1, R_2 ... R_n. The n solutions of SR_1, SR_2 ... SR_n are mixed, and split again for the next step. Mixtures containing up to 52 million different peptides have been prepared in this way and the probability of finding an active molecule is enormously increased. Certain techniques of peptide synthesis such as the construction of sequences on polystyrene beads, make it possible to program, at the same time of this construction, final operations of physical separation and analysis of active molecules (Nielsen 1994; Janda 1994).

The nature of sugars makes the manufacturing of a library of oligosaccharides clearly more difficult. For instance, glycosidation of unprotected disaccharide glycoside Galβ-(1$\rightarrow$3)-GlcNAcβ-(1-OR) with tri-*O*-benzylfucopyranosyl trichloroacetimidate gave a mixture of the six possible α-fucosylated trisaccharides in proportions varying from 8% for the α(1$\rightarrow$4) to 23% for the α(1$\rightarrow$3). One of these was the methyl glycoside of the Lewis$^{\alpha}$ blood group trisaccharide **16.4** (Kanie 1995).

References

Bock, G. and Harnett, S. (ed.) (1989), *Carbohyrate recognition in cellular function*, Ciba Foundation Symposium 145, Wiley, New York.

Casu, B. (1985), *Adv. Carbohydr. Chem. Biochem.*, **43**, 51–134.

Casu, B. (1989), *Ann. N. Y. Acad. Sci.*, **556**, 1–17.

Choay, J., Petitou, M., Lormeau, J.-C., Sinaÿ, P., Casu, B., and Gatti, G. (1983), *Biochem. Biophys. Res. Commun.*, **116**, 492–499.

Feizi, T. (1985*a*), *Nature*, **314**, 53–57.

Feizi, T. (1985*b*), *Cancer Surveys*, **4**, 245–269.

Feizi, T. (1991), *Trends Biochem. Sci.*, **16**, 84–86.

Feizi, T. and Childs, R. A. (1985), *T.I.B.S.*, **10**, 24–29.

Feizi, T. and Childs, R. A. (1987), *Biochem. J.*, **245**, 1–11.

Gooi, H. C., Feizi, T., Kapadia, A., Knowles, B. B., Solter, D., and Evans, M. J. (1981), *Nature*, **292**, 156–158.

Green, P. J., Tamatani, T., Watanabe, T., Miyasaka, M., Hasegawa, A., Kiso, M. *et al.* (1992), *Biochem. Biophys. Res. Commun.*, **188**, 244–251.

Höök, M., Björk, I., Hopwood, J., and Lindahl, U. (1976), *FEBS Lett.*, **66**, 90–93.

Jacquinet, J.-C., Petitou, M., Duchaussoy, P., Lederman, I., Choay, J., Torri, G., and Sinaÿ, P. (1984), *Carbohydr. Res.*, **130**, 221–241.

Janda, K. D. (1994), *Proc. Natl. Acad. Sci. USA*, **91**, 10779–10785.

Kanie, O., Barresi, F., Ding, Y., Labbe, J., Otter, A., Forsberg, L. S. *et al.* (1995), *Angew. Chem., Int. Ed. Engl.*, **34**, 2720–2722.

Kapadia, A., Feizi, T., and Evans, M. J. (1981), *Exp. Cell Res.*, **131**, 185–195.

Lubineau, A., Le Gallic, J., and Lemoine, R. (1994), *Bioorg. Med. Chem.*, **2**, 1143–1151.

Nicolaou, K. C., Bocovich, N. J., Carcanague, D. R., Hummel, C. W., and Even, L. F. (1992), *J. Am. Chem. Soc.*, **114**, 8701–8702.

Nielsen, J. (1994), *Chemisty and Industry*, 902–905.

Ørntoft, T. F. and Bech, E. (1995), *Glycoconj. J.*, **12**, 200–205.

Paulson, J. C. (1992), In *Adhesion. Its role in inflammatory disease* (ed. J. M. Harlan and D. Y. Liu), pp. 19–42, W. H. Freeman, New York.

Petitou, M., Duchaussoy, P., Lederman, I., Choay, J., Sinaÿ, P., Jacquinet, J.-C., and Torri, G. (1986), *Carbohydr. Res.*, **147**, 221–236.

Roche, P., Debellé, F., Maillet, F., Lerouge, P., Faucher, C., Truchet, G., Dénarié, J., and Promé, J.-C. (1991), *Cell*, **67**, 1131–1143 (and other publications from the same group in 1990–1991).

Varki, A. (1993), *Glycobiology*, **3**, 97–130.

Yuen, C.-T., Lawson, A. M., Chai, W., Larkin, M., Stoll, M. S., Sutart, A. C. *et al.* (1992), *Biochemistry*, **31**, 9126–9131.

18 Oligosaccharides as ligands to DNA

18.1 Recognition

The recognition of oligosaccharides by DNA was brought to the fore by a family of antibiotics, the calicheamicins (Lee *et al.* 1992) and esperamicins (Golik *et al.* 1987*a*; 1987*b*), which exhibit potent cytotoxicity and antitumoural properties. This chapter will be devoted essentially to calicheamicin $\gamma_1{}^{\mathrm{I}}$, the more studied of the two, which we will simply refer to as calicheamicin. In structure **18.1** the reader will recognize several exceptional elements. It is first and foremost a glycoside, whose bicyclic aglycon, the calicheamicinone, contains a ten-membered enediyne ring system. It also has a rare trisulfide sequence. The oligosaccharide part is also very unusual. The **B** residue is glycosidated by a hydroxylamine, so that the bond between **A** and **B** is a –NH–O bridge. The **B** and **D** residues are linked by a completely substituted benzene, **C**.

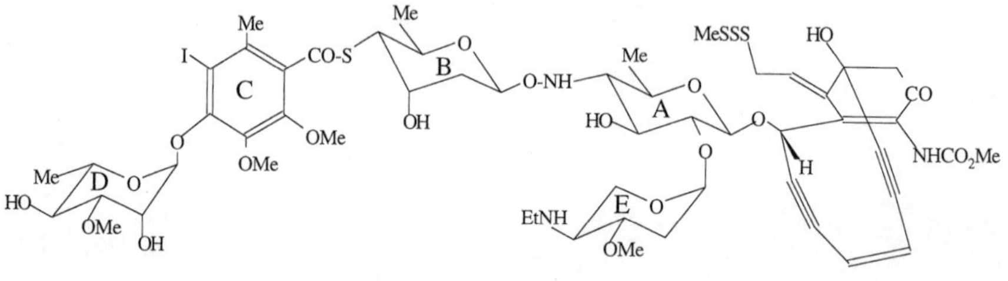

18.1

The cytotoxicity of calicheamicin and its congeners is due to the cleavage, apparently irreparable, of the double helix of DNA, the result of a remarkable transformation of the aglycon. It can be observed in the absence of DNA. Treatment of calicheamicin in CD_2Cl_2 solution by triphenylphosphine gives compound **18.2** (in which R represents the oligosaccharide residue), with a deuterium atom at each of the *x* and *y* positions. This can be interpreted as follows: the reagent attacks the central sulfur of the trisulfide group thus separating the sulfur atom from the side of the aglycon, which is now found in the form of thiolate. Michael addition of the thiolate to the cyclohexenone gives dihydrothiophen **18.3**. In this intermediate, the enediyne system is destabilized and cyclizes to the benzene diradical **18.2**, where *x* and *y* represent, this time, single electrons, mainly located at these two sites. This diradical abstracts two deuterium atoms from the solvent.

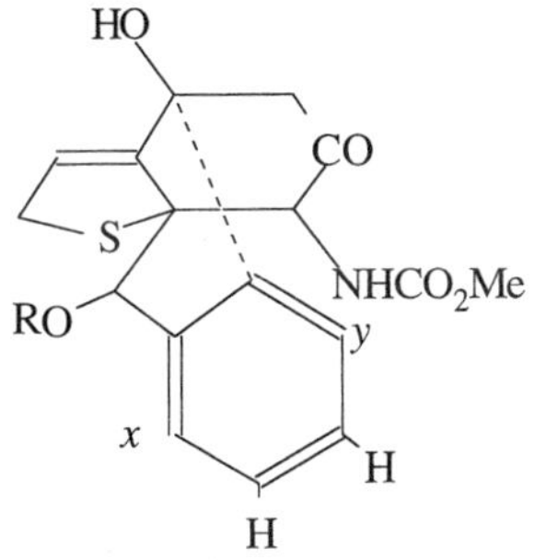

18.2 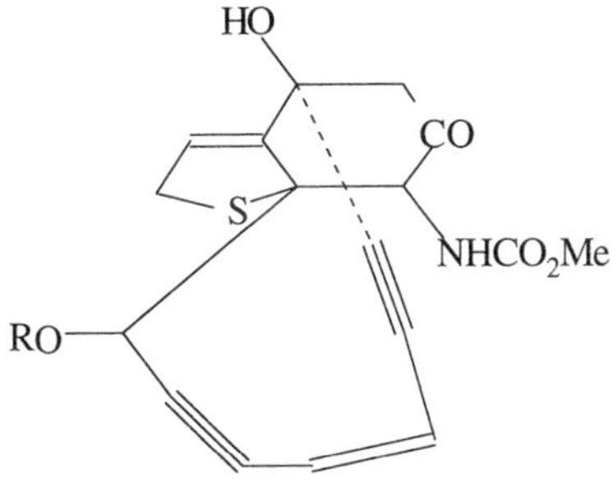 **18.3**

Before looking at the action on DNA, we should recall a few of the writing conventions in this area. The capital letters A, G, C, and T represent the monophosphates of 2′-deoxyadenosine, 2′-deoxyguanosine, 2′-deoxycytidine, and thymidine, respectively (see Section 3.6.3). The symbol TCCT represents a DNA fragment of four nucleotides in which each nucleoside is linked to its neighbour by a phophodiester bridge between the alcohol function at the 5′ position of one of the sugars and the 3′ position of the other. The primes have been reserved for the numbers of the deoxyribose carbon atoms. The 5′TCCT symbol indicates that the phosphodiester bridge links the alcohol function at the 3′ position of a nucleoside to the alcohol function at the 5′ position of the following one, going from left to right. This practice can also be applied to any length or composition of the polynucleotide. In the double helix of DNA, two polynucleotide chains are associated, with the well-known complementary bases opposite each other as adenine and thymine or guanine and cytosine. To the TCCT sequence corresponds the AGGA sequence in the complementary strand. The two chains are positioned in reverse order so that on the last sequence, read in this order, the phosphodiester bridge links the alcohol function at the 5′ position of a nucleoside to the alcohol function at the 3′ position of the one which follows. Consequently, the complete symbol would be 3′AGGA. Figure 18.1 shows a double helical DNA fragment in which are depicted the complementary tetranucleotides, 5′TCCT and 3′AGGA, which play a role in the phenomena described below. The identity of other nucleotides, N, does not need to be clarified for the moment. The aglycon of calicheamicin binds to the minor groove between the two polynucleotide chains, indicated by the wavy line. The addition of a reducing thiol induces the cleavage of two polynucleotide chains, at the level of the cytidine marked by an arrow on one strand, and at the level of a non-specified nucleoside on the other strand, located two nucleotide units beyond the adenosine towards the 3′ side. These cleavages are initiated by the abstraction of a hydrogen atom of the deoxyribose by the diradical **18.2** (*x* and *y*, single electrons). The cleavage at the level of the cytidine could be shown directly with a simpler system. The 'duplex' of dodecanucleotides **18.4** was prepared synthetically. The cytidyl unit C^7 is marked by deuterium at the 5′ position

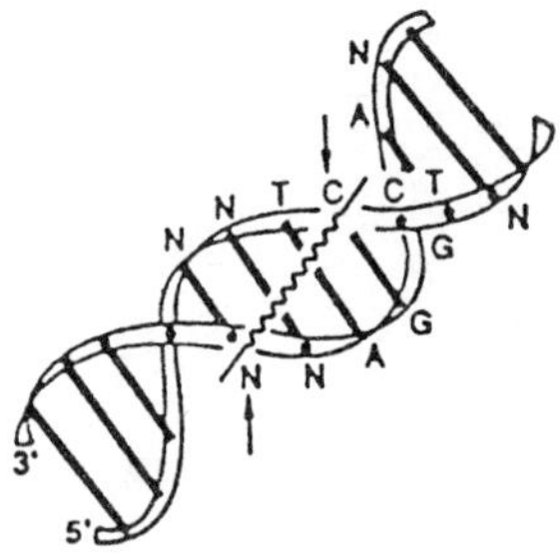

Fig. 18.1 Location of calicheamicin in the minor groove of the double helix (from Lee *et al.* 1991) (reproduced with kind permission from *Accounts of Chemical Research*; © 1991 American Chemical Society).

on the deoxyribose as represented by the partial formula **18.5**. Following the reaction of calicheamicin on the DNA fragment **18.4**, the transformation product **18.2** was recovered with deuterium incorporation at postion $x(x = D, y = H)$. The corresponding carbon at x on the enediyne system is thus the one which is found close to the cytidine in the association of the double helix with calicheamicin. Other experiments confirm that non-exchangeable protons of DNA are the source of hydrogens introduced at positions x and y, but they do not allow their location on the polynucleotides to be defined. In a deuterated buffer, transformation of calicheamicin, initiated by the deuterated thiol $DSCH_2COOCH_3$, gives compound **18.2** (20%), with more than 98% deuterium incorporation at positions x and y. The same experiment, *in the presence of DNA*, gives **18.2** (65%), with only hydrogen at the same positions. Thus the hydrogen atoms come from non-exchangeable sites of DNA. Cleavage, after the removal of a hydrogen atom at the 5′ position, probably leads to aldehyde **18.6**, since the remaining fragment loses a derivative of the cytosine in a basic medium, undoubtedly by β-elimination. In contrast, no isotope was transferred to the aglycon when deoxycytidine C^7 was labelled with D at the 1′ or 4′ position.

To determine the principal site of hydrogen abstraction from the AGGA strand of dodecamer **18.4**, similar experiments were conducted with samples labeled at deoxycytidine C^{21} with [1′-^{2}H]-, [4′-^{2}H]-, and [5′-^{2}H]-deoxycytidine. When calicheamicin was incubated with the dodecamer labelled at the 4′ position of C^{21}, deuterium incorporation was observed at only position y to give **18.2** ($x = H$, $y = D$). No significant isotope enrichment was observed with the other two labelled substrates.

Finally, the two labelled molecules, (5′-*R*)- and (5′-*S*)-[5′-^{2}H]-deoxycytidine obtained by synthesis were incorporated at position 7 in dodecamer **18.4**. Only the (5′-*S*)-diastereomer transferred deuterium (89 ± 1%), exclusively to position x of **18.2**. In conclusion, all the hydrogen abstracted originates from position 5′-*S* of C^7 and 4′ of C^{21}. 'These hydrogen atoms point into the minor groove at comparatively exposed sites near the exterior of the helix with proper alignment for reaction with calicheamicin' (Hangeland *et al.* 1992). However, it appears that H-4′ of deoxycytidine C^{21} is not very close to calicheamicin (Paloma *et al.* 1994).

5' C C C G G T C[7] C T A A G 3'
3' G G G C[21]C A G G A T T C 5'

18.4

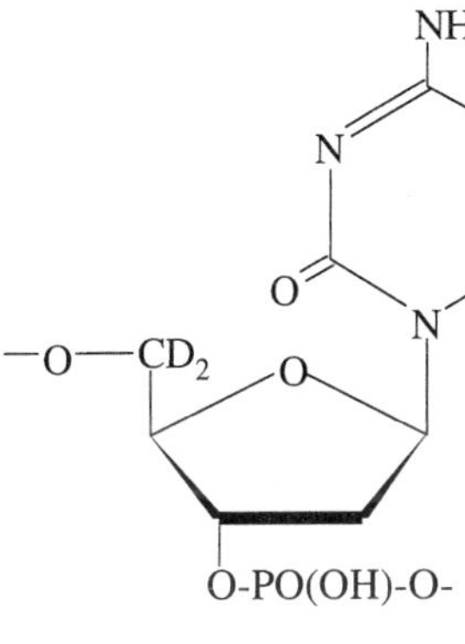

18.5

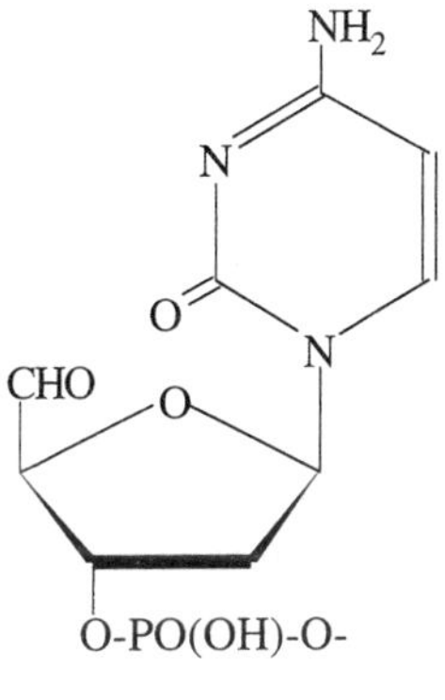

18.6

This general cleavage mechanism does not explicitly involve the pseudo-oligosaccharide and the aglycon alone should have the same properties. Calicheamicin **18.7** can not be separated from the oligosaccharide without degrading it, but the racemic form was obtained by a rather laborious total synthesis (Halcomb *et al.* 1992). This synthesis is outside the framework of this book. Calicheamicinone is indeed capable of cutting a DNA double helix, but in a statistical fashion. Other differences in the behaviour of the natural product and the aglycon show that the pseudo-oligosaccharide part is very important. This will-be examined in closer detail in Section 18.3, whereas Section 18.2 will deal with the synthesis of the unusual pseudo-oligosaccharide.

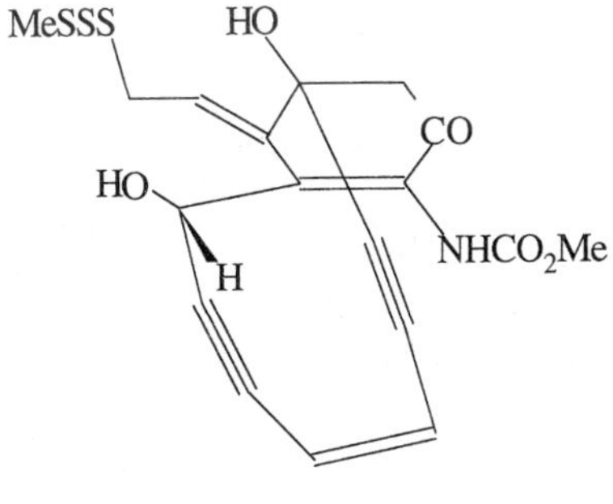

18.7

18.2 Synthesis of the pseudo-oligosaccharide of calicheamicin

18.2.1 The DCBAE route (Halcomb *et al.* 1992)

The five units **ABCDE** of the tetrasaccharide domain are assembled as follows:

$$\mathbf{D + C \rightarrow DC}$$

$$\mathbf{DC + B \rightarrow DCB}$$

$$\mathbf{A + E \rightarrow AE}$$

The precursors of the monosaccharide units **A**, **B**, **D**, and **E** are glycal derivatives, but each is used in a different way. The hydroxylamine linkage between the **A** and **B** residues is protected in the form of a urethane derivative to prevent rearrangement of the **A** residue which we will explain further on. The precursor of the **D** unit is di-*O*-acetyl-L-rhamnal **18.8** which gives, in the presence of benzylic alcohol and BF_3. Et_2O, the benzyl glycoside **18.9** via Ferrier rearrangement (see Section 7.5.1). The final functionalization of **18.9**, to give **18.10**, requires four steps, among which is the *bis*-hydroxylation of the double bond with the reagent OsO_4/*N*-methylmorpholine-*N*-oxide/water. After acetylation and catalytic debenzylation, the sugar is activated by conversion to trichloroacetimidate **18.11**. The precursor of the **C** ring is diiodo phenol **18.12**, according to an earlier suggested scheme (Nicolaou *et al.* 1990; 1992*a*). Coupling with trichloroacetimidate **18.11** gives the phenyl α-L-glycoside **18.13**, because of the participation of the acetate function at position 2. The latter is replaced by a *t*-butyldimethylsilyl group, then the methoxycarbonyl group is introduced by treating the diiodo phenol with a mixture of methanol and carbon monoxide in a reaction catalysed by palladium acetate in the presence of diphosphine $Ph_2P(CH_2)_3PPh_2$ and triethylamine. The methoxycarbonyl derivative **18.14** is then transformed to acid chloride **18.15**. This is how the **CD** fragment is obtained in a form ready for the next coupling.

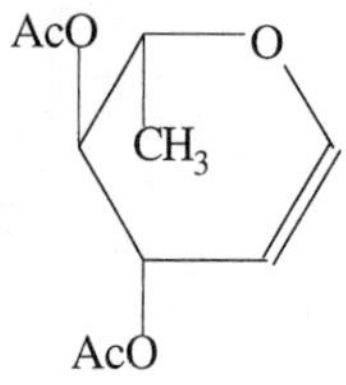

AcO O CH$_3$ AcO

18.8

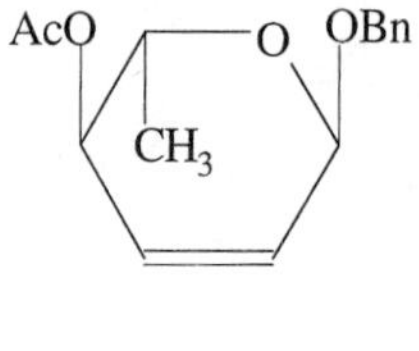

AcO O OBn CH$_3$

18.9

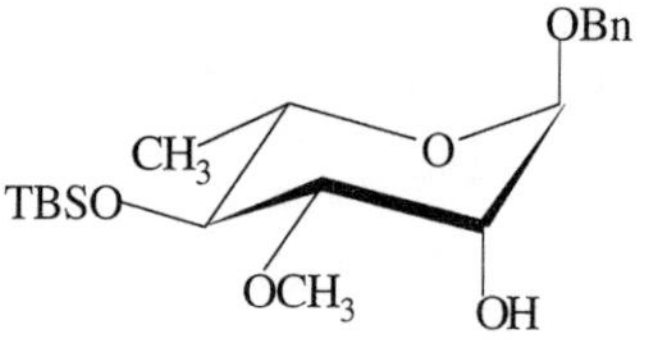

OBn CH$_3$ O TBSO OCH$_3$ OH

18.10

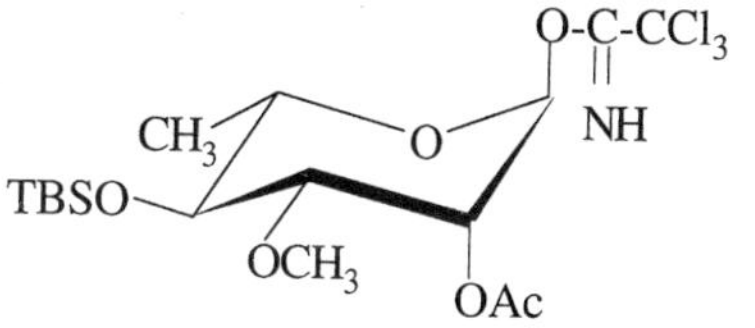

18.11

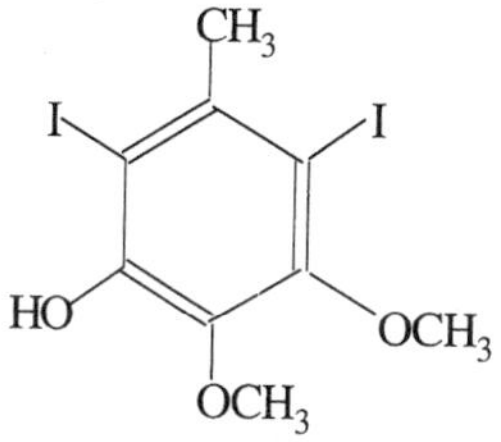

18.12

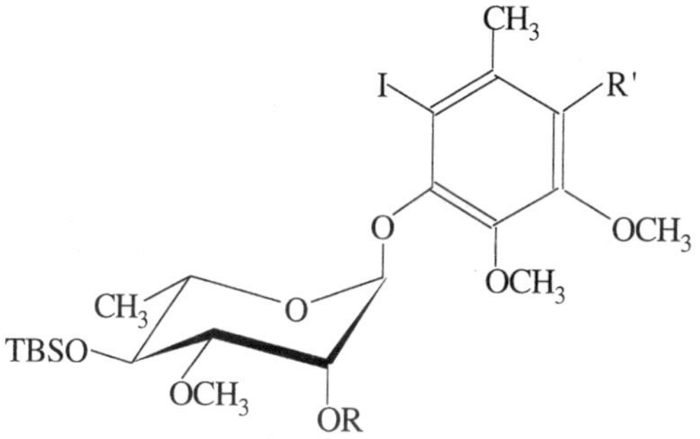

18.13 R = Ac, R' = I
18.14 R = TBS, R' = CO_2Me
18.15 R = TBS, R' = COCl

The precursor of unit **B**, glycal **18.16**, possesses a thiol function protected by a dinitrophenyl group (DNP) and a hydroxyl group protected by a *t*-butyldimethylsilyl group. Addition of substituted hydroxylamine $Me_3SiCH_2CH_2OCONHOH$ on the double bond in the presence of catalytic amounts of $Ph_3P{\cdot}HBr$ gives urethane **18.17**. The thiol function is then liberated to give **18.18** which is condensed without difficulty to acid chloride **18.15**. The protected version of the **DCB** fragment **18.19** is thus obtained, including the nitrogen atom of unit **A**.

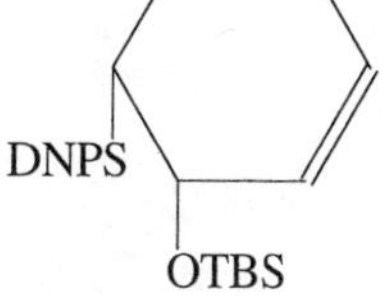

18.16

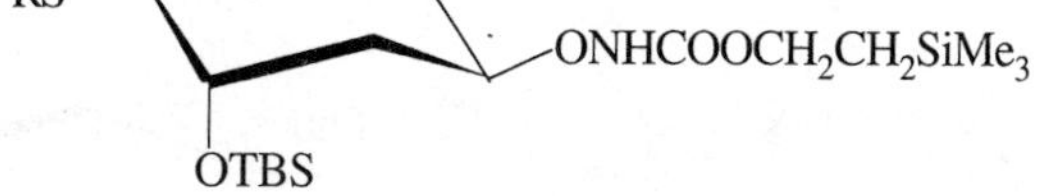

18.17 R = DNP
18.18 R = H

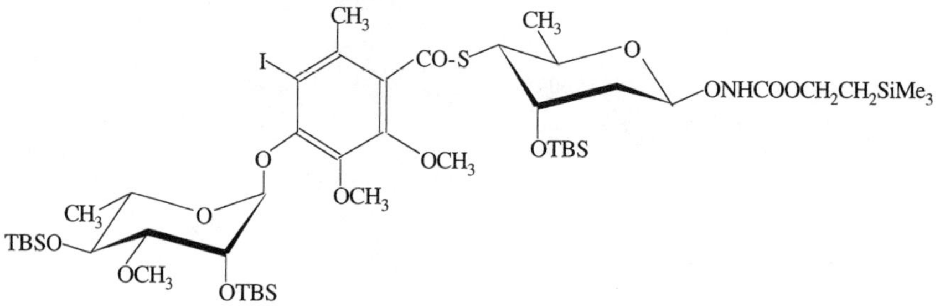

18.19

The precursor of unit **A** is glycal **18.20**, prepared from D-fucose whose one hydroxyl is protected by a *p*-methoxybenzyl group (PMB). Treatment with 2,2-dimethyldioxirane gave oxirane **18.21** (a Brigl anhydride) which is easily converted to the *trans*-glycoside **18.22** by simple mixing with *p*-methoxybenzyl alcohol at room temperature. Of the two free hydroxyl groups of this glycoside, only the equatorial one reacts in the consecutive coupling.

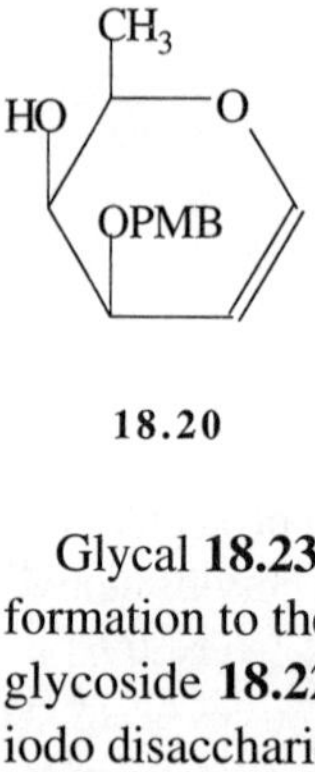

18.20 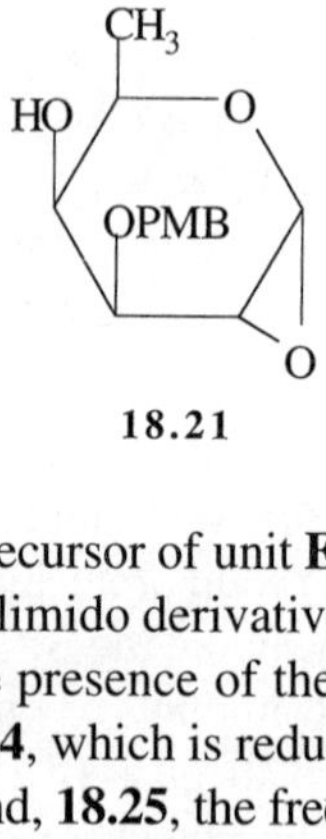 **18.21** 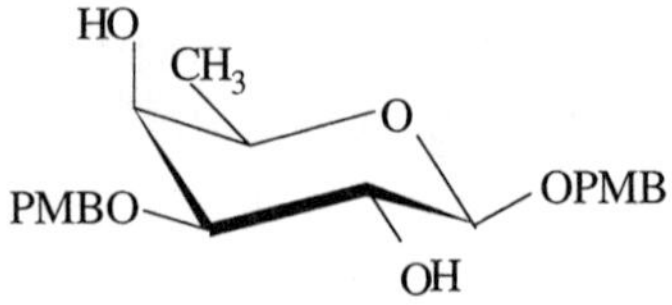 **18.22**

Glycal **18.23**, the precursor of unit **E**, has an amino function protected by transformation to the phthalimido derivative (abbreviated to Pht). It is condensed with glycoside **18.22** in the presence of the complex $I^+ClO_4^-$ (*sym*-collidine)$_2$ to give iodo disaccharide **18.24**, which is reduced to compound **18.25** with triphenylstannane. On this compound, **18.25**, the free hydroxyl group is activated by conversion to trifluoromethanesulfonate **18.26**. Thus was prepared the **AE** fragment.

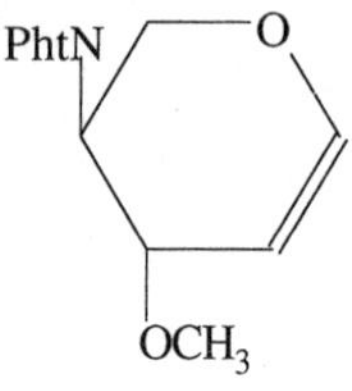

18.23

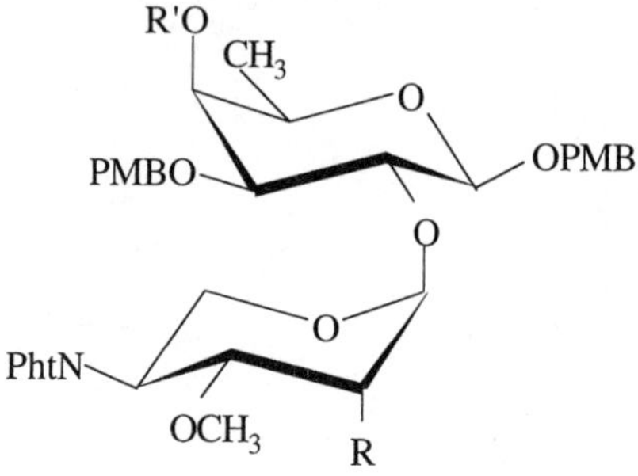

18.24 R = I, R' = H
18.25 R = R' = H
18.26 R = H, R' = Tf

The very unusual –N–O-bridge between fragments **AE** and **DCB** is built by reaction of trifluoromethanesulfonate **18.26** with the sodium salt of urethane **18.19** in *N,N*-dimethylformamide. This terminates the assembling of the pseudo-tetrasaccharide. The removal of the *p*-methoxybenzyl groups by oxidation still gives a partially protected compound, **18.27**, with a free hemi-acetal function for possible coupling to the aglycon.

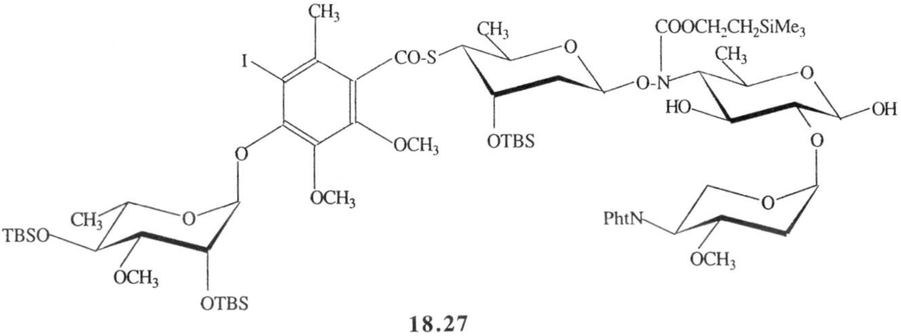

18.27

The protection of nitrogen by a urethane derivative deserves a few comments. Unit **A** is only stable when glycosylated. If the anomeric hydroxyl group is liberated, this unit is transformed to pyrrolidinose according to reaction (18.1). Derivation to urethane, inspired from an earlier study (Yang *et al.* 1991) is useful in two ways: the proximity of the carbonyl group facilitates the deprotonation of the nitrogen, and the 4-amino-4-deoxyhexose, free at the anomeric centre, is protected on nitrogen, which avoids its transformation to pyrrolidinose. After the hemi-acetal is engaged in a glycosidic bond, deprotection of the hydroxylamine nitrogen atom is carried out nearly quantitatively with tetrabutylammonium fluoride.

(18.1)

18.2.2 The EABCD route (Nicolaou *et al.* 1990; 1992*a*, 1992*b*)

Precursor **18.28** of unit **E** is prepared from L-serine. The secondary amino function is protected by derivation to 9-fluorenylmethylcarbamate (R = $C_{13}H_9CH_2OCO$) which can be cleaved in moderate alkaline conditions. Coupling of the pyranosyl fluoride **18.28** with **18.29** uses $AgClO_4$ (2 eq) and $SnCl_2$ (2 eq) as promotors (α/β = 4.5:1). In this way the **AE** fragment is obtained. The *cis*-diol system is liberated which is selectively oxidized to hydroxy ketone **18.30**. The **BA** linkage is then carried out by oximation of this ketone by a glycopyranosylhydroxylamine to give **18.31**. Because of the before-mentioned instability of the non-glycosidated unit **A**, it is necessary to conserve the oxime function until after the coupling takes place. It is then reduced by cyanoborohydride at pH 3. Coming back to **18.31**, the product was prepared for the introduction of sulfur at position 4 of the **EAB** trisaccharide. The aroyl group is replaced by a thiocarbonylimidazole and thionocarbonate **18.32** is isomerized by thermolysis to the thioester at position 4 (98%), by a [3,3]-sigmatropic rearrangement. The final assembling takes place as in the preceding synthesis by coupling of an acid chloride attached to the **CD** fragment of the free thiol linked to unit **EAB**.

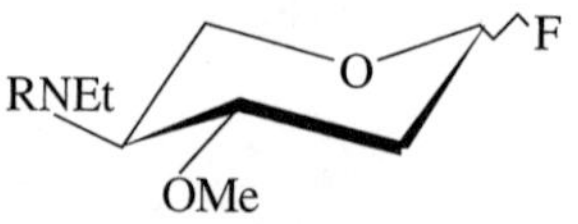

18.28

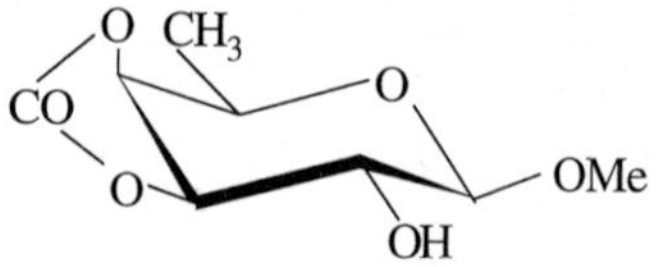

18.29

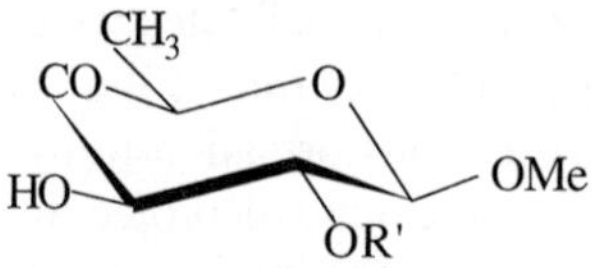

18.30

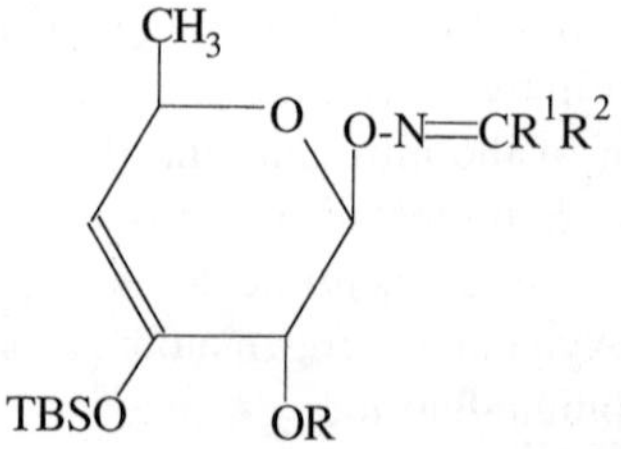

18.31 R = Aroyl

18.32 R = CS–N (imidazolyl)

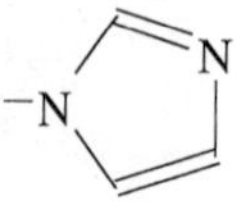

18.2.3 Alternative methods (Bamhaoud *et al.* 1992; Dupradeau *et al.* 1993; Da Silva *et al.* 1994)

The glycosidation of oxime **18.33** by a phenyl pyranosyl sulfoxide G–SO–Ph, in the presence of trimethylsilyl triflate, takes place on O-2 with isomerization of the acetal to give **18.34**. Reduction of **18.34** by cyanoborohydride gives only the α-D-*gluco* configuration **18.35**, hydrolysable to the free hydroxylamine. Nitrone **18.36**, obtained by treating the hydroxylamine with *p*-anisaldehyde, is glycosidated by a pyranosyl halide exclusively on the oxygen atom (80%).

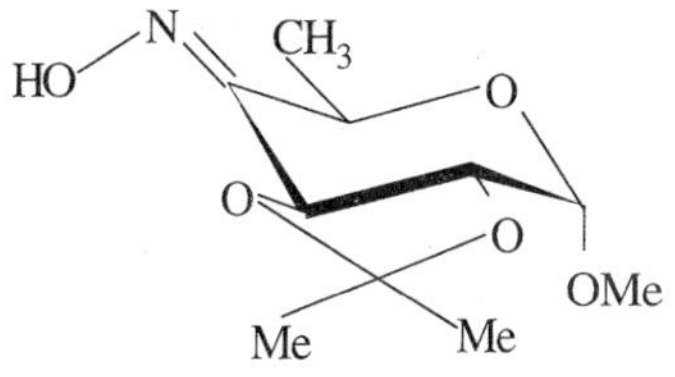

Simple access to unit **B** begins with the glycal of D-fucose which gives glycoside **18.37** by treatment with methanol in the presence of acid. The latter is selectively silylated at position 3 via stannylene to give **18.38**, then converted to trifluoromethanesulfonate **18.39**. The sulfur is introduced at position 4 with inversion of configuration by nucleophilic substitution with PhCOSK in *N,N*-dimethylformamide at 0°C (80%), **18.40**. After hydrolysis of the silyl ether, **18.40** is converted to trifluoromethanesulfonate **18.41**. Solvolysis with water takes place with participation of the benzoyl group to give the desired configuration, **18.42**. These alternative methods have allowed the preparation of the analogue of trisaccharide **ABE** present in esperamicin A_1 (Da Silva *et al.* 1994).

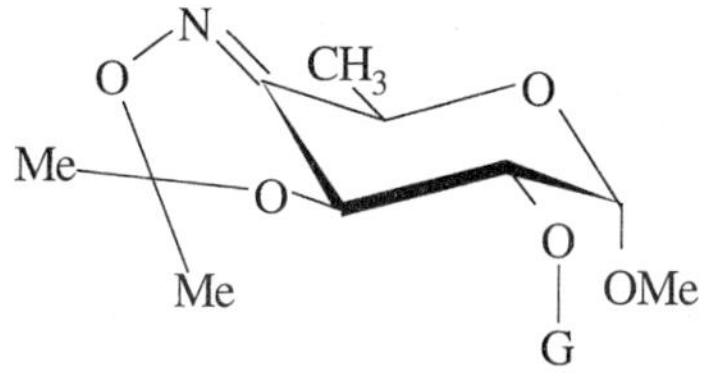

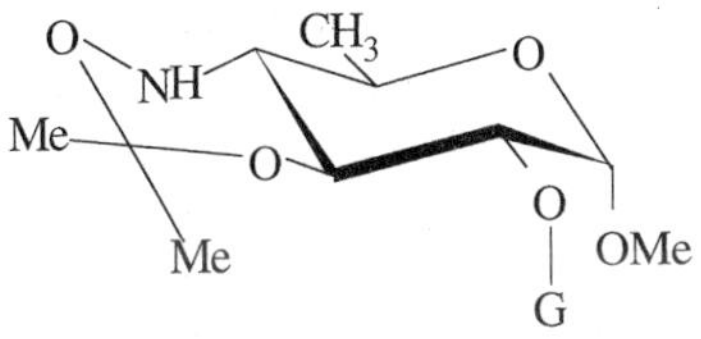

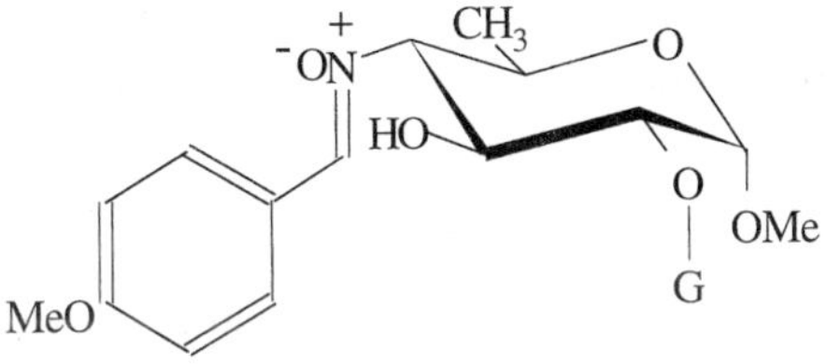

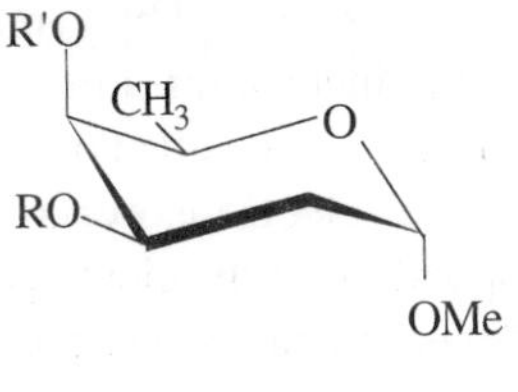

18.37 R = R' = H
18.38 R = t-BuSiMe$_2$, R'=H
18.39 R = t-BuSiMe$_2$, R'=Tf

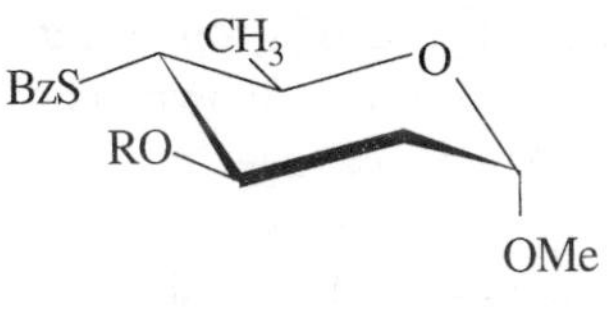

18.40 R = H
18.41 R = Tf

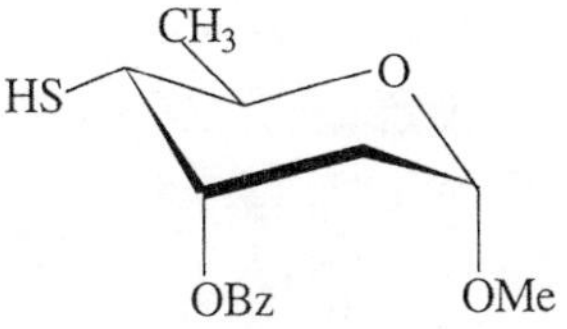

18.42

18.3 Recognition of the pseudo-tetrasaccharide by DNA

The reaction of calicheamicin with the double helix of DNA is significantly different from that with the isolated aglycon (Drak *et al.* 1991).

18.3.1 Cleavage of one or two strands

The substrate is a circular 'supercoiled' DNA, 'form I', in which the double helix is closed on itself by covalent bonds. One of the large DNA rings used was 9000 nucleotides long. Cleavage of a single strand produces 'form II', and cleavage at close sites of both strands, a linear double strand, 'form III' (Fig. 18.2). It is possible to separate and estimate these three forms by gel electrophoresis. With calicheamicin, at the beginning of the reaction, the ratio of double-stranded to a single-stranded cleavage is 1:2. With calicheamicinone, this ratio falls to 1:30 with the cleavage of a single strand being predominant.

18.3.2 Comparing efficiencies

The same substrate and the same technique make it possible to compare the efficiency of calicheamicin and its separated aglycon. In the presence of a reducing thiols, calicheamicin causes degradation at concentrations as low as 0.7 nM. At a concentration of 1.5 μM, only small nucleotides remain. Calicheamicinone causes cleavage observable at a concentration of 13 μM.

18.3.3 Sequence-selective cleavage

It was immediately recognized that chain cleavage by calicheamicin takes place at certain tetranucleotide sequences such as TCCG, TCCT, and GCCT. In order to study this, polynucleotide **18.43**, with 53 base pairs, was prepared by synthesis (only one strand is represented). This polynucleotide incorporates the presumed cleavage sites, shown in bold letters. It was radioactively labelled on both strands. After cleavage, the pieces are separated by gel electrophoresis and located by autoradiography. Indeed the pieces corresponding to the cleavage at TCCT are observed, but other fragmentation points are not as easily interpreted.

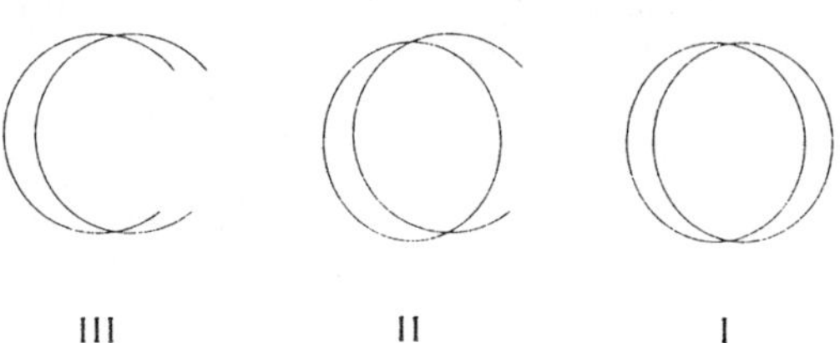

Fig. 18.2 Schematic representation of forms I, II, and III of DNA.

In other experiments (Walker *et al.* 1992) using another type of polynucleotide sequence, preferential fragmentations were seen at other sites, thus among others at sequence 5′TTTT. In both cases, the preferential cuts limit the number of cleaved polynucleotides and gel electrophoresis only revealed a small number of compounds. On the other hand, cleavage is statistical with the aglycon and this explains the presence of a good number of spots of uniform intensity.

5'-TTTAACCGATCAGAAT**TCCG**GTGCATGC**TCCT**AAGTGTAC**GCCT**AAGCTTCTT

18.43

18.3.4 Details of the association of the oligosaccharide with a DNA sequence

These experiments were carried out with decamer **18.44** and calicheamicin $\theta_1{}^{I}$, a close analogue of **18.1** which only differs by its side chain, $SCOCH_3$ instead of $SSSCH_3$, and which is more stable during NMR studies. This modification should have no effect on the association with sugar. Three NMR spectra examined were from decamer **18.44**, the complex with calicheamicin $\theta_1{}^{I}$, and the complex with the methyl glycoside of the synthetic pseudo-oligosaccharide **ABCDE**. One method for assessing the interactions consists in locating variations in the chemical shift of certain protons (H-6 for pyrimidines, H-8 for purines) of the polynucleotide when complexed. These variations are found to be similar with the most intense effect taking place for the C^5G^{16} pair. Likewise, analysing Overhauser effects found in the spectra of the two complexes showed that their intermolecular contacts were virtually identical. Thus the association mode of pseudo-oligosaccharides is largely independent of the presence of an aglycon, which implies that sequence specificity in the complexation with calicheamicin is essentially due to the pseudo-oligosaccharide. The Overhauser effects showed that all residues of the pseudo-oligosaccharide, with the exception of unit **E**, were in direct contact with the polynucleotide. In that respect, among the protons involved, H-4′, H-5′, H-5″ of the pentose and H-2 of the adenine of the sequences A^3-C^6 and G^{15}-A^{17} stand out, all located in the minor groove of the double helix.

5' G C A^3 T^4 C^5 C^6 T A G C 3'
3' C G T^{18} A^{17} G^{16} G^{15} A T C G 5'

18.44

Although molecular modelling gives a great number of details on structures (Paloma 1994), we will only give a few here. The minor groove's width normally measures 5–6 Å (the distance between the phosphorus atoms) but enlarges to

7.1–8.5 Å in order to accept the ligand. The oligosaccharide chain is slightly curved in order to adapt itself into the curvature of the minor groove. The aglycon is in contact with the A^3T^{18} and T^4A^{17} pairs. Sugars **A** and **B** are close to the T^4A^{17} and C^5G^{16} pairs. Sugar **E** is outside the minor groove. The aromatic ring is located between the guanine residues G^{15} and G^{16}, closer to AGGA. Carbons C-1, C-2, and the ring oxygen of residue **D** are close to the phosphates of the AGGA sequence. In general, the combination is stabilized by van der Waals forces, resulting from the lipophilic character of the ligand, four hydrogen bonds, and perhaps the formation of a salt bridge between the ethylamino group and the phosphate of C^5. The signals of the two amino protons of guanine G^{15} are relatively narrow, and highly shifted downfield. This can be explained by noting that they are neighbours of the iodine atom whose bulkiness slows down their rotation, while the aromatic ring has the deshielding effect.

The proximity of this iodine and the amino group of residue G^{15} seems to be an important factor in the interaction. The affinity of various ligands can be compared by the apparent dissociation constant, K_{obs}, with their complexes at a single site, TCCT·AGGA in a 20-mer DNA duplex. The results, K_{obs} 0.135 μM for calicheamicin and 4.1 μM for the pseudo-oligosaccharide indicates that over a total bond energy of 9.7 kcal mol^{-1}, the interactions at the oligosaccharide level correspond to 7.6 kcal mol^{-1}. But the affinity of the pseudo-oligosaccharide is 45 times lower if the iodine atom is replaced by a hydrogen atom (Li *et al.* 1994). This interaction can also be cancelled by replacing guanine by inosine in a sequence 5′AIGA. This lessens considerably the bond with calicheamicin and the consecutive cleavage (Li *et al.* 1994). The reactivities of a 160 base pair DNA have also been compared with its analogues in which guanine was replaced by inosine (I-DNA), adenine by 2,6-diaminopurine (DAP-DNA), and both guanine and adenine are replaced by inosine and 2,6-diaminopurine (I-DAP-DNA), that suppresses, increases, or displaces the amino groups. A decrease in specificity is observed, and with the modification (I-DAP-DNA), new sites, extremely sensitive to cleavage, appear (Bailly 1995).

18.3.5 Conclusion

These results suggest a selective recognition between certain nucleotide sequences of the double helix and the pseudo-oligosaccharide. The latter intervenes in order to transport the reactive aglycon towards a particular site and, perhaps, at this site, give it the most favourable and precise orientation for double-stranded cleavage. The methyl glycoside of the pseudo-oligosaccharide, by occupying a particular site on the double helix, protects it against outside agressions, as for example the cutting by calicheamicinone (Aiyar *et al.* 1992) and deoxyribonuclease, a general hydrolytic degradation enzyme of DNA (Nicolaou *et al.* 1992*b*). We can wonder what are the fundamental elements in the structure of the pseudo-oligosaccharide. A few results can be indicative. The specificity of the scission remains unchanged if residues **D** or **E** are omitted. On the other hand, calicheamicin T, an artificial product in which only sugars **A** and

E remain, only reacts with a much higher concentration than does the natural product, in a non-selective manner. It is now certain that the iodine atom plays an important role in this interaction (Li *et al.* 1994).

References

Aiyar, J., Danishefsky, S. J., and Crothers, D. M. (1992), *J. Am. Chem. Soc.*, **114**, 7552–7554.

Bamhaoud, T., Lancelin, J.-M., and Beau, J.-M. (1992), *J. Chem. Soc., Chem. Commun.*, 1494–1496.

Bailly, C. and Waring, M. J. (1995), *J. Am. Chem. Soc.*, **117**, 7311–7316.

Da Silva, E., Prandi, J., and Beau, J.-M. (1994), *J. Chem. Soc., Chem. Commun.*, 2127–2128.

Drak, J., Iwasawa, N., Crothers, D. M., Danishefsky, S. J. (1991), *Proc. Natl. Acad. Sci. USA*, **88**, 7464–7468.

Dupradeau, F.-Y., Allaire, S., Prandi, J., and Beau, J.-M. (1993), *Tetrahedron Lett.*, **34**, 4513–4516.

Golik, J., Clardy, J., Dubay, G., Groenewold, G., Kawaguchi, H., Konishi, M. *et al.* (1987*a*), *J. Am. Chem. Soc.*, **109**, 3461–3462.

Golik, J., Dubay, G., Groenewold, G., Kawaguchi, H., Konishi, M., Krishnan, B. *et al.* (1987*b*), *J. Am. Chem. Soc.*, **109**, 3462–3464.

Halcomb, R. L., Boyer, S. H. and Danishefsky, S. J. (1992), *Angew. Chem., Int. Ed. Engl.*, **31**, 338–342.

Hangeland, J. J., De Voss, J. J., Heath, J. A., Townsend, C. A., Ding, W.-d., Ashcroft, J. S., and Ellestad, G. A. (1992), *J. Am. Chem. Soc.*, **114**, 9200–9202.

Lee, M. D., Dunne, T. S., Chang, C. C., Siegel, M. M., Morton, G. O., Ellestad, G. A. *et al.* (1992), *J. Am. Chem. Soc.*, **114**, 985–997.

Lee, M. D., Ellestad, G. A., and Borders, D. B. (1991), *Acc. Chem. Res.*, **24**, 235–243.

Li, T., Zeng, Z., Estevez, V. A., Baldenius, K. U., Nicolaou, K. C., and Joyce, G. F. (1994), *J. Am. Chem. Soc.*, **116**, 3709–3715.

Nicolaou, K. C., Groneberg, P. D., Miyazaki, T., Stylianides, N. A., Schulze, T. J., and Stahl, W. (1990), *J. Am. Chem. Soc.*, **112**, 8193–8195.

Nicolaou, K. C., Schreiner, E. P., Iwabuchi, Y., and Suzuki, T. (1992*a*), *Angew. Chem., Int. Ed. Engl.*, **31**, 340–342.

Nicolaou, K. C., Tsay, S.-C., Suzuki, T., and Joyce, G. F. (1992*b*), *J. Am. Chem. Soc.*, **114**, 7555–7557.

Paloma, L. G., Smith, J. A., Chazin, W. J., and Nicolaou, K. C. (1994), *J. Am. Chem. Soc.*, **116**, 3697–3708.

Walker, S., landovitz, R., Ding, W.-D., Ellestad, G. A., and Kahne, D., (1992), *Proc. Natl. Acad. Sci. USA*, **89**, 4608–4612.

Yang, D., Kim, S. H., and Kahne, D. (1991), *J. Am. Chem. Soc.*, **113**, 4715–4716.

Index

Note: Figures, formulae, and Tables are indicated (in this index) by *italic page numbers*